An Introduction to Design and Construction of Dams

J. Paul Guyer, P.E., R.A.

Editor

The Clubhouse Press
El Macero, California

CONTENTS

CHAPTER 1: ARCH DAMS

1.1 INTRODUCTION. Unlike gravity dams that use the weight of the concrete for stability, arch dams utilize the strength of the concrete to resist the hydrostatic loads. Therefore, the concrete used in arch dams must meet very specific strength requirements. In addition to meeting strength criteria, concrete used in arch dams must meet the usual requirements for durability, permeability, and workability. Like all mass concrete structures, arch dams must keep the heat of hydration to a minimum by reducing the cement content, using low-heat cement, and using pozzolans. This chapter discusses the material investigations and mixture proportioning requirements necessary to assure the concrete used in arch dams will meet each of these special requirements. This chapter also discusses the testing for structural and thermal properties that relates to the design and analysis of arch dams, and it provides recommended values which may be used prior to obtaining test results.

1.2 MATERIAL INVESTIGATIONS. The material discussed in the next few paragraphs is intended to provide guidance in the investigations that should be performed for arch dams.

1.1.1 CEMENT. Under normal conditions the cementitious materials used in an arch dam will simply be a Type II portland cement (with heat of hydration limited to 70 cal/gm) in combination with a pozzolan. However, Type II cement may not be available in all project areas. The lack of Type II cement does not imply that massive concrete structures, such as arch dams, are not constructable. It will only be necessary to investigate how the available materials and local conditions can be utilized. For example, the heat of hydration for a Type I cement can be reduced by modifying the cement grinding process to provide a reduced fineness. Most cement manufacturers should be willing to do this since it reduces their cost in grinding the cement. However, there would not necessarily be a cost savings to the Government, since separate silos would be required to store the specially ground cement. In evaluating the cement sources, it is preferable to test each of the available cements at various fineness to determine the heat generation characteristics of each. This information is useful in performing parametric thermal studies.

1.1.2 POZZOLANS. Pozzolans are siliceous or siliceous and aluminous materials which in themselves possess little or no cementitious value; however, pozzolans will chemically react, in finely divided form and in the presence of moisture, with calcium hydroxide at ordinary temperatures to form compounds possessing cementitious properties. There are three classifications for pozzolans: Class N, Class F, and Class C.

1.1.2.1 CLASS N POZZOLANS are naturally occurring pozzolans that must be mined and ground before they can be used. Many natural pozzolans must also be calcined at high temperatures to activate the clay constituent. As a result, Class N pozzolans are not as economical as Classes F and C, if these other classes are readily available.

1.1.2.2 CLASSES F AND C are fly ashes which result from burning powdered coal in boiler plants, such as electric generating facilities. The ash is collected to prevent it from entering the atmosphere. Being a byproduct of another industry, fly ash is usually much cheaper than cement. However, some of the savings in material may be offset by the additional material handling and storage costs. Fly ash particles are spherical and are about the same fineness as cement. The spherical shape helps reduce the water requirement in the concrete mix.

1.1.2.3 IT IS IMPORTANT THAT THE POZZOLAN source produce consistent material properties, such as constant fineness and constant carbon content. Otherwise, uniformity of concrete will be affected. Therefore, an acceptable pozzolan source must be capable of supplying the total project needs.

1.1.2.4 IN MASS CONCRETE, pozzolans are usually used to replace a portion of the cement, not to increase the cementitious material content. This will reduce the amount of portland cement in the mixture proportions. Not only will this cement reduction lower the heat generated within the mass, but the use of pozzolans will improve workability, long-term strength, and resistance to attack by sulphates and other destructive agents. Pozzolans can also reduce bleeding and permeability and control alkali-aggregate reaction.

1.1.2.5 HOWEVER, WHEN POZZOLANS are used as a cement replacement, the time rate of strength gain will be adversely affected; the more pozzolan replacement, the lower the early strength. As a result, an optimization study should be performed to determine the appropriate amount of pozzolan to be used. The optimization study must consider both the long-term and the early-age strengths. The early-age strength is important because of the need for form stripping, setting of subsequent forms, and lift joint preparation. The practical limits on the percentage of pozzolan that should be used in mass concrete range from 15 to 50 percent. During the planning phases and prior to performing the optimization study, a value of 25 to 35 percent can be assumed.

1.1.3 AGGREGATES. Aggregates used in mass concrete will usually consist of natural sand, gravel, and crushed rock. Natural sands and gravels are the most common and are used whenever they are of satisfactory quality and can be obtained economically in sufficient quantity. Crushed rock is widely used for the larger coarse aggregates and occasionally for smaller aggregates including sand when suitable materials from natural deposits are not economically available. However, production of workable concrete from sharp, angular, crushed fragments usually requires more vibration and cement than that of concrete made with well-rounded sand and gravel.

1.1.3.1 SUITABLE AGGREGATE SHOULD be composed of clean, uncoated, properly shaped particles of strong, durable materials. When incorporated into concrete, it should satisfactorily resist chemical or physical changes such as cracking, swelling, softening, leaching, or chemical alteration. Aggregates should not contain contaminating substances which might contribute to deteriorating or unsightly appearance of the concrete.

1.1.3.2 THE DECISION TO DEVELOP an on-site aggregate quarry versus hauling from an existing commercial quarry should be based on an economic feasibility study of each. The study should determine if the commercial source(s) can produce aggregates of the size and in the quantity needed. If an on-site quarry is selected, then the testing of the on-site source should include a determination of the effort required to produce the aggregate. This information should be included in the contract documents for the prospective bidders.

1.1.4 WATER. All readily available water sources at the project site should be investigated during the design phase for suitability for mixing and curing water. The purest available water should be used. When a water source is of questionable quality, it should be tested in accordance with approved standards. When testing the water, the designer may want to consider including ages greater than the 7 and 28 days required in the typical specification. This is especially true when dealing with a design age of 180 or 360 days, because the detrimental effects of the water may not become apparent until the later ages. Since there are usually differences between inplace concrete using an on-site water source and lab mix designs using ordinary tap water, the designer may want to consider having the lab perform the mix designs using the anticipated on-site water.

1.1.5 ADMIXTURES. Admixtures normally used in arch dam construction include air-entraining, water-reducing, retarding, and water-reducing/ retarding admixtures. During cold weather, accelerating admixtures are sometimes used. Since each of these admixtures is readily available throughout the United States, no special investigations are required.

1.3 MIX DESIGNS. The mixture proportions to be used in the main body of the dam should be determined by a laboratory utilizing materials that are representative of those to be used on the project. The design mix should be the most economical one that will produce a concrete with the lowest practical slump that can be efficiently consolidated, the largest maximum size aggregate that will minimize the required cementitious materials, adequate early-age and later-age strength, and adequate long-term durability. In addition, the mix design must be consistent with the design requirements discussed in the other chapters of this manual. A mix design study should be performed to include various mixture proportions that would account for changes in material properties that might reasonably be expected to occur during the construction of the project. For example, if special requirements are needed for the cement (such as a reduced fineness), then a mix with the cement normally available should be developed to account for the possibility that the special cement may not always be obtainable. This would provide valuable information during construction that could avoid prolonged delays.

1.3.1 COMPRESSIVE STRENGTH. The required compressive strength of the concrete is determined during the static and dynamic structural analyses. The mixture proportions for the concrete should be selected so that the average compressive strength (fcr) exceeds the required compressive strength (f'c) by a specified amount. The amount that fcr should exceed f'c depends upon the classification of the concrete (structural or nonstructural) and the availability of test records from the concrete production facility. Concrete for an arch dam meets the requirements for both structural concrete and nonstructural (mass) concrete. That is, arch dams rely on the strength of the concrete in lieu of its mass, which would classify it as a structural concrete. However, the concrete is unreinforced mass concrete, which would classify it as a nonstructural concrete. For determining the required compressive strength (fcr) for use in the mix designs, the preferred method would be the method for nonstructural concrete. Assuming that no test records are available from the concrete plant, this would require the mix design to be based on the following relationship:

$$f_{cr} = f'_c + 600 \text{ psi} \qquad (9-1)$$

This equation will be valid during the design phase, but it can change during construction of the dam as test data from the concrete plant becomes available.

1.3.2 WATER-CEMENT RATIO. Table 9-1 shows the maximum water-cement ratios recommended for concrete used in most arch dams. These water-cement ratios are based on minimum durability requirements. The strength requirements in the preceding paragraph may dictate an even lower value. In thin structures (thicknesses less than approximately 30 feet) it may not be practical to change mixes within the body of the structure, so the lowest water-cement ratio should be used throughout the dam. In thick dams (thicknesses greater than approximately 50 feet), it may be practical to use an interior class of concrete with a water-cement ratio as high as 0.80. However, the concrete in the upstream and downstream faces should each extend into the dam a minimum of 15 to 20 feet before transitioning to the interior mixture. In addition, the interior concrete mix should meet the same strength requirements of the surface mixes.

Location in Dam	Severe or Moderate Climate	Mild Climate
Upstream face above minimum pool	0.47	0.57
Interior (for thick dams only)	0.80	0.80
Downstream face and upstream face below minimum pool	0.52	0.57

TABLE 9-1

Maximum Permissible Water-Cement Ratio

1.3.3 MAXIMUM SIZE AGGREGATE. Typical specifications recommend 6 inches as the nominal maximum size aggregate (MSA) for use in massive sections of dams. However, a 6-inch MSA may not always produce the most economical mixture proportion. Figure 9-1 shows that for a 90-day compressive strength of 5,000 psi, a 3-inch MSA would require less cement per cubic yard than a 6-inch MSA. Therefore, the selection of a MSA should be based on the size that will minimize the cement requirement. Another consideration in the selection of the MSA is the availability of the larger sizes and the cost of handling additional sizes. However, if the various sizes are available, the savings in cement and the savings in temperature control measures needed to control heat generation should offset the cost of handling the additional aggregate sizes.

1.3.4 DESIGN AGE. The design age for mass concrete is usually set between 90 days and 1 year. This is done to limit the cement necessary to obtain the desired strength. However, there are early-age strength requirements that must also be considered: form removal, resetting of subsequent forms, lift line preparation, construction loading, and impoundment of reservoir. There are also some difficulties that must be considered when selecting a very long design age. These include: the time required to develop the mixture proportions and perform the necessary property testing and the quality assurance evaluation of the mixture proportions during construction

1.3.4.1 DURING THE DESIGN PHASE, selection of a 1-year design strength would normally require a minimum of 2 years to complete the mix design studies and then perform the necessary testing for structural properties. This assumes that all the material investigations discussed earlier have been completed and that the laboratory has an adequate supply of representative material to do all required testing. In many cases, there may not be sufficient time to perform these functions in sequence, thus they must be done concurrently with adjustments made at the conclusion of the testing.

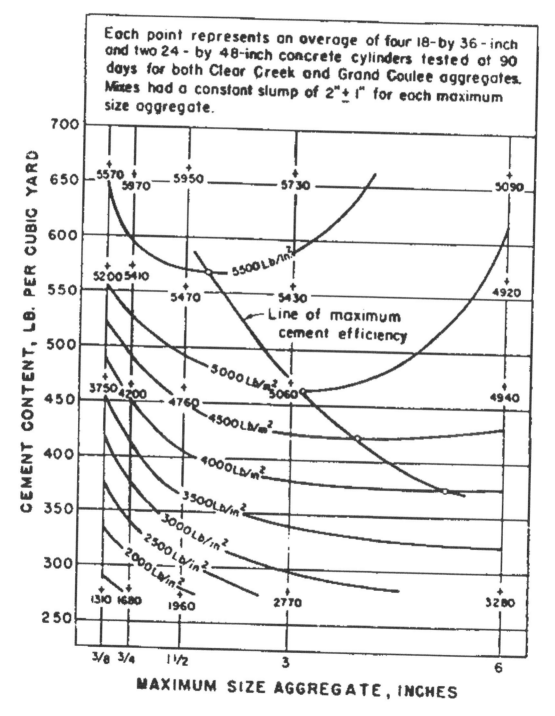

Figure 9-1

Variation of cement content with maximum size aggregate for various compressive strengths. Chart shows that compressive strength varies inversely with maximum size aggregate for minimum cement content

1.3.4.2 DURING THE CONSTRUCTION PHASE of the project, the problems with extended design ages become even more serious. The quality assurance program requires that the contracting officer be responsible for assuring that the strength requirements of the project are met. With a Government-furnished mix design, the Contracting Officer must perform strength testing to assure the adequacy of the mixture proportions and make adjustments in the mix proportions when necessary. The problem with identifying variability of concrete batches with extended design ages of up to 1 year are obvious. As a result, the laboratory should develop a relationship between accelerated tests and standard cured specimens. EM 1110-2-2000 requires that, during construction, two specimens be tested in accordance with the accelerated test procedures, one specimen be tested at an information age, and two specimens be tested at the design age. The single-information-age specimen should coincide with form stripping or form resetting schedules. For mass concrete construction where the design age usually exceeds 90 days, it is recommended that an additional information-age specimen be tested at 28 days.

1.4 TESTING DURING DESIGN. During the design phase of the project, test information is needed to adequately define the expected properties of the concrete. For the purposes of this manual, the type of tests required are divided into two categories: structural properties testing and thermal properties testing. The number of tests and age at which they should be performed will vary depending on the type of analyses to be performed. However, Table 9-2 should be used in developing the overall testing program. If several mixes are to be investigated by the laboratory for possible use, then the primary mix should be tested in accordance with Table 9-2 and sufficient isolated tests performed on the secondary mixes to allow for comparisons to the primary mix results. ACI STP 169B (1978) and Neville (1981) provide additional information following tests and their significance.

1.4.1 STRUCTURAL PROPERTIES TESTING.

1.4.1.1 COMPRESSIVE STRENGTH. Compressive strength testing at various ages will be available from the mix design studies. However, additional companion compressive tests at various ages may be required for correlation to other properties, such as tensile strength, shear strength, modulus of elasticity, Poisson's ratio, and dynamic compressive strength. Most of the compression testing will be in accordance with a standard which is a uniaxial test. However, if the stress analysis is to consider a biaxial stress state, then additional biaxial testing may need to be performed.

Test	\multicolumn{7}{c}{Age of Specimens (days)}
	1	3	7	28	90	180	1 yr.
Compression test	*	*	*	*	*	*	*
Modulus of rupture	*	*	*	*	*	*	*
Splitting tensile test				*			*
Shear test				*			*
Modulus of elasticity	*	*	*	*	*	*	*
Poisson's ratio	*	*	*	*	*	*	*
Dynamic tests							*
Creep	*	*	*	*	*	*	*
Strain capacity		*	*	*			*
Coefficient of thermal expansion	*	*	*	*			*
Specific heat	*	*	*	*			*
Thermal diffusivity	*	*	*	*			*

TABLE 9-2

Recommended Testing Program

1.4.1.2 TENSILE STRENGTH. The limiting factor in the design and analysis of arch dams will usually be the tensile strength of the concrete. Currently there are three accepted methods of obtaining the concrete tensile strength: direct tension test; the splitting tension test; and the modulus of rupture test. The direct tension test can give the truest indication of the tensile strength but is highly susceptible to problems in handling, sample preparation, surface cracking due to drying, and testing technique; therefore, it can often give erratic results. The splitting tension test also provides a good indication of the true tensile strength of the concrete and it has the advantage of being the easiest tension test to perform. In addition, the splitting tensile test compensates for any surface cracking and gives consistent and repeatable test results. However, when using the splitting tension test results as a criteria for determining the acceptability of a design, the designer should be aware that these results represent nonlinear performance that is normally being compared to a tensile stress computed from a linear analysis. The modulus of rupture test gives a value that is calculated based on assumed linear elastic behavior of the concrete. It gives consistent results and has the advantage of also being more consistent with the assumption of linear elastic behavior used in the design. A more detailed discussion of the importance of these tests in the evaluation of concrete dams is presented the ACI Journal (Raphael 1984). In testing for tensile strength for arch dam projects, the testing should include a combination of splitting tension tests and modulus of rupture tests.

1.4.1.3 SHEAR STRENGTH. The shear strength of concrete results from a combination of internal friction (which varies with the normal compressive stress) and cohesive strength (zero normal load shear strength). Companion series of shear strength tests should be conducted at several different normal stress values covering the range of normal stresses to be expected in the dam. These values should be used to obtain a curve of shear strength versus normal stress. Shear strength is determined in accordance with approved standards.

1.4.1.4 MODULUS OF ELASTICITY. When load is applied to concrete it will deform. The amount of deformation will depend upon the magnitude of the load, the rate of loading, and the total time of loading. In the analysis of arch dams, three types of deformations must be considered: instantaneous modulus of elasticity; sustained modulus of elasticity; and dynamic modulus of elasticity. Dynamic modulus of elasticity will be discussed in a separate section.

1.4.1.4.1 THE INSTANTANEOUS MODULUS OF ELASTICITY is the static modulus of elasticity. The modulus of elasticity in tension is usually assumed to be equal to that in compression. Therefore, no separate modulus testing in tension is required. Typical values for instantaneous (static) modulus of elasticity will range from 3.5×10^6 psi to 5.5×10^6 psi at 28 days and from 4.3×10^6 psi to 6.8×10^6 psi at 1 year. (b) The sustained modulus of elasticity includes the effects of creep, and can be obtained directly from creep tests. This is done by dividing the sustained load on the test specimen by the total deformation. The age of the specimen at the time of loading and the total time of loading will affect the result. It is recommended that the age of a specimen at the time of loading be at least 1 year and that the total time under load also be 1 year. The sustained modulus under these conditions will typically be approximately two-thirds that of the instantaneous modulus of elasticity.

1.4.1.5 DYNAMIC PROPERTIES. Testing for concrete dynamic properties should include compressive strength, modulus of rupture or splitting tensile strength, and modulus of elasticity. The dynamic testing can be performed at any age for information but is only required at the specified design age. The rate of loading used in the testing should reflect the actual rate with which the dam will be stressed from zero to the maximum value. This rate should be available from a preliminary dynamic analysis. If the rate of loading is not available, then several rates should be used covering a range that can be reasonably expected. For example, a range of rates that would cause failure at 20 to 150 milliseconds could be used.

1.4.1.6 POISSON'S RATIO. Poisson's ratio is defined in American Society for Testing and Materials (ASTM) E6 (ASTM 1992) as "the absolute value of the ratio of transverse strain to the corresponding axial strain below the proportional limit of the material." In simplified terms, it is the ratio of lateral strain to axial strain. Poisson's ratio for mass concrete will typically range from 0.15 to 0.20 for static loads, and from 0.24 to 0.25 for dynamic loads.

1.4.1.7 CREEP. Creep is time-dependent deformation due to sustained load. Creep can also be thought of as a relaxation of stress under a constant strain. In addition to using the creep test to determine the sustained modulus of elasticity (discussed previously), creep is extremely important in the thermal studies. However, unlike the sustained modulus of elasticity, the thermal studies need early age creep information.

1.4.1.8 STRAIN CAPACITY. Analyses that are based on tensile strain capacity rather than tensile strength will require some information on strain capacity. Examples of these types of analyses include the closure temperature and NISA as discussed in Chapter 8. Strain capacity can be measured in accordance with approved standards or can be estimated from the results of the modulus of elasticity, modulus of rupture, and specific creep tests.

1.4.2 THERMAL PROPERTIES. Understanding the thermal properties of concrete is vital in planning mass concrete construction. The basic properties involved include coefficient of thermal expansion, specific heat, thermal conductivity, and thermal diffusivity.

1.4.2.1 COEFFICIENT OF THERMAL EXPANSION. Coefficient of thermal expansion is the change in linear dimension per unit length divided by the temperature change. The coefficient of thermal expansion is influenced by both the cement paste and the aggregate. Since these materials have dissimilar thermal expansion coefficients, the coefficient for the concrete is highly dependent on the mix proportions, and since aggregate occupies a larger portion of the mix in mass concrete, the thermal expansion coefficient for mass concrete is more influenced by the aggregate. Coefficient of thermal expansion is expressed in terms of inch per inch per degree Fahrenheit (in./in./ F). In many cases, the length units are dropped and the quantities are expressed in terms of the strain value per F. This abbreviated form is completely acceptable. Typical values for mass concrete range from 3.0 to 7.5×10^{-6} /F. Testing for coefficient of thermal expansion should be in accordance with approved standards. However, the test should be modified to account for the temperature ranges to which the concrete will be subjected, including the early-age temperatures.

1.4.2.2 SPECIFIC HEAT. Specific heat is the heat capacity per unit temperature. It is primarily influenced by moisture content and concrete temperature. Specific heat is typically expressed in terms of Btu/pound· degree Fahrenheit (Btu/lb-F). Specific heat for mass concrete typically ranges from 0.20 to 0.25 Btu/lb-F. Testing for specific heat should be in accordance with approved standards.

1.4.2.3 THERMAL CONDUCTIVITY. Thermal conductivity is a measure of the ability of the material to conduct heat. It is the rate at which heat is transmitted through a material of unit area and thickness when there is a unit difference in temperature between the two faces. For mass concrete, thermal conductivity is primarily influenced by aggregate type and water content, with aggregate having the larger influence. Within the normal ambient temperatures, conductivity is usually constant. Conductivity is typically expressed in terms of Btu-inch per hour-square foot-degree Fahrenheit (Btu-in./ hr-ft2-F). Thermal conductivity for mass concrete typically ranges from 13 to 24 Btu-in./hr-ft2-F. It can be determined in accordance with approved standards or it can be calculated by the following equation:

$$k = \sigma c \rho \qquad (9\text{-}2)$$

where

k = thermal conductivity
σ = thermal diffusivity
c = specific heat
ρ = density of concrete

1.4.2.4 THERMAL DIFFUSIVITY. Thermal diffusivity is a measure of the rate at which temperature changes can take place within the mass. As with thermal conductivity, thermal diffusivity is primarily influenced by aggregate type and water content, with aggregate having the larger influence. Within the normal ambient temperatures, diffusivity is usually constant. Diffusivity is typically expressed in terms of square feet/hour (ft^2/hr). For mass concrete, it typically ranges from 0.02 to 0.06 ft^2/hr .

1.4.2.5 ADIABATIC TEMPERATURE RISE. The adiabatic temperature rise should be determined for each mix anticipated for use in the project. The adiabatic temperature rise is determined using appropriate standards.

1.5 PROPERTIES TO BE ASSUMED PRIOR TO TESTING. During the early stages of design analysis it is not practical to perform in-depth testing. Therefore, the values shown in Tables 9-3, 9-4, and 9-5 can be used as a guide during the early design stages and as a comparison to assure that the test results fall within reasonable limits.

Compressive strength (f'_c)	$\geq 4{,}000$ psi
Tensile strength (f'_t)	10% of f'_c

Shear strength (v)
    Cohesion
    Coef. of internal friction
	10% of f'_c
	1.0 ($\phi = 45°$)

Instantaneous modulus of elasticity (E_i)	4.5×10^6 psi
Sustained modulus of elasticity (E_s)	3.0×10^6 psi
Poisson's ratio (μ_s)	0.20
Unit weight of concrete (ρ_c)	150 pcf

Strain capacity (ε_c)
    rapid load
    slow load
	100×10^{-6} in/in
	150×10^{-6} in/in

TABLE 9-3

Static Values (Structural)

Compressive strength (f'$_{cd}$)	130% f'$_c$
Tensile strength (f'$_{td}$)	130% f'$_t$
Modulus of elasticity (E$_D$)	5.5×10^6 psi
Poisson's ratio (μ_d)	0.25

TABLE 9-4

Dynamic Values (Structural)

Coefficient of thermal expansion (e)	5.0×10^{-6} per °F
Specific heat (c)	0.22 Btu/lb-°F
Thermal conductivity (k)	16 Btu-in./hr-ft^2-°F
Thermal diffusivity (σ)	0.04 ft^2/hr

TABLE 9-5

Thermal Values

1.2 DESIGN

1.2.1 INTRODUCTION. The design and analysis of any dam assumes that the construction will provide a suitable foundation (with minimal damage) and uniform quality of concrete. The design also depends on the construction to satisfy many of the design assumptions, including such items as the design closure temperature. The topics in this chapter will deal with the construction aspects of arch dams as they relate to the preparation of planning, design, and construction documents. Contract administration aspects are beyond the scope of this discussion.

1.2.2 DIVERSION.

1.2.2.1 GENERAL. With arch dams, as with all concrete dams, flooding of the working area is not as serious an event as it is with embankment dams, since flooding should not cause serious damage to the completed portions of the dam and will not cause the works to be abandoned. For this reason, the diversion scheme can be designed for floods of relatively high frequencies corresponding to a 25-, 10-, or 5-year event. In addition, if sluices and low blocks are incorporated into the diversion scheme, the design flood frequency is reduced and more protection against flooding is added as the completed concrete elevation increases.

1.2.2.2 THE DIVERSION SCHEME. The diversion scheme will usually be a compromise between the cost of diversion and the amount of risk involved. The proper diversion plan should minimize serious potential flood damage to the construction site while also minimizing expense. The factors that the designer should consider in the study to determine the best diversion scheme include: streamflow characteristics; size and frequency of diversion flood; available regulation by existing upstream dams; available methods of diversion; and environmental concerns. Since streamflow characteristics, existing dams, and environmental concerns are project-specific items, they will not be addressed in this discussion.

1.2.2.2.1 SIZE AND FREQUENCY OF DIVERSION FLOOD. In selecting the flood to be used in any diversion scheme, it is not economically feasible to plan on diverting the largest flood that has ever occurred or may be expected to occur at the site. Consideration should be given to how long the work will be under construction (to determine the number of flood seasons which will be encountered), the cost of possible damage to work completed or still under construction if it is flooded, the delay to completion of the work, and the safety of workers and the downstream inhabitants caused by the diversion works. For concrete dams, the cost of damage to the project would be limited to loss of items such as formwork and stationary equipment. Delays to the construction would be primarily the cleanup of the completed work, possible replacement of damaged equipment, and possible resupply of construction materials (Figure 13-1). It is doubtful that there would be any part of the completed work that would need to be removed and reworked. The risk to workers and downstream residents should also be minimal since the normal loads on the dam are usually much higher than the diversion loads, and overtopping should not cause

Figure 13-1

View from right abutment of partially complete Monticello Dam in California showing water flowing over low blocks

serious undermining of the dam. And, as the concrete construction increases in elevation, the size of storm that the project can contain is substantially increased. Based on these considerations, it is not uncommon for the diversion scheme to be designed to initially handle a flood with a frequency corresponding to a 5- or 10-year event.

1.2.2.2.2 PROTECTION OF THE DIVERSION WORKS. The diversion scheme should be designed to either allow floating logs, ice and other debris to pass through the diversion works without jamming inside them and reducing their capacity, or prevent these items from entering the diversion works (International Commission on Large Dams 1986).

Systems such as trash structures or log booms can be installed upstream of the construction site to provide protection by collecting floating debris.

1.2.2.2.3 OTHER CONSIDERATIONS. If partial filling of the reservoir before completing the construction of the dam is being considered, then the diversion scheme must take into account the partial loss of flood storage. In addition, partial filling will require that some method (such as gates, valves, etc.) be included in the diversion scheme so that the reservoir level can be controlled.

1.2.2.3 METHODS OF DIVERSION. Typical diversion schemes for arch dams include tunnels, flumes or conduits, sluices, low blocks, or a combination of any of these. Each of these will require some sort of cofferdam across the river upstream of the dam to dewater the construction site. A cofferdam may also be required downstream. The determination of which method or methods of diversion should be considered will rest on the site conditions.

1.2.2.3.1 TUNNELS AND CULVERTS. Tunnels are the most common method of diversion used in very narrow valleys (Figure 13-2). The main advantage to using a diversion tunnel is that it eliminates some of the need for the staged construction required by other diversion methods. They also do not interfere with the foundation excavation or dam construction. The main disadvantage of a tunnel is that it is expensive, especially when a lining is required. As a result, when a tunnel is being evaluated as part of the diversion scheme, an unlined tunnel should be considered whenever water velocities and the rock strengths will allow. At sites where a deep crevice exists at the stream level and restitution concrete is already required to prevent excessive excavation, a culvert may be more economical than a tunnel (Figure 13-3). All tunnels and culverts will require a bulkhead scheme at their intakes to allow for closure and will need to be plugged with concrete once the dam is completed. The concrete plug will need to have a postcooling and grouting system (Figure 13-4).

1.2.2.3.2 CHANNELS AND FLUMES. Channels and flumes are used in conjunction with sluices and low blocks to pass the stream flows during the early construction phases.

Channels and flumes should be considered in all but very narrow sites (Figures 13-5 and 13-6). Flumes work well where flows can be carried around the construction area or across low blocks. Channels will require that the construction be staged so that the dam can be constructed along one abutment while the river is being diverted along the base of the other abutment (Figure 13-6). Once the construction reaches sufficient height, the water is diverted through sluices in the completed monoliths (Figure 13-7). A similar situation applies for flumes. However, flumes can be used to pass the river through the construction site over completed portions of the dam. If the flume has adequate clearance, it is possible to perform limited work under the flume.

1.2.2.3.3 SLUICES AND LOW BLOCKS. Sluices and low blocks are used in conjunction with channels and flumes. Sluices and low blocks are used in the intermediate and later stages of construction when channels and flumes are no longer useful. Sluices become very economical if they can be incorporated into the appurtenant structures, such as part of the outlet works or power penstock (Figure 13-7). It is acceptable to design this type of diversion scheme so that the permanent outlet works or temporary low-level outlets handle smaller floods, with larger floods overtopping the low blocks (Figure 13-8). The low blocks are blocks which are purposely lagged behind the other blocks. As with the tunnel option, sluices not incorporated into the appurtenant structures will need to be closed off and concrete backfilled at the end of the construction (Figure 13-9).

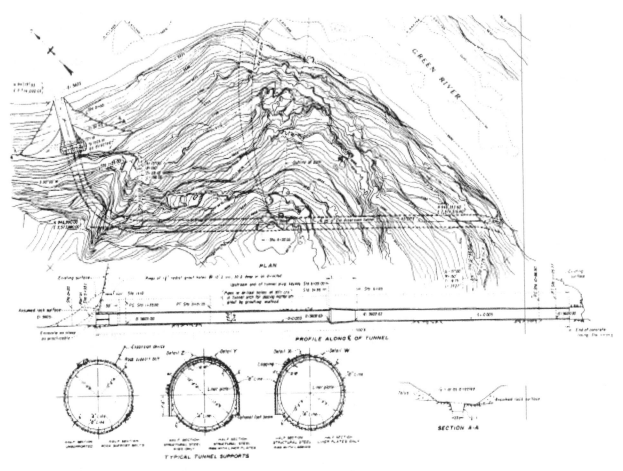

Figure 13- 2

Diversion tunnel for Flaming Gorge Dam

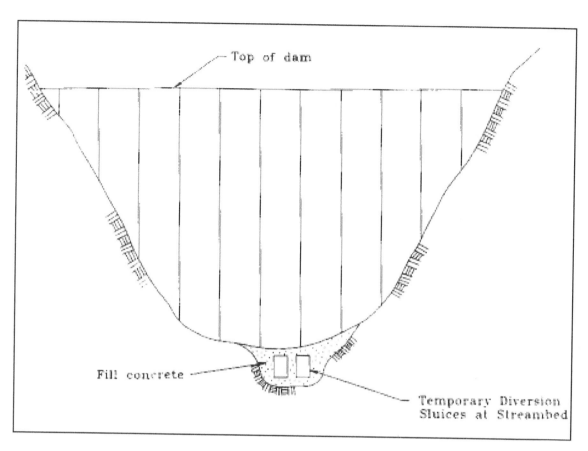

Figure 13-3

Diversion sluices through concrete pad at the base of a dam

1.2.3 FOUNDATION EXCAVATION. Excavation of the foundation for an arch dam will require that weak areas be removed to provide a firm foundation capable of withstanding the loads applied to it. Sharp breaks and irregularities in the excavation profile should be avoided (U.S. Committee on Large Dams 1988). Excavation should be performed in such manner that a relatively smooth foundation contact is obtained without excessive damage due to blasting. However, it is usually not necessary to over-excavate to produce a symmetrical site. A symmetrical site can be obtained by the use of restitutional concrete such as dental concrete, pads, thrust blocks, etc. Foundations that are subject to weathering and that will be left exposed for a period of time prior to concrete placement should be protected by leaving a sacrificial layer of the foundation material in place or by protecting the final foundation grade with shotcrete that would be removed prior to concrete placement.

1.2.4 CONSOLIDATION GROUTING AND GROUT CURTAIN. The foundation grouting program should be the minimum required to consolidate the foundation and repair any damage done to the foundation during the excavation and to provide for seepage control. The final grouting plan must be adapted to suit field conditions at each site.

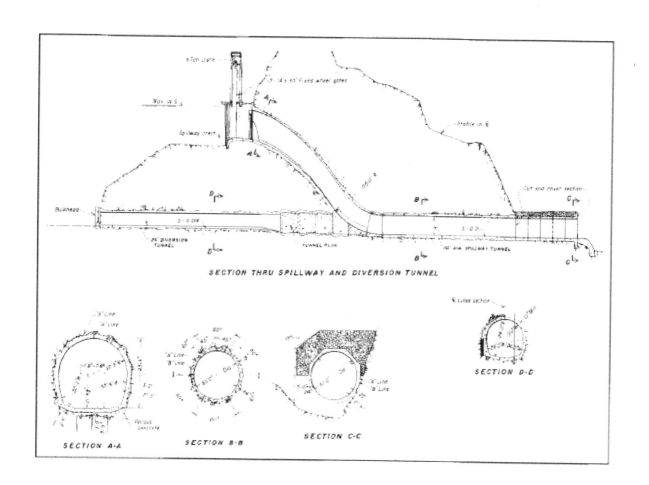

Figure 13-4

Typical diversion tunnel plug

Figure 13-5

Complete diversion flume at Canyon Ferry dam site in use for first stage diversion

1.2.4.1 CONSOLIDATION GROUTING. Consolidation grouting consists of all grouting required to fill voids, fracture zones, and cracks at and slightly below the excavated foundation. Consolidation grouting is performed before any other grouting is done and is usually accomplished from the excavated surface using low pressures. In cases of very steep abutments, the grouting can be performed from the top of concrete placements to prevent "slabbing" of the rock. Holes should be drilled normal to the excavated surface, unless it is desired to intersect known faults, shears, fractures, joints, and cracks. Depths vary from 20 to 50 feet depending on the local conditions. A split spacing process should be used to assure that all groutable voids, fracture zones, and cracks have been filled.

1.2.4.2 GROUT CURTAIN. The purpose of a grout curtain is to control seepage. It is installed using higher pressures and grouting to a greater depth than the consolidation grouting. The depth of the grout curtain will depend on the foundation characteristics but will typically vary from 30 to 70 percent of the hydrostatic head. To permit the higher pressures, grout curtains are usually installed from the foundation gallery in the dam or from the top of concrete placements. If lower pressures in the upper portion can be used, the grout curtain can also be placed at foundation grade prior to concrete placement and supplemented by a short segment of grouting from the gallery to connect the dam and the previously installed curtain. Grout curtains can also be installed from adits or tunnels that extend into the abutments. The grout curtain should be positioned along the footprint of the dam in a region of zero or minimum tensile stress. It should extend into the foundation so that the base of the grout curtain is located at the vertical projection of the heel of the dam. To assist in the drilling operations, pipes are embedded in the concrete. Once the concrete placement has reached an elevation that the grouting pressure will not damage or lift the concrete, the grout holes are drilled through these pipes and into the foundation. Grout curtain operations should be performed before concrete placement starts or after the monolith joints have been grouted to prevent damage to the embedded grout stops and to prevent leakage into the monolith joints.

Figure 13-6

Channel used in the Phase 1 Diversion for Smith Mountain Dam

Figure 13·7

Sluice used in the Phase II diversion for Smith Mountain dam

Figure 13-8

Aerial view of partially completed New Bullards Bar Dam in California showing water

flowing over the low blocks

1.2.5 CONCRETE OPERATIONS. The concrete operations discussed in this section are the special concrete construction requirements for arch dams which are not covered in other chapters of this and other manuals.

1.2.5.1 FORMWORK. Although modern arch dams are almost always curved in both plan and section, the forms used in constructing these dams are usually comprised of short, plane segments. The length of these segments correspond to the length of form panels, which should not exceed 8 feet. The detailed positioning of the formwork is usually supplied to the contractor as part of the contract documents. The computer program Arch Dam Construction Program (ADCOP) has been developed to generate the information needed to establish the location of the formwork along the upstream and downstream faces of each lift. ADCOP is a PC based computer program which requires arch dam geometry information similar to the ADSAS program and generates data which will control layout of formwork for each monolith in the body of an arch dam. ADCOP generates contraction joints which are spiral in nature.

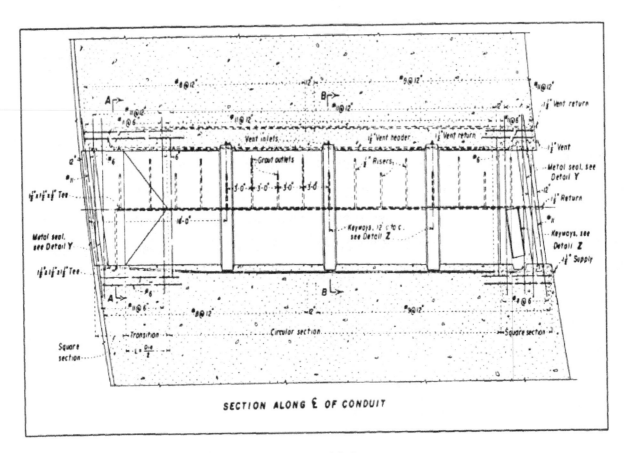

Figure 13-9

Diversion conduit through Morrow Point Dam

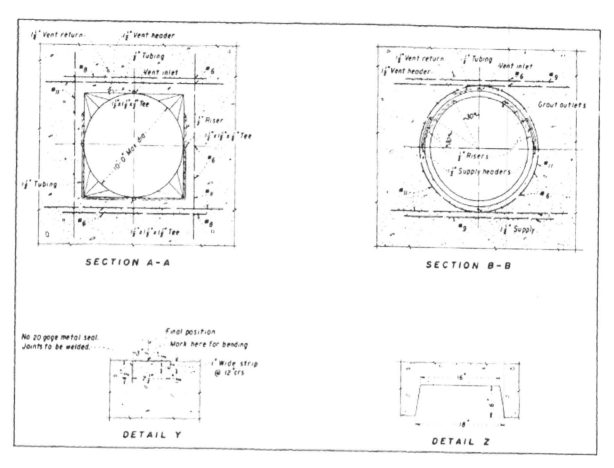

Figure 13-9 (continued)

Diversion conduit through Morrow Point Dam

1.2.5.2 HEIGHT DIFFERENTIAL. The maximum height differential between adjacent monoliths should be no more than 40 feet and no more than 70 to 100 feet between highest and lowest blocks in the dam. For dams with a pronounced overhang, limits should be placed on the placement of concrete with respect to ungrouted monoliths to avoid undesirable deflection and coincident tensile stresses.

1.2.5.3 CONCRETE QUALITY. The current trend in mass concrete construction is to minimize the use of cement by varying the concrete mixes used in different parts of the structure. This is usually done between the upstream and downstream faces and in areas of localized high stresses. For large dams, this practice can be very beneficial. For smaller dams, where the thicknesses are less than 30 feet, it may not be practical to vary the mix from the upstream to downstream faces. The benefits of adjusting the concrete mix for localized areas should be addressed, but the minimum requirements discussed in

Chapter 9 should not be relaxed. Also, the effects of the differing cement contents created by these adjustments must be considered in the construction temperature studies.

1.2.5.4 JOINT CLEANUP. A good bond between horizontal construction joints is essential if cantilever tensile load transfer is to be achieved and leakage to the downstream face is to be minimized. Bond strength between lift lines can be almost as strong as the in-place concrete if the joints are properly prepared and the freshly placed concrete is well consolidated. Vertical contraction (monolith) joints that are to be grouted should be constructed so that no bond exists between the blocks in order that effective joint opening can be accomplished during the final cooling phase without cracking the mass concrete.

1.2.5.5 EXPOSED JOINT DETAILS. Joints should be chamfered at the exposed faces of the dam to give a desirable appearance and to minimize spalling.

1.2.6 MONOLITH JOINTS.

1.2.6.1 SPACING OF MONOLITH JOINTS. The width of a monolith is the distance between monolith joints as measured along the axis of the dam. The determination of the monolith width will be part of the results of the closure temperature analysis discuss.. Once set, the monolith joint spacing should be constant throughout the dam if possible. In general, if the joints are to be grouted, monolith widths will be set at a value that allows for sufficient contraction to open the joints for grouting. However, there are limits to monolith widths. In recent construction, monolith widths have been commonly set at approximately 50 feet but with some structures having monoliths ranging from 30 to 80 feet. The USBR (1977) recommends that the ratio of the longer to shorter dimensions of a monolith be between 1 and 2. Therefore, a dam with a base thickness of 30 feet could have a maximum monolith width of 60 feet. However, the thickness of the base of a dam will vary significantly as the dam progresses from the crown cantilever up the abutment contacts. Near the upper regions of the abutment contacts, the base thickness of the dam will drop to a value close to the thickness of the crest. If a dam has a base thickness of 30 feet at the crown cantilever and a crest thickness of only 10 feet, then the limit on the monolith width at the crown cantilever will be 60 feet while the limit near the ends of the dam would be 20 feet. Since monolith spacing must be uniform throughout the dam, the maximum limit can be set at the higher value determined for the crown cantilever, with the understanding that some cracking could be expected along the abutment contacts near the ends of the dam. In any case, the joint spacing should never exceed 80 feet.

1.2.6.2 LIFT HEIGHTS. Lift heights are typically set at 5, 7.5, or 10 feet. This height should be uniform as the construction progresses from the base.

1.2.6.3 WATER STOPS AND GROUT STOPS. The terms water stop and grout stop describe the same material used for different purposes. A water stop is used to prevent seepage from migrating through a joint, typically from the upstream face of the dam to the downstream face. A grout stop is used to confine the grout within a specified area of a joint. General layout for the water stops/grout stops for a grouted dam is shown in Figure 13-10. Figure 13-20 also shows a typical layout at the upstream face for ungrouted cases where a joint drain is added. Proper installation and protection of the water stop/grout stop during construction is as important as the shape and material.

1.2.6.4 SHEAR KEYS. Shear keys are installed in monolith joints to provide shearing resistance between monoliths. During construction they also help maintain alignment of the blocks, and they may help individual blocks "bridge" weak zones within the foundation. Vertical shear keys, similar to those shown in Figure 13-11, will also increase the seepage path between the upstream and downstream faces. Other shapes of shear key, such as the dimple, waffle, etc., shown in Figures 13-12 and 13-13, have also been used. The main advantage in these other shaped keys is the use of standard forms and, in the case of the dimple shape, less interference with the grouting operations. However, these other shapes provide additional problems in concrete placement and are not as effective in extending the seepage path as are the vertical keys.

1.2.6.5 VERTICAL VERSUS SPIRAL JOINTS. Arch dams can be constructed with either vertical or spiral radial contraction joints. Vertical radial joints are created when all monolith joints, at each lift, are constructed on a radial line between the dam axis and the axis center on the line of centers (see Chapter 5 for definition of dam axis and axis center). Therefore, when looking at a plan view of a vertical joint, the joints created at each construction lift will fall on the same radial line. Spiral joints are created when the monolith joints, at each lift, are constructed on a radial line between the dam axis and with the intrados line of centers for that lift elevation. Therefore, when looking at a plan view of a spiral joint, each joint created at each construction lift will be slightly rotated about the dam axis. Spiral radial joints will also provide some keying of the monoliths, while vertical radial joints will not.

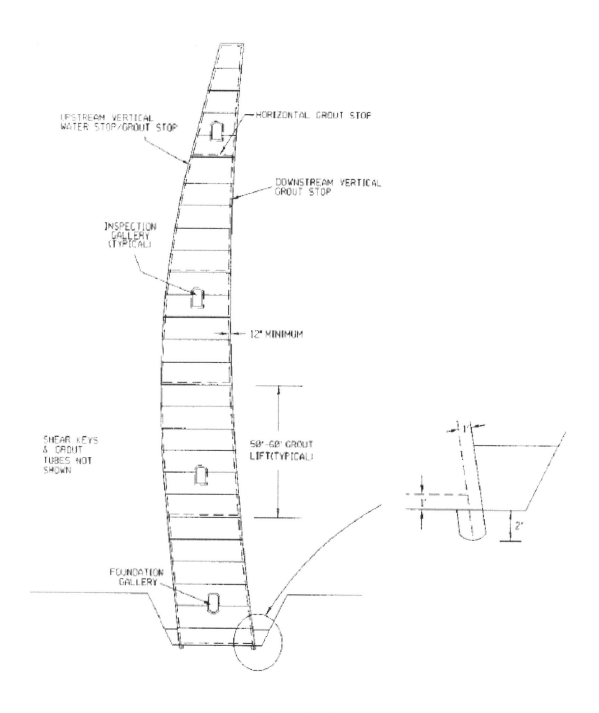

Figure 13-10

Water stop and grout stop details

Figure 13-11

Vertical keys at monolith joints of Monar Dam, Scotland

a)

b)

Figure 13-12

Dimple shear keys

1.2.6.6 MONOLITH JOINT GROUTING. The purpose of grouting monolith joints is to provide a monolithic structure at a specified "closure temperature." To accomplish this, grout is injected into the joint by a means of embedded pipes similar to those shown in Figures 13-14 and 13-15. To ensure complete grouting of the joint prior to grout set, and to prevent excessive pressure on the seals and the blocks, grout lifts are usually limited to between 50 and 60 feet.

1.2.6.6.1 JOINT OPENING FOR GROUTING. The amount that a joint will open during the final cooling period will depend upon the spacing of the monolith joints, the thermal properties of the concrete, and the temperature drop during the final cooling period. The typical range of joint openings needed for grouting is 1/16 to 3/32 inch. With arch dams, where it is desirable to have a monolithic structure that will transfer compressive loads across the monolith joints, it is better to have an opening that will allow as thick a grout as practical to be placed. A grout mixture with a water-cement ratio of 0.66 (1 to 1 by volume) will usually require an opening on the higher side of the range given (Water Resources Commission 1981). In the initial design process, a required joint opening of 3/32 inch should be assumed.

1.2.6.6.2 LAYOUT DETAILS. The supply loop at the base of the grout lift provides a means of delivering the grout to the riser pipes. Grout can be injected into one end of the looped pipe. The vertical risers extend from the supply loop at 6-foot intervals. Each riser contains outlets spaced 10 feet on centers, which are staggered to provide better grout coverage to the joint. The risers are discontinued near the vent loop (groove) provided at the top of the grout lift. Vent pipes are positioned at each end of the vent groove to permit air, water, and thin grout to escape in either direction. Normally, these systems are terminated at the downstream face. Under some conditions, they can be terminated in galleries. The ends of the pipes have nipples that can be removed after completion of the grouting and the remaining holes filled with dry pack. Typical grout outlet and vent details are shown in Figures 13-14 and 13-15.

Figure 13-13

Waffle keys at the vertical monolith joints of Gordon Dam, Australia

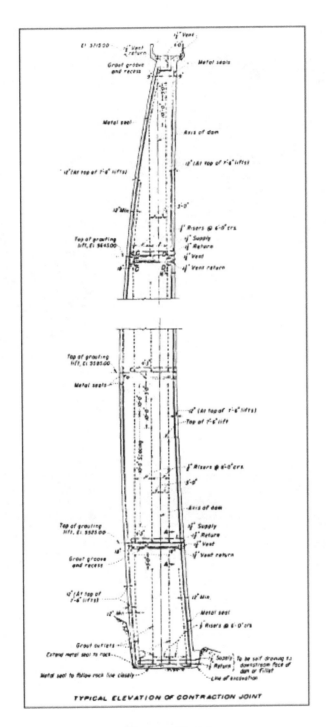

Figure 13-14

Contraction joint and grouting system for East Canyon Dam

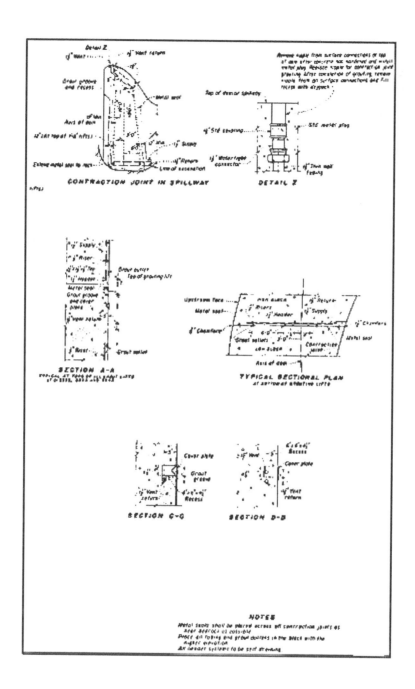

Figure 13-14 (continued)

Contraction joint and grouting system for East Canyon Dam

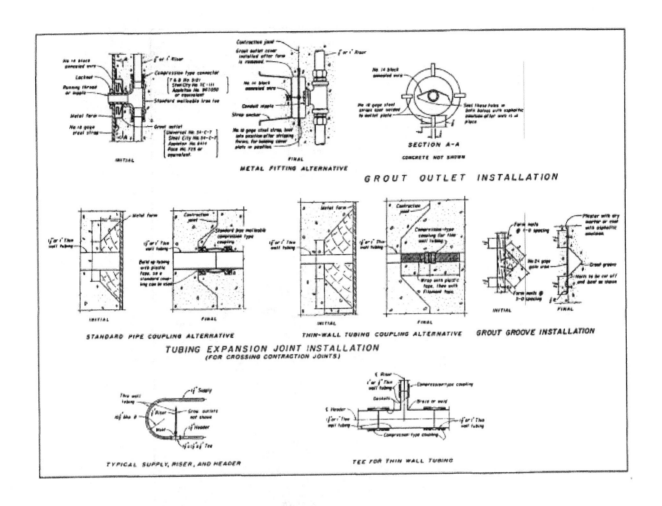

Figure 13-15

Example of grouting system details

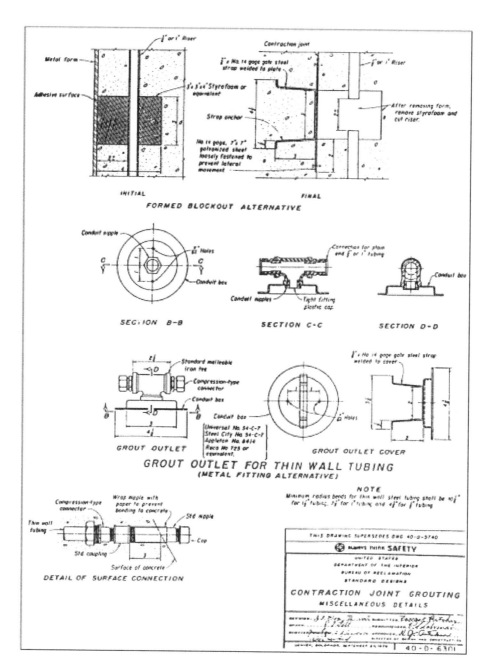

Figure 13-15 (continued)

Example of grouting system details

1.2.6.6.3 PREPARATION FOR GROUTING. Prior to grouting, the system is tested to assure that obstructions do not exist. The monolith joint is then cleaned with air and water under pressure. The joint is filled with water for a period of 24 hours. The water is drained from the joint to be grouted. Joints of two or more ungrouted joints on either side are filled with water, but not pressurized. Once the grout reaches the top of the grout lift, the lift above is filled with water to protect the upper grout stop. Immediately after a grouting operation is completed, the water is drained from the joints in the lift above. Water in the adjacent ungrouted joints should remain in place for at least 6 hours after the grouting operation is completed.

1.2.6.6.4 GROUT MIX. The grout should consist of the thickest mix that will enter the joint, fill all of the small voids, and travel to the vent. Grout mixes usually vary from water-cement ratios of 2 to 1 by volume (1.33 by weight) at the start of the grouting operation to thicker mixes (1 to 1 by volume or 0.66 by weight) as the operation progresses. If the joints are sufficiently open to accept a thicker grout, then mixes with ratios of 0.70 to 1 by volume (0.46 by weight) should be used to finish the job.

1.2.6.6.5 GROUTING OPERATION. Grout is injected in the supply loop at the bottom of the grout lift so that grout first comes in contact with the riser farthest from the supply portal. This will allow for the most favorable expulsion of air, water, and diluted grout as the grouting operation progresses. Once the grout appears at the return end of the supply loop, the return end is closed and the grout is forced up the risers and into the joint. The grouting must proceed at a rate fast enough that the grout will not set before the entire joint is filled with a thick grout. However, the rate of grouting must also be slow enough to allow the grout to settle into the joint. When thick grout reaches one end of the vent loop at the top of the grout lift, grouting operations are stopped for a short time (5 to 10 minutes) to allow the grout to settle in the joint. After three to five repetitions of a showing of thick grout, the valves are closed and the supply pressure increased to the allowable limit (usually 30 to 50 psi) to force grout into all small openings of the joint and to force the excess water into the pores of the concrete, leaving a grout film of lower water-cement ratio and higher density in the joint. The limiting pressure is set at a value that will avoid excessive deflection in the block and joint opening in the grouted portions below the joint.

The system is sealed off when no more grout can be forced into the joint as the pressure is maintained.

1.2.7. GALLERIES AND ADITS.

1.2.7.1 GENERAL. Adits are near horizontal passageways that extend from the surface into the dam or foundation. Galleries are the internal passageways within the dam and foundation and can be horizontal, vertical, or sloped. Chambers or vaults are created when galleries are enlarged to accommodate equipment. Galleries serve a variety of purposes. During construction, they can provide access to manifolds for the concrete postcooling and grouting operations. The foundation gallery also provides a work space for the installation of the grout and drainage curtains. During operation, galleries provide access for inspection and for collection of instrumentation data. They also provide a means to collect the drainage from the face and gutter drains and from the foundation drains. Galleries can also provide access to embedded equipment such as gates or valves. However, with all the benefits of galleries, there are also many problems. Galleries interfere with the construction operations and, therefore, increase the cost of construction. They provide areas of potential stress concentrations, and they may interfere with the proper performance of the dam. Therefore, galleries, as well as other openings in the dam, should be minimized as much as possible.

1.2.7.2 LOCATION AND SIZE. Typical galleries are 5 feet wide by 7.5 feet high. Figure 13-16 shows the most typical shapes of galleries currently being used. Foundation galleries are somewhat larger to allow for the drilling and grouting operations required for the grout and drainage curtains. Foundation galleries can be as large as 6 feet wide by 8.5 feet high. Personnel access galleries, which provide access only between various features within the dam, can be as small as 3 feet wide by 7 feet high. Spiral stairs should be 6 feet 3 inches in diameter to accommodate commercially available metal stairs.

1.2.7.3 LIMITATIONS OF DAM THICKNESS. Galleries should not be put in areas where the thickness of the dam is less than five times the width of the gallery.

1.2.7.4 REINFORCEMENT REQUIREMENTS. Reinforcement around galleries is not recommended unless the gallery itself will produce localized high tensile stresses or if it is positioned in an area where the surrounding concrete is already in tension due to the other external loads being applied to it. Even under these conditions, reinforcement is only required if the cracking produced by these tensions is expected to propagate to the reservoir. Reinforcement will also be required in the larger chambers formed to accommodate equipment.

1.2.7.5 LAYOUT DETAILS (Figure 13-17). Gallery and adit floors should be set at the top of a placement lift for ease of construction. Galleries should be at a slope comfortable for walking. Ramps can be used for slopes up to 10 degrees without special precautions and up to 15 degrees if nonslip surfaces are provided. Stairs can be used for slopes up to 40 degrees. Spiral stairs or vertical ladders can be used in areas where slopes exceed these limits. Landings should be provided approximately every 12 vertical feet when spiral stairs or ladders are used. Landings should also be provided in stairways if at all possible. Handrails should be provided in all galleries where the slope is greater than 10 degrees. There should be a minimum of 5 feet between the floor of the foundation gallery and the rock interface. There should also be a minimum 5-foot spacing between a gallery or adit and the monolith joints and external faces. The preferable location of the galleries is near the center of the monolith to minimize its impact on the section modulus of the cantilever. As a minimum, galleries should be located away from the upstream face at a distance that corresponds to 5 percent of the hydrostatic head.

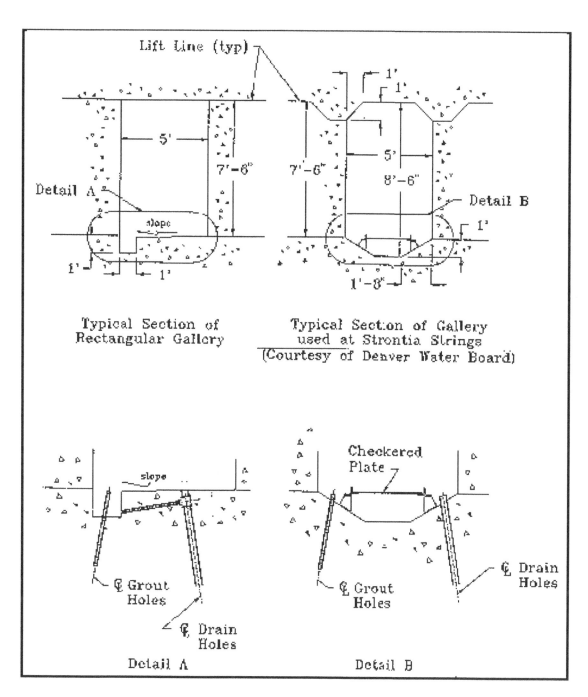

Figure 13-16

Typical gallery details

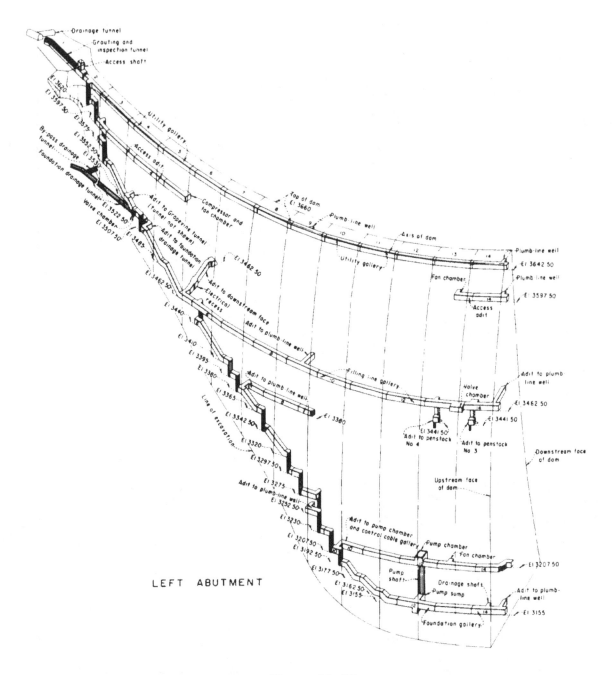

Figure 13-17

Left abutment gallery system for Yellowtail Dam

1.2.7.6 UTILITIES. Water and air lines should be embedded in the concrete to help in future maintenance operations. Lighting and ventilation should also be provided for the convenience and safety of personnel working in the galleries. Telephones should be located in gate rooms or chambers within the dam, as well as scattered locations throughout the galleries, for use in emergencies and for convenience of operation and maintenance personnel.

1.2.8 DRAINS. Drains fall into two categories: foundation drains and embedded drains. Embedded drains include face drains, gutter drains, and joint drains. All arch dams should include provisions for foundation drains and for face and gutter drains. Joint drains are not recommended if the monolith joints are to be grouted because they can interfere with the grouting process. Providing water/grout stops on each side of the drain help alleviate that problem, but the addition of the drain and the grout stop reduces the available contact area between the grout and mass concrete and thereby reduces the area for load transfer between adjacent monoliths.

1.2.8.1 FOUNDATION DRAINS. Foundation drains provide a way to intercept the seepage that passes through and around the grout curtain and thereby prevents excessive hydrostatic pressures from building up within the foundation and at the dam/foundation contact. The depth of the foundation drains will vary depending on the foundation conditions but typically ranges from 20 to 40 percent of the reservoir depth and from 35 to 75 percent of the depth of the grout curtain. Holes are usually 3 inches in diameter and are spaced on 10- foot centers. Holes should not be drilled until after all foundation grouting in the area has been completed. Foundation drains are typically drilled from the foundation gallery, but if no foundation gallery is provided, they can be drilled from the downstream face. Foundation drains can also be installed in adits or tunnels that extend into the abutments.

1.2.8.2 FACE DRAINS. Face drains are installed to intercept seepage along the lift lines or through the concrete. They help minimize hydrostatic pressure within dam as well as staining on the downstream face. Face drains extend from the crest of the dam to the foundation gallery. If there is no foundation gallery, then the drains are extended to the downstream face and connected to a drain pipe in the downstream fillet. They should be 5 or 6 inches in diameter and located 5 to 10 feet from upstream face. If the crest of the dam is thin, the diameter of the drains can be reduced and/or the distance from the upstream face can be reduced as they approach the crest. Face drains should be evenly spaced along the face at approximately 10 feet on centers (Figure 13-18).

1.2.8.3 GUTTER DRAINS. Gutter drains are drains that connect the gutters of the individual galleries to provide a means of transporting seepage collected in the upper galleries to the foundation gallery and eventually to the downstream face or to a sump. These drains are 8-inch-diameter pipes and extend from the drainage gutter in one gallery to the wall or drainage gutter in the next lower gallery. These drains are located in approximately every fourth monolith (Figure 13-19).

1.2.8.4 JOINT DRAINS. As noted earlier, joint drains are not normally installed in arch dams because of the joint grouting operations. However, if the monolith joints are not to be grouted, or if drains are required in grouted joints and can effectively be installed, then a 5- or 6-inch drain similar to the face drains should be used. The drain should extend from the crest of the dam to the foundation and should be connected to the foundation gallery (Figure 13-20).

1.2.9 APPURTENANT STRUCTURES. The appurtenant structures should be kept as simple as possible to minimize interference with the mass concrete construction. Outlet works should be limited to as few monoliths as possible. Conduits should be aligned horizontally through the dam and should be restricted to a single construction lift. Vertical sections of the conduits can be placed outside the main body of the dam.

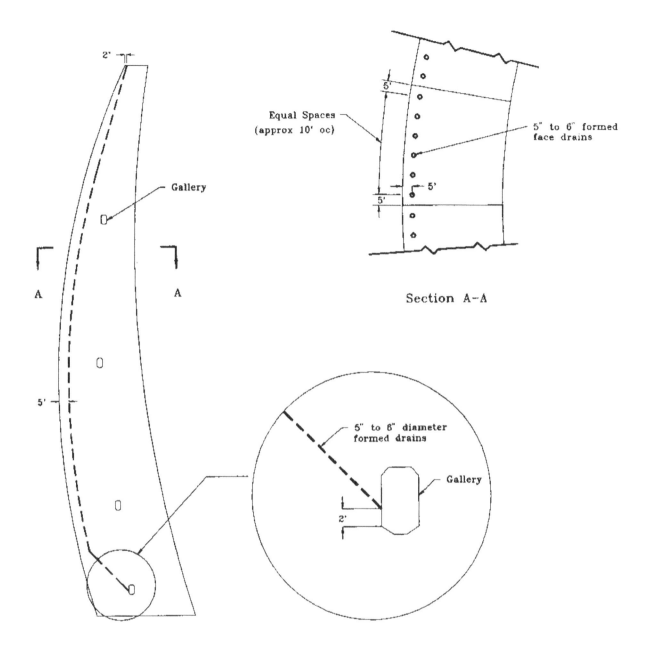

2'

Gallery

A   A

5'

Equal Spaces
(approx 10' oc)

5'

5'

5"
5" to 6" formed
face drains

Section A-A

5" to 6" diameter
formed drains

Gallery

2'

Figure 13-18

Details of face drains

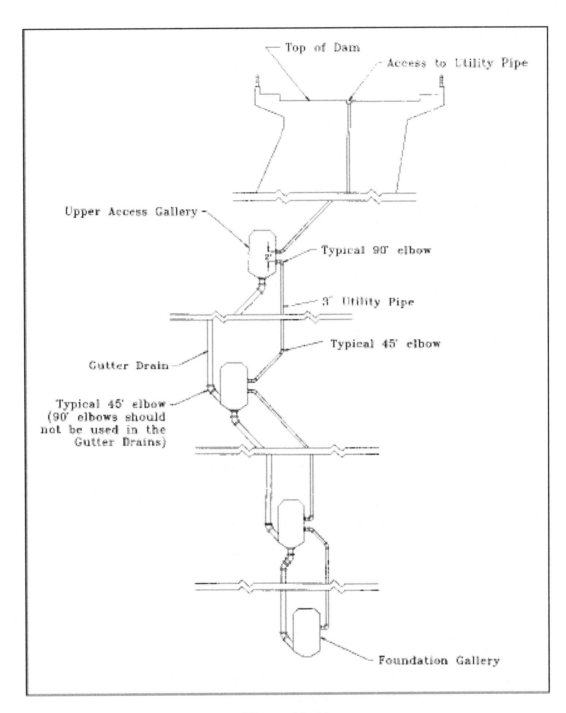

Figure 13-19

Details of gutter drains and utility piping

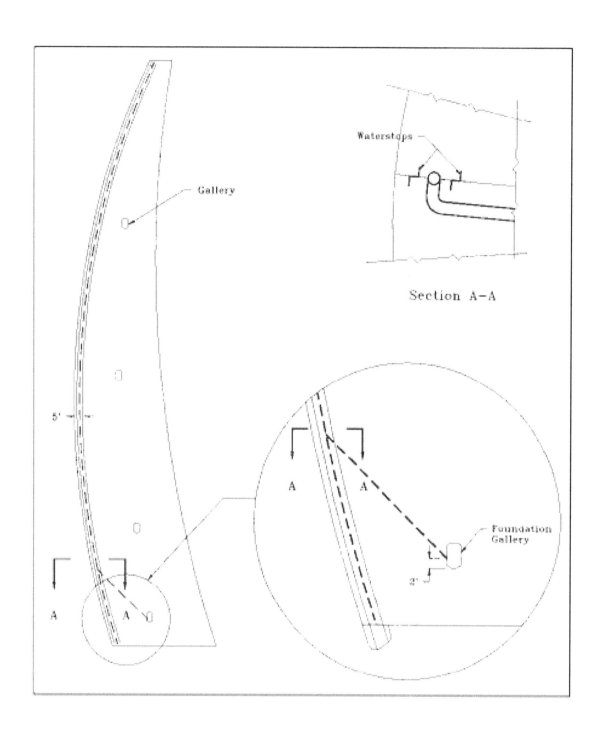

Figure 13-20

Details of joint drains

CHAPTER 2: GRAVITY DAMS

2.1 GENERAL DESIGN CONSIDERATIONS

2.1.1 TYPES OF CONCRETE GRAVITY DAMS. Basically, gravity dams are solid concrete structures that maintain their stability against design loads from the geometric shape and the mass and strength of the concrete. Generally, they are constructed on a straight axis, but may be slightly curved or angled to accommodate the specific site conditions. Gravity dams typically consist of a nonoverflow section(s) and an overflow section or spillway. The two general concrete construction methods for concrete gravity dams are conventional placed mass concrete and roller compacted concrete (RCC).

2.1.1 CONVENTIONAL CONCRETE DAMS.

2.1.1.1 CONVENTIONALLY PLACED MASS CONCRETE dams are characterized by construction using materials and techniques employed in the proportioning, mixing, placing, curing, and temperature control of mass concrete (American Concrete Institute (ACI) 207.1 R-87). Typical overflow and nonoverflow sections are shown on Figures 2-1 and 2-2. Construction incorporates methods that have been developed and perfected over many years of designing and building mass concrete dams. The cement hydration process of conventional concrete limits the size and rate of concrete placement and necessitates building in monoliths to meet crack control requirements. Generally using large-size coarse aggregates, mix proportions are selected to produce a low-slump concrete that gives economy, maintains good workability during placement, develops minimum temperature rise during hydration, and produces important properties such as strength, impermeability, and durability. Dam construction with conventional concrete readily facilitates installation of conduits, penstocks, galleries, etc., within the structure.

2.1.1.2 CONSTRUCTION PROCEDURES INCLUDE batching and mixing, and transportation, placement, vibration, cooling, curing, and preparation of horizontal construction joints between lifts. The large volume of concrete in a gravity dam normally justifies an onsite batch plant, and requires an aggregate source of adequate quality and quantity, located at or within an economical distance of the project. Transportation from the batch plant to the dam is generally performed in buckets ranging in size from 4 to 12 cubic yards carried by truck, rail, cranes, cableways, or a combination of these methods. The maximum bucket size is usually restricted by the capability of effectively spreading and vibrating the concrete pile after it is dumped from the bucket. The concrete is placed in lifts of 5- to 10-foot depths. Each lift consists of successive layers not exceeding 18 to 20 inches. Vibration is generally performed by large one-man, air-driven, spud-type vibrators. Methods of cleaning horizontal construction joints to remove the weak laitance film on the surface during curing include green cutting, wet sand-blasting, and high-pressure air-water jet.

2.1.1.3 THE HEAT GENERATED AS CEMENT HYDRATES requires careful temperature control during placement of mass concrete and for several days after placement. Uncontrolled heat generation could result in excessive tensile stresses due to extreme gradients within the mass concrete or due to temperature reductions as the concrete approaches its annual temperature cycle. Control measures involve precooling and postcooling techniques to limit the peak temperatures and control the temperature drop. Reduction in the cement content and cement replacement with pozzolans have reduced the temperature-rise potential. Crack control is achieved by constructing the conventional concrete gravity dam in a series of individually stable monoliths separated by transverse contraction joints. Usually, monoliths are approximately 50 feet wide.

2.1.2 ROLLER-COMPACTED CONCRETE (RCC) GRAVITY DAMS. The design of RCC gravity dams is similar to conventional concrete structures. The differences lie in the construction methods, concrete mix design, and details of the appurtenant structures. Construction of an RCC dam is a relatively new and economical concept. Economic advantages are achieved with rapid placement using construction techniques that are similar to those employed for embankment dams. RCC is a relatively dry, lean, zero slump concrete material containing coarse and fine aggregate that is consolidated by external vibration using vibratory rollers, dozer, and other heavy equipment. In the hardened condition, RCC has similar properties to conventional concrete. For effective consolidation, RCC must be dry enough to support the weight of the construction equipment, but have a consistency wet enough to permit adequate distribution of the past binder throughout the mass during the mixing and vibration process and, thus, achieve the necessary compaction of the RCC and prevention of undesirable segregation and voids. The consistency requirements have a direct effect on the mixture proportioning requirements (ACI 207.1 R-87).

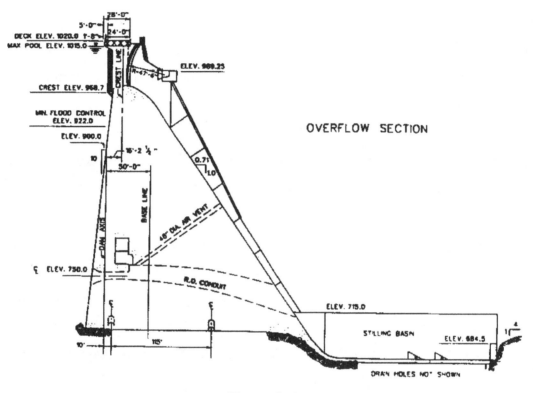

Figure 2-1

Typical dam overflow section

2.1.2 COORDINATION BETWEEN DISCIPLINES. A fully coordinated team of structural, material, and geotechnical engineers, geologists, and hydrological and hydraulic engineers should ensure that all engineering and geological considerations are properly integrated into the overall design. Some of the critical aspects of the analysis and design process that require coordination are:

2.1.2.1 PRELIMINARY ASSESSMENTS OF GEOLOGICAL data, subsurface conditions, and rock structure. Preliminary designs are based on limited site data. Planning and evaluating field explorations to make refinements in design based on site conditions should be a joint effort of structural and geotechnical engineers.

2.1.2.2 SELECTION OF MATERIAL PROPERTIES, DESIGN parameters, loading conditions, loading effects, potential failure mechanisms, and other related features of the analytical models. The structural engineer should be involved in these activities to obtain a full understanding of the limits of uncertainty in the selection of loads, strength parameters, and potential planes of failure within the foundation.

2.1.2.3 EVALUATION OF THE TECHNICAL AND ECONOMIC feasibility of alternative type structures. Optimum structure type and foundation conditions are interrelated. Decisions on alternative structure types to be used for comparative studies need to be made jointly with geotechnical engineers to ensure the technical and economic feasibility of the alternatives.

2.1.2.4 PARTICIPATION IN CONSTRUCTIBILITY REVIEWS is necessary to ensure that design assumptions and methods of construction are compatible. Constructibility reviews should be followed by a memorandum concerning special design considerations and scheduling of construction visits by design engineers during crucial stages of construction.

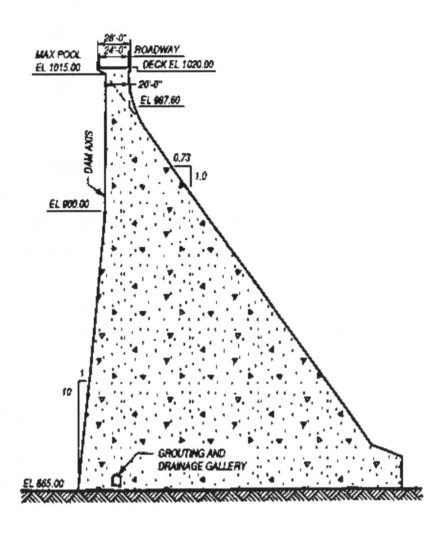

Figure 2-2

Nonoverflow section

2.1.2.5 REFINEMENT OF THE PRELIMINARY structure configuration to reflect the results of detailed site explorations, materials availability studies, laboratory testing, and numerical analysis. Once the characteristics of the foundation and concrete materials are defined, the founding levels of the dam should be set jointly by geotechnical and structural engineers, and concrete studies should be made to arrive at suitable mixes, lift thicknesses, and required crack control measures.

2.1.2.6 COFFERDAM AND DIVERSION LAYOUT, design, and sequencing requirements. Planning and design of these features will be based on economic risk and require the joint effort of hydrologists and geotechnical, construction, hydraulics, and structural engineers. Cofferdams must be set at elevations which will allow construction to proceed with a minimum of interruptions, yet be designed to allow controlled flooding during unusual events.

2.1.2.7 SIZE AND TYPE OF OUTLET WORKS AND SPILLWAY. The size and type of outlet works and spillway should be set jointly with all disciplines involved during the early stages of design. These features will significantly impact on the configuration of the dam and the sequencing of construction operations. Special hydraulic features such as water quality control structures need to be developed jointly with hydrologists and mechanical and hydraulics engineers.

2.1.2.8 MODIFICATION TO THE STRUCTURE configuration during construction due to unexpected variations in the foundation conditions. Modifications during construction are costly and should be avoided if possible by a comprehensive exploration program during the design phase. However, any changes in foundation strength or rock structure from those upon which the design is based must be fully evaluated by the structural engineer.

2.1.3 CONSTRUCTION MATERIALS. The design of concrete dams involves consideration of various construction materials during the investigations phase. An assessment is required on the availability and suitability of the materials needed to manufacture concrete qualities meeting the structural and durability requirements, and of adequate quantities for the volume of concrete in the dam and appurtenant structures. Construction materials include fine and coarse aggregates, cementitious materials, water for washing aggregates, mixing, curing of concrete, and chemical admixtures. One of the most important factors in determining the quality and economy of the concrete is the selection of suitable sources of aggregate. In the construction of concrete dams, it is important that the source have the capability of producing adequate quantitives for the economical production of mass concrete. The use of large aggregates in concrete reduces the cement content.

2.1.4 SITE SELECTION

2.1.4.1 GENERAL. During the feasibility studies, the preliminary site selection will be dependent on the project purposes within the Corps' jurisdiction. Purposes applicable to dam construction include navigation, flood damage reduction, hydroelectric power generation, fish and wildlife enhancement, water quality, water supply, and recreation. The feasibility study will establish the most suitable and economical location and type of structure. Investigations will be performed on hydrology and meteorology, relocations, foundation and site geology, construction materials, appurtenant features, environmental considerations, and diversion methods.

2.1.4.1.1 A CONCRETE DAM REQUIRES A SOUND bedrock foundation. It is important that the bedrock have adequate shear strength and bearing capacity to meet the necessary stability requirements. When the dam crosses a major fault or shear zone, special design features (joints, monolith lengths, concrete zones, etc.) should be incorporated in the design to accommodate the anticipated movement. All special features should be designed based on analytical techniques and testing simulating the fault movement. The foundation permeability and the extent and cost of foundation grouting, drainage, or other seepage and uplift control measures should be investigated. The reservoir's suitability from the aspect of possible landslides needs to be thoroughly evaluated to assure that pool fluctuations and earthquakes would not result in any mass sliding into the pool after the project is constructed.

2.1.4.1.2 THE TOPOGRAPHY IS AN important factor in the selection and location of a concrete dam and its appurtenant structures. Construction as a site with a narrow canyon profile on sound bedrock close to the surface is preferable, as this location would minimize the concrete material requirements and the associated costs.

2.1.4.1.3 THE CRITERIA SET FORTH FOR THE SPILLWAY, powerhouse, and the other project appurtenances will play an important role in site selection. The relationship and adaptability of these features to the project alignment will need evaluation along with associated costs.

2.1.4.1.4 ADDITIONAL FACTORS OF LESSER IMPORTANCE that need to be included for consideration are the relocation of existing facilities and utilities that lie within the reservoir and in the path of the dam. Included in these are railroads, powerlines, highways, towns, etc. Extensive and costly relocations should be avoided.

2.1.4.1.5 THE METHOD OR SCHEME OF DIVERTING flows around or through the damsite during construction is an important consideration to the economy of the dam. A concrete gravity dam offers major advantages and potential cost savings by providing the option of diversion through alternate construction blocks, and lowers risk and delay if overtopping should occur.

2.1.5 DETERMINING FOUNDATION STRENGTH PARAMETERS

2.1.5.1 GENERAL. Foundation strength parameters are required for stability analysis of the gravity dam section. Determination of the required parameters is made by evaluation of the most appropriate laboratory and/or in situ strength tests on representative foundation samples coupled with extensive knowledge of the subsurface geologic characteristics of a rock foundation. In situ testing is expensive and usually justified only on very large projects or when foundation problems are know to exist. In situ testing would be appropriate where more precise foundation parameters are required because rock strength is marginal or where weak layers exist and in situ properties cannot be adequately determined from laboratory testing of rock samples.

2.1.5.2 FIELD INVESTIGATION. The field investigation must be a continual process starting with the preliminary geologic review of known conditions, progressing to a detailed drilling program and sample testing program, and concluding at the end of construction with a safe and operational structure. The scope of investigation and sampling should be based on an assessment of homogeneity or complexity of geological structure. For example, the extent of the investigation could vary from quite limited (where the foundation material is strong even along the weakest potential failure planes) to quite extensive and detailed (where weak zones or seams exist). There is a certain minimum level of investigation necessary to determine that weak zones are not present in the foundation. Field investigations must also evaluate depth and severity of weathering, ground-water conditions (hydrogeology), permeability, strength, deformation characteristics, and excavatability. Undisturbed samples are required to determine the engineering properties of the foundation materials, demanding extreme care in application and sampling methods. Proper sampling is a combination of science and art; many procedures have been standardized, but alteration and adaptation of techniques are often dictated by specific field procedures.

2.1.5.3 STRENGTH TESTING. The wide variety of foundation rock properties and rock structural conditions preclude a standardized universal approach to strength testing. Decisions must be made concerning the need for in situ testing. Before any rock testing is initiated, the geotechnical engineer, geologist, and designer responsible for formulating the testing program must clearly define what the purpose of each test is and who will supervise the testing. It is imperative to use all available data, such as results from geological and geophysical studies, when selecting representative samples for testing. Laboratory testing must attempt to duplicate the actual anticipated loading situations as closely as possible. Compressive strength testing and direct shear testing are normally required to determine design values for shear strength and bearing capacity. Tensile strength testing in some cases as well as consolidation and slakeability testing may also be necessary for soft rock foundations. Rock testing procedures are discussed in the International Society of Rock Mechanics, "Suggested Methods for Determining Shear Strength," (International Society of Rock Mechanics 1974). These testing methods may be modified as appropriate to fit the circumstances of the project.

2.1.5.4 DESIGN SHEAR STRENGTHS. Shear strength values used in sliding analyses are determined from available laboratory and field tests and judgment. For preliminary designs, appropriate shear strengths for various types of rock may be obtained from numerous available references. It is important to select the types of strength tests to be performed based upon the probable mode of failure. Generally, strengths on rock discontinuities would be used for the active wedge and beneath the structure. A combination of strengths on discontinuities and/or intact rock strengths would be used for the passive wedge when included in the analysis. Strengths along preexisting shear planes (or faults) should be determined from residual shear tests, whereas the strength along other types of discontinuities must consider the strain characteristics of the various materials along the failure plane as well as the effect of asperities.

2.2 DESIGN DATA

2.2.1 CONCRETE PROPERTIES

2.2.1.1 GENERAL. The specific concrete properties used in the design of concrete gravity dams include the unit weight, compressive, tensile, and shear strengths, modulus of elasticity, creep, Poisson's ratio, coefficient of thermal expansion, thermal conductivity, specific heat, and diffusivity. These same properties are also important in the design of RCC dams. Investigations have generally indicated RCC will exhibit properties equivalent to those of conventional concrete. Values of the above properties that are to be used by the designer in the reconnaissance and feasibility design phases of the project are available in ACI 207.1R-87 or other existing sources of information on similar materials. Follow-on laboratory testing and field investigations should provide the values necessary in the final design.

2.2.1.2 STRENGTH.

2.2.1.2.1 CONCRETE STRENGTH VARIES with age; the type of cement, aggregates, and other ingredients used; and their proportions in the mixture. The main factor affecting concrete strength is the water-cement ratio. Lowering the ratio improves the strength and overall quality. Requirements for workability during placement, durability, minimum temperature rise, and overall economy may govern the concrete mix proportioning. Concrete strengths should satisfy the early load and construction requirements and the stress criteria described in Chapter 4. Design compressive strengths at later ages are useful in taking full advantage of the strength properties of the cementitious materials and lowering the cement content, resulting in lower ultimate internal temperature and lower potential cracking incidence. The age at which ultimate strength is required needs to be carefully reviewed and revised where appropriate.

2.2.1.2.2 COMPRESSIVE STRENGTHS are determined from the standard unconfined compression test excluding creep effects (American Society for Testing and Materials (ASTM) C 39, "Test Method for Compressive Strength of Cylindrical Concrete Specimens"; C 172, "Method of Sampling Freshly Mixed Concrete"; ASTM C 31, "Method of Making and Curing Concrete Test Specimens in the Field").

2.2.1.2.3 THE SHEAR STRENGTH ALONG construction joints or at the interface with the rock foundation can be determined by the linear relationship $T = C + d \tan f$ in which C is the unit cohesive strength, d is the normal stress, and $\tan f$ represents the coefficient of internal friction.

2.2.1.2.4 THE SPLITTING TENSION TEST (ASTM C 496) or the modulus of rupture test (ASTM C 78) can be used to determine the strength of intact concrete. Modulus of rupture tests provide results which are consistent with the assumed linear elastic behavior used in design. Spitting tension test results can be used; however, the designer should be aware that the results represent nonlinear performance of the sample. A more detailed discussion of these tests is presented in the ACI Journal (Raphael 1984).

2.2.1.3 ELASTIC PROPERTIES.

2.2.1.3.1 THE GRAPHICAL STRESS-STRAIN relationship for concrete subjected to a continuously increasing load is a curved line. For practical purposes, however, the modulus of elasticity is considered a constant for the range of stresses to which mass concrete is usually subjected.

2.2.1.3.2 THE MODULUS OF ELASTICITY and Poisson's ratio are determined by the ASTM C 469, "Test Method for Static Modulus of Elasticity and Poisson's Ratio of Concrete in Compression."

2.2.1.3.3 THE DEFORMATION RESPONSE of a concrete dam subjected to sustained stress can be divided into two parts. The first, elastic deformation, is the strain measured immediately after loading and is expressed as the instantaneous modulus of elasticity. The other, a gradual yielding over a long period, is the inelastic deformation or creep in concrete. Approximate values for creep are generally based on reduced values of the instantaneous modulus. When design requires more exact values, creep should be based on the standard test for creep of concrete in compression (ASTM C 512).

2.2.1.4 THERMAL PROPERTIES. Thermal studies are required for gravity dams to assess the effects of stresses induced by temperature changes in the concrete and to determine the temperature controls necessary to avoid undesirable cracking. The thermal properties required in the study include thermal conductivity, thermal diffusivity, specific heat, and the coefficient of thermal expansion.

2.2.1.5 DYNAMIC PROPERTIES.

2.2.1.5.1 THE CONCRETE PROPERTIES REQUIRED for input into a linear elastic dynamic analysis are the unit weight, Young's modulus of elasticity, and Poisson's ratio. The concrete tested should be of sufficient age to represent the ultimate concrete properties as nearly as practicable. One-year-old specimens are preferred. Usually, upper and lower bound values of Young's modulus of elasticity will be required to bracket the possibilities.

2.2.1.5.2 THE CONCRETE PROPERTIES NEEDED to evaluate the results of the dynamic analysis are the compressive and tensile strengths. The standard compression test (see paragraph 3-1b) is acceptable, even though it does not account for the rate of loading, since compression normally does not control in the dynamic analysis. The splitting tensile test or the modulus of rupture test can be used to determine the tensile strength. The static tensile strength determined by the splitting tensile test may be increased by 1.33 to be comparable to the standard modulus of rupture test.

2.2.1.5.3 THE VALUE DETERMINED BY the modulus of rupture test should be used as the tensile strength in the linear finite element analysis to determine crack initiation within the mass concrete. The tensile strength should be increased by 50 percent when used with seismic loading to account for rapid loading. When the tensile stress in existing dams exceeds 150 percent of the modulus of rupture, nonlinear analyses will be required in consultation with CECW-ED to evaluate the extent of cracking. For initial design investigations, the modulus of rupture can be calculated from the following equation:

$$f_t = 2.3 f_c'^{2/3} \qquad\qquad (3\text{-}1)$$

where

f_t = tensile strength, psi (modulus of rupture)

f_c' = compressive strength, psi

2.2.2 FOUNDATION PROPERTIES

2.2.2.1 DEFORMATION MODULUS. The deformation modulus of a foundation rock mass must be determined to evaluate the amount of expected settlement of the structure placed on it. Determination of the deformation modulus requires coordination of geologists and geotechnical and structural engineers. The deformation modulus may be determined by several different methods or approaches, but the effect of rock inhomogeneity (due partially to rock discontinuities) on foundation behavior must be accounted for. Thus, the determination of foundation compressibility should consider both elastic and inelastic (plastic) deformations. The resulting "modulus of deformation" is a lower value than the elastic modulus of intact rock. Methods for evaluating foundation moduli include in situ (static) testing (plate load tests, dilatometers, etc.); laboratory testing (uniaxial compression tests, ASTM C 3148; and pulse velocity test, ASTM C 2848); seismic field testing; empirical data (rock mass rating system, correlations with unconfined compressive strength, and tables of typical values); and back calculations using compression measurements from instruments such as a borehole extensometer. The foundation deformation modulus is best estimated or evaluated by in situ testing to more accurately account for the natural rock discontinuities. Laboratory testing on intact specimens will yield only an "upper bound" modulus value. If the foundation contains more than one rock type, different modulus values may need to be used and the foundation evaluated as a composite of two or more layers.

2.2.2.2 STATIC STRENGTH PROPERTIES. The most important foundation strength properties needed for design of concrete gravity structures are compressive strength and shear strength. Allowable bearing capacity for a structure is often selected as a fraction of the average foundation rock compressive strength to account for inherent planes of weakness along natural joints and fractures. Most rock types have adequate bearing capacity for large concrete structures unless they are soft sedimentary rock types such as mudstones, clayshale, etc.; are deeply weathered; contain large voids; or have wide fault zones. Foundation rock shear strength is given as two values: cohesion (c) and internal friction (f). Design values for shear strength are generally selected on the basis of laboratory direct shear test results. Compressive strength and tensile strength tests are often necessary to develop the appropriate failure envelope during laboratory testing. Shear strength along the foundation rock/structure interface must also be evaluated. Direct shear strength laboratory tests on composite grout/rock samples are recommended to assess the foundation rock/structure interface shear strength. It is particularly important to determine strength properties of discontinuities and the weakest foundation materials (i.e., soft zones in shears or faults), as these will generally control foundation behavior.

2.2.2.3 DYNAMIC STRENGTH PROPERTIES.

2.2.2.3.1 WHEN THE FOUNDATION IS INCLUDED in the seismic analysis, elastic moduli and Poisson's ratios for the foundation materials are required for the analysis. If the foundation mass is modeled, the rock densities are also required.

2.2.2.3.2 DETERMINING THE ELASTIC moduli for a rock foundation should include several different methods or approaches.

2.2.2.3.3 POISSON'S RATIOS SHOULD be determined from uniaxial compression tests, pulse velocity tests, seismic field tests, or empirical data. Poisson's ratio does not vary widely for rock materials.

2.2.2.3.4 THE RATE OF LOADING EFFECT on the foundation modulus is considered to be insignificant relative to the other uncertainties involved in determining rock foundation properties, and it is not measured.

2.2.2.3.5 TO ACCOUNT FOR THE UNCERTAINTIES, a lower and upper bound for the foundation modulus should be used for each rock type modeled in the structural analysis.

2.3 LOADS

2.3.1 GENERAL. In the design of concrete gravity dams, it is essential to determine the loads required in the stability and stress analysis. The following forces may affect the design:

- Dead load.
- Headwater and tailwater pressures.
- Uplift.
- Temperature.
- Earth and silt pressures.
- Ice pressure.
- Earthquake forces.
- Wind pressure.
- Subatmospheric pressure.
- Wave pressure.
- Reaction of foundation.

2.3.2 DEAD LOAD. The unit weight of concrete generally should be assumed to be 150 pounds per cubic foot until an exact unit weight is determined from the concrete materials investigation. In the computation of the dead load, relatively small voids such as galleries are normally not deducted except in low dams, where such voids could create an appreciable effect upon the stability of the structure. The dead loads considered should include the weight of concrete, superimposed backfill, and appurtenances such as gates and bridges.

2.3.3 HEADWATER AND TAILWATER.

2.3.3.1 GENERAL. The headwater and tailwater loadings acting on a dam are determined from the hydrology, meteorology, and reservoir regulation studies. The frequency of the different pool levels will need to be determined to assess which will be used in the various load conditions analyzed in the design.

2.3.3.2 HEADWATER.

2.3.3.2.1 THE HYDROSTATIC PRESSURE against the dam is a function of the water depth times the unit weight of water. The unit weight should be taken at 62.5 pounds per cubic foot, even though the weight varies slightly with temperature.

2.3.3.2.2 IN SOME CASES THE JET of water on an overflow section will exert pressure on the structure. Normally such forces should be neglected in the stability analysis except as noted.

2.3.3.3 TAILWATER.

2.3.3.3.1 FOR DESIGN OF NONOVERFLOW SECTIONS. The hydrostatic pressure on the downstream face of a nonoverflow section due to tailwater shall be determined using the full tailwater depth.

2.3.3.3.2 FOR DESIGN OF OVERFLOW SECTIONS. Tailwater pressure must be adjusted for retrogression when the flow conditions result in a significant hydraulic jump in the downstream channel, i.e. spillway flow plunging deep into tailwater. The forces acting on the downstream face of overflow sections due to tailwater may fluctuate significantly as energy is dissipated in the stilling basin. Therefore, these forces must be conservatively estimated when used as a stabilizing force in a stability analysis. Studies have shown that the influence of tailwater retrogression can reduce the effective tailwater depth used to calculate pressures and forces to as little as 60 percent of the full tailwater depth. The amount of reduction in the effective depth used to determine tailwater forces is a function of the degree of submergence of the crest of the structure and the backwater conditions in the downstream channel.

2.3.3.3.3 TAILWATER SUBMERGENCE. When tailwater conditions significantly reduce or eliminate the hydraulic jump in the spillway basin, tailwater retrogression can be neglected and 100 percent of the tailwater depth can be used to determine tailwater forces.

2.3.3.3.4 UPLIFT DUE TO TAILWATER. Full tailwater depth will be used to calculate uplift pressures at the toe of the structure in all cases, regardless of the overflow conditions.

2.3.4 UPLIFT. Uplift pressure resulting from headwater and tailwater exists through cross sections within the dam, at the interface between the dam and the foundation, and within the foundation below the base. This pressure is present within the cracks, pores, joints, and seams in the concrete and foundation material. Uplift pressure is an active force that must be included in the stability and stress analysis to ensure structural adequacy. These pressures vary with time and are related to boundary conditions and the permeability of the material. Uplift pressures are assumed to be unchanged by earthquake loads.

2.3.4.1 ALONG THE BASE.

2.3.4.1.1 GENERAL. The uplift pressure will be considered as acting over 100 percent of the base. A hydraulic gradient between the upper and lower pool is developed between the heel and toe of the dam. The pressure distribution along the base and in the foundation is dependent on the effectiveness of drains and grout curtain, where applicable, and geologic features such as rock permeability, seams, jointing, and faulting. The uplift pressure at any point under the structure will be tailwater pressure plus the pressure measured as an ordinate from tailwater to the hydraulic gradient between upper and lower pool.

2.3.4.1.2 WITHOUT DRAINS. Where there have not been any provisions provided for uplift reduction, the hydraulic gradient will be assumed to vary, as a straight line, from headwater at the heel to zero or tailwater at the toe. Determination of uplift, at any point on or below the foundation, is demonstrated in Figure 3-1.

2.3.4.1.3 WITH DRAINS. Uplift pressures at the base or below the foundation can be reduced by installing foundation drains. The effectiveness of the drainage system will depend on depth, size, and spacing of the drains; the character of the foundation; and the facility with which the drains can be maintained. This effectiveness will be assumed to vary from 25 to 50 percent, and the design memoranda should contain supporting data for the assumption used. If foundation testing and flow analysis provide supporting justification, the drain effectiveness can be increased to a maximum of 67 percent with approval. This criterion deviation will depend on the pool level operation plan instrumentation to verify and evaluate uplift assumptions and an adequate drain maintenance program. Along the base, the uplift pressure will vary linearly from the undrained pressure head at the heel, to the reduced pressure head at the line of drains, to the undrained pressure head at the toe, as shown in Figure 3-2. Where the line of drains intersects the foundation within a distance of 5 percent of the reservoir depth from the upstream face, the uplift may be assumed to vary as a single straight line, which would be the case if the drains were exactly at the heel. This condition is illustrated in Figure 3-3. If the drainage gallery is above tailwater elevation, the pressure of the line of drains should be determined as though the tailwater level is equal to the gallery elevation.

2.3.4.1.4 GROUT CURTAIN. For drainage to be controlled economically, retarding of flow to the drains from the upstream head is mandatory. This may be accomplished by a zone of grouting (curtain) or by the natural imperviousness of the foundation. A grouted zone (curtain) should be used wherever the foundation is amenable to grouting. Grout holes shall be oriented to intercept the maximum number of rock fractures to maximize its effectiveness. Under average conditions, the depth of the grout zone should be two-thirds to three-fourths of the headwater-tailwater differential and should be supplemented by foundation drain holes with a depth of at least two-thirds that of the grout zone (curtain).

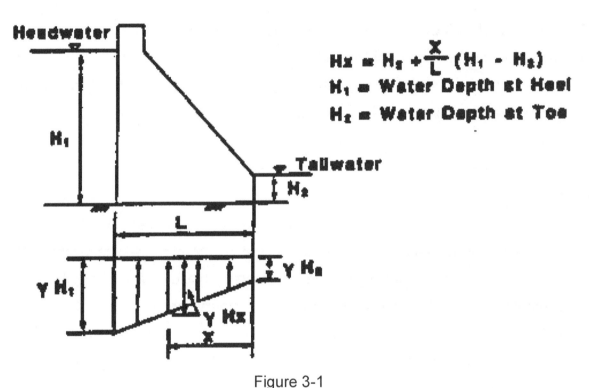

Figure 3-1

Uplift distribution without foundation drainage

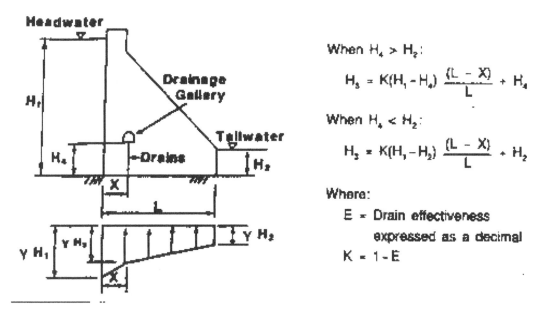

When $H_4 > H_2$:

$$H_3 = K(H_1 - H_4) \frac{(L - X)}{L} + H_4$$

When $H_4 < H_2$:

$$H_3 = K(H_1 - H_2) \frac{(L - X)}{L} + H_2$$

Where:

E = Drain effectiveness expressed as a decimal

K = 1 - E

Figure 3-2

Uplift distribution with drainage gallery

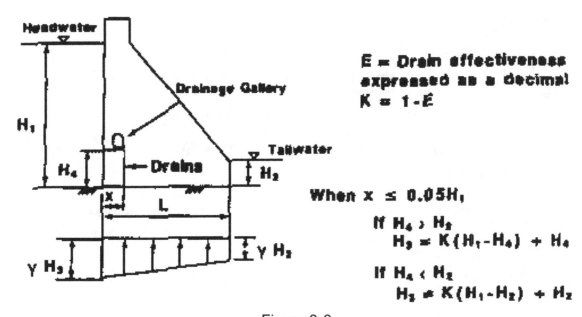

E = Drain effectiveness expressed as a decimal
$K = 1-E$

When $x \leq 0.05H_1$

If $H_4 > H_2$
$$H_3 = K(H_1-H_4) + H_4$$

If $H_4 < H_2$
$$H_3 = K(H_1-H_2) + H_2$$

Figure 3-3

Uplift distribution with foundation drains near upstream face

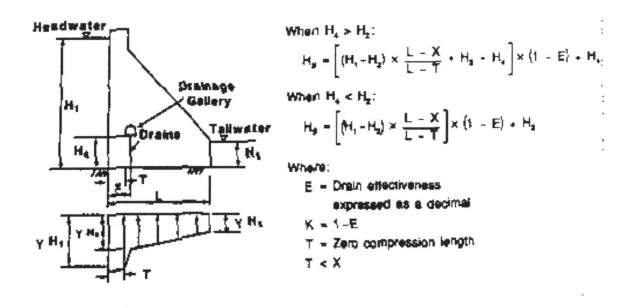

When $H_4 > H_2$:
$$H_3 = \left[(H_1-H_2) \times \frac{L-X}{L-T} + H_3 - H_4 \right] \times (1-E) + H_4$$

When $H_4 < H_2$:
$$H_3 = \left[(H_1-H_2) \times \frac{L-X}{L-T} \right] \times (1-E) + H_2$$

Where:

E = Drain effectiveness expressed as a decimal

$K = 1-E$

T = Zero compression length

$T < X$

Figure 3-4

Uplift distribution cracked base with drainage, zero compression zone not extending beyond drains

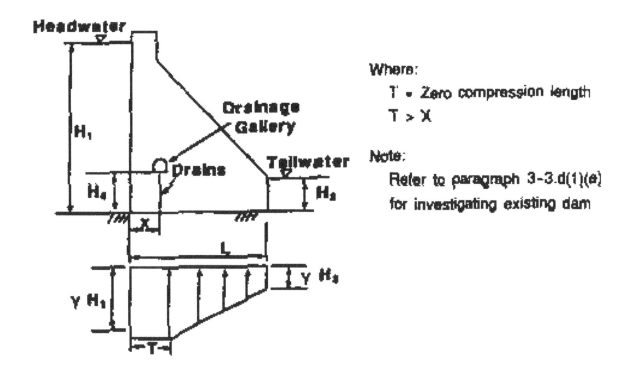

Figure 3-5

Uplift distribution cracked base with drainage, zero compression zone extending beyond drains

Where the foundation is sufficiently impervious to retard the flow and where grouting would be impractical, an artificial cutoff is usually unnecessary. Drains, however, should be provided to relieve the uplift pressures that would build up over a period of time in a relatively impervious medium. In a relatively impervious foundation, drain spacing will be closer than in a relatively permeable foundation.

2.3.4.1.5 ZERO COMPRESSION ZONES. Uplift on any portion of any foundation plane not in compression shall be 100 percent of the hydrostatic head of the adjacent face, except where tension is the result of instantaneous loading resulting from earthquake forces. When the zero compression zone does not extend beyond the location of the drains, the uplift will be as shown in Figure 3-4. For the condition where the zero compression zone extends beyond the drains, drain effectiveness shall not be considered. This uplift condition is shown in Figure 3-5. When an existing dam is being investigated, the design office should submit a request for a deviation if expensive remedial measures are required to satisfy this loading assumption.

2.3.4.2 WITHIN DAM.

2.3.4.2.1 CONVENTIONAL CONCRETE. Uplift within the body of a conventional concrete-gravity dam shall be assumed to vary linearly from 50 percent of maximum headwater at the upstream face to 50 percent of tailwater, or zero, as the case may be, at the downstream face. This simplification is based on the relative impermeability of intact concrete which precludes the buildup of internal pore pressures. Cracking at the upstream face of an existing dam or weak horizontal construction joints in the body of the dam may affect this assumption. In these cases, uplift along these discontinuities should be determined as described.

2.3.4.2.2 RCC CONCRETE. The determination of the percent uplift will depend on the mix permeability, lift joint treatment, the placements, techniques specified for minimizing segregation within the mixture, compaction methods, and the treatment for watertightness at the upstream and downstream faces. A porous upstream face and lift joints in conjunction with an impermeable downstream face may result in a pressure gradient through a cross section of the dam considerably greater than that outlined above for conventional concrete. Construction of a test section during the design phase shall be used as a means of determining the permeability and, thereby, the exact uplift force for use by the designer.

2.3.4.2.3 IN THE FOUNDATION. Sliding stability must be considered along seams or faults in the foundation. Material in these seams or faults may be gouge or other heavily sheared rock, or highly altered rock with low shear resistance. In some cases, the material in these zones is porous and subject to high uplift pressures upon reservoir filling. Before stability analyses are performed, engineering geologists must provide information regarding potential failure planes within the foundation. This includes the location of zones of low shear resistance, the strength of material within these zones, assumed potential failure planes, and maximum uplift pressures that can develop along the failure planes. Although there are no prescribed uplift pressure diagrams that will cover all foundation failure plane conditions, some of the most common assumptions made are illustrated in Figures 3-6 and 3-7. These diagrams assume a uniform head loss along the failure surface from point "A" to tailwater, and assume that the foundation drains penetrate the failure plane and are effective in reducing uplift on that plane. If there is concern that the drains may be ineffective or partially effective in reducing uplift along the failure plane, then uplift distribution as represented by the dashed line in Figures 3-6 and 3-7 should be considered for stability computations. Dangerous uplift pressures can develop along foundation seams or faults if the material in the seams or faults is pervious and the pervious zone is intercepted by the base of the dam or by an impervious fault. These conditions are described in Casagrande (1961) and illustrated by Figures 3-8 and 3-9. Every effort is made to grout pervious zones within the foundation prior to constructing the dam. In cases where grouting is impractical or ineffective, uplift pressure can be reduced to safe levels through proper drainage of the pervious zone. However, in those circumstances where the drains do not penetrate the pervious zone or where drainage is only partially effective, the uplift conditions shown in Figures 3-8 and 3-9 are possible.

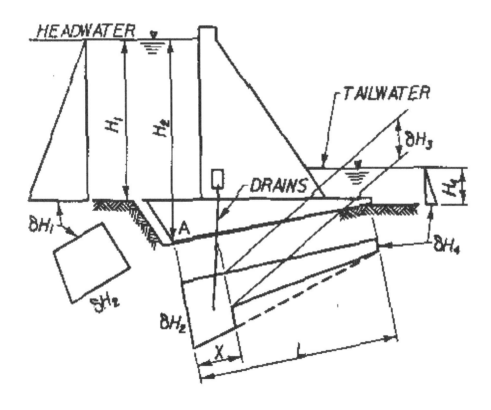

Figure 3-6.

Uplift pressure diagram. Dashed line represents uplift distribution to be considered for stability computations

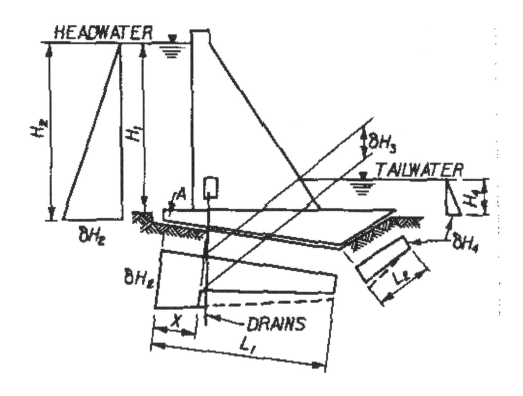

Figure 3-7

Dashed line in uplift pressure diagram represents uplift distribution to be considered for

stability computations

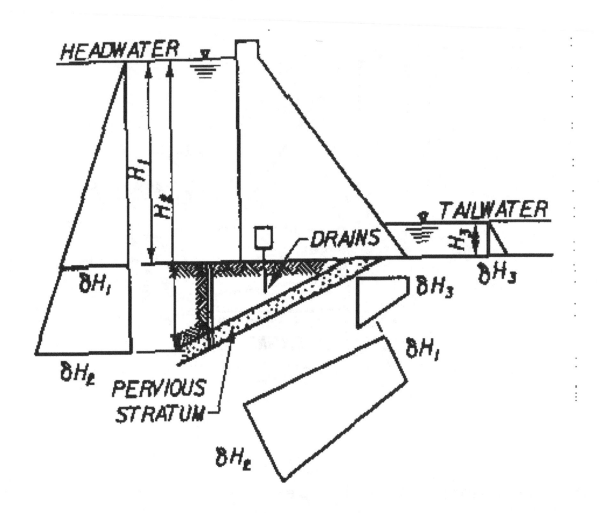

Figure 3-8

Development of dangerous uplift pressure along foundation seams or faults

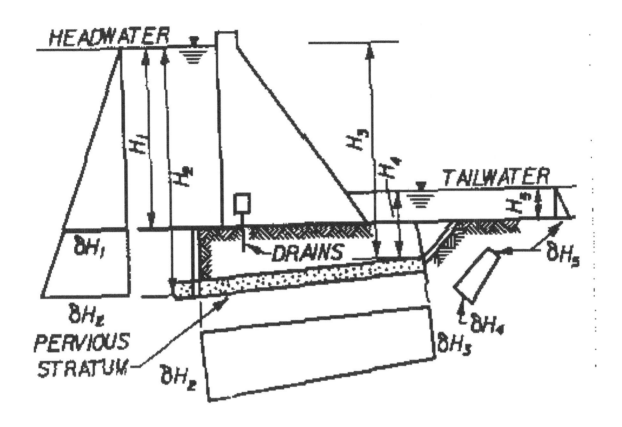

Figure 3-9

Effect along foundation seams or faults if material is pervious and pervious zone is intercepted by base of dam or by impervious fault

2.3.5 TEMPERATURE.

2.3.5.1 A MAJOR CONCERN IN CONCRETE DAM CONSTRUCTION is the control of cracking resulting from temperature change. During the hydration process, the temperature rises because of the hydration of cement. The edges of the monolith release heat faster than the interior; thus the core will be in compression and the edges in tension. When the strength of the concrete is exceeded, cracks will appear on the surface. When the monolith starts cooling, the contraction of the concrete is restrained by the foundation or concrete layers that have already cooled and hardened. Again, if this tensile strain exceeds the capacity of the concrete, cracks will propagate completely through the monolith. The principal concerns with cracking are that it affects the watertightness, durability, appearance, and stresses throughout the structure and may lead to undesirable crack propagation that impairs structural safety.

2.3.5.2 IN CONVENTIONAL CONCRETE DAMS, various techniques have been developed to reduce the potential for temperature cracking (ACI 224R-80). Besides contraction joints, these include temperature control measures during construction, cements for limiting heat of hydration, and mix designs with increased tensile strain capacity.

2.3.5.2 IF AN RCC DAM IS BUILT WITHOUT VERTICAL contraction joints, additional internal restraints are present. Thermal loads combined with dead loads and reservoir loads could create tensile strains in the longitudinal axis sufficient to cause transverse cracks within the dam.

2.3.6 EARTH AND SILT. Earth pressures against the dam may occur where backfill is deposited in the foundation excavation and where embankment fills abut and wrap around concrete monoliths. The fill material may or may not be submerged. Silt pressures are considered in the design if suspended sediment measurements indicate that such pressures are expected. Whether the lateral earth pressures will be in an active or an at-rest state is determined by the resulting structure lateral deformation.

2.3.7 ICE PRESSURE. Ice pressure is of less importance in the design of a gravity dam than in the design of gates and other appurtenances for the dam. Ice damage to the gates is quite common while there is no known instance of any serious ice damage occurring to the dam. For the purpose of design, a unit pressure of not more than 5,000 pounds per square foot should be applied to the contact surface of the structure. For dams in this country, the ice thickness normally will not exceed 2 feet. Climatology studies will determine whether an allowance for ice pressure is appropriate.

2.3.8 EARTHQUAKE.

2.3.8.1 GENERAL.

2.3.8.1.1 THE EARTHQUAKE LOADINGS USED in the design of concrete gravity dams are based on design earthquakes and site-specific motions determined from seismological evaluation. As a minimum, a seismological evaluation should be performed on all projects located in seismic zones 2, 3, and 4.

2.3.8.1.2 THE SEISMIC COEFFICIENT METHOD OF ANALYSIS should be used in determining the resultant location and sliding stability of dams.. In strong seismicity areas, a dynamic seismic analysis is required for the internal stress analysis.

2.3.8.1.3 EARTHQUAKE LOADINGS SHOULD BE checked for horizontal earthquake acceleration and, if included in the stress analysis, vertical acceleration. While an earthquake acceleration might take place in any direction, the analysis should be performed for the most unfavorable direction.

2.3.8.2 SEISMIC COEFFICIENT. The seismic coefficient method of analysis is commonly known as the pseudostatic analysis. Earthquake loading is treated as an inertial force applied statically to the structure. The loadings are of two types: inertia force due to the horizontal acceleration of the dam and hydrodynamic forces resulting from the reaction of the reservoir water against the dam (see Figure 3-10). The magnitude of the inertia forces is computed by the principle of mass times the earthquake acceleration. Inertia forces are assumed to act through the center of gravity of the section or element. The seismic coefficient is a ratio of the earthquake acceleration to gravity; it is a dimensionless unit, and in no case can it be related directly to acceleration from a strong motion instrument. The coefficients used are considered to be the same for the foundation and are uniform for the total height of the dam.

2.3.8.2.1 INERTIA OF CONCRETE FOR HORIZONTAL EARTHQUAKE
ACCELERATION. The force required to accelerate the concrete mass of the dam is determined from the equation:

$$Pe_x = Ma_x = \frac{W}{g}\, \alpha g = W\alpha \qquad (3\text{-}2)$$

where

Pe_x = horizontal earthquake force

M = mass of dam

a_x = horizontal earthquake acceleration = g

W = weight of dam

g = acceleration of gravity

α = seismic coefficient

2.3.8.2.2 INERTIA OF RESERVOIR FOR HORIZONTAL EARTHQUAKE
ACCELERATION. The inertia of the reservoir water induces an increased or decreased pressure on the dam concurrently with concrete inertia forces. Figure 3-10 shows the pressures and forces due to earthquake by the seismic coefficient method. This force may be computed by means of the Westergaard formula using the parabolic approximation:

$$Pew = \frac{2}{3} Ce \ (\alpha) \ y \ (\sqrt{hy}) \qquad (3\text{-}3)$$

where

Pew = additional total water load down to depth y (kips)

Ce = factor depending principally on depth of water and the earthquake vibration period, t_e, in seconds

h = total height of reservoir (feet)

Westergaard's approximate equation for Ce, which is sufficiently accurate for all usual conditions, in pound-second feet units is:

$$Ce = \frac{51}{\sqrt{1 - 0.72 \left(\dfrac{h}{1.000 \ t_e}\right)^2}} \qquad (3\text{-}4)$$

where t_e is the period of vibration.

2.3.8.3 DYNAMIC LOADS. The first step in determining wind setup are usually important factors in determining earthquake induced loading involves a geological and seismological investigation of the damsite. The objectives of the investigation are to establish the controlling maximum credible earthquake (MCE) and operating basis earthquake (OBE) and the corresponding ground motions for each, and to assess the possibility of earthquake produced foundation dislocation at the site.

The ground motions are characterized by the site-dependent design response spectra and, when necessary in the analysis, acceleration-time records. The dynamic method of analysis determines the structural response using either a response spectrum or acceleration-time records. The dynamic method of analysis determines the structural response using either a response spectrum or acceleration-time records for the dynamic input.

2.3.8.3.1 SITE-SPECIFIC DESIGN RESPONSE spectrum is a plot of the maximum values of acceleration, velocity, and/or displacement of single-degree-of-freedom systems subjected to an earthquake. The maximum response values are expressed as a function of natural period for a given damping value. The site-specific response spectra is developed statistically from response spectra of strong motion records of earthquakes that have similar source and propagation path properties or from the controlling earthquakes and that were recorded on a similar foundation.

2.3.8.3.2 ACCELERATION--TIME RECORDS. Accelerograms, used for input for the dynamic analysis, provide a simulation of the actual response of the structure to the given seismic ground motion through time. The acceleration-time records should be compatible with the design response spectrum.

2.3.9 SUBATMOSPHERIC PRESSURE. At the hydrostatic head for which the crest profile is designed, the theoretical pressures along the downstream face of an ogee spillway crest approach atmospheric pressure. For heads higher than the design head, subatmospheric pressures are obtained along the spillway. When spillway profiles are designed for heads appreciably less than the probable maximum that could be obtained, the magnitude of these pressures should be determined and considered in the stability analysis.

2.3.10 WAVE PRESSURE. While wave pressures are of more importance in their effect upon gates and appurtenances, they may, in some instances, have an appreciable effect upon the dam proper. The height of waves, runup, and wind setup are usually important factors in determining the required freeboard of any dam. Wave dimensions and forces depend on the extent of water surface or fetch, the wind velocity and duration, and other factors.

2.3.11 REACTION OF FOUNDATIONS. In general, the resultant of all horizontal and vertical forces including uplift must be balanced by an equal and opposite reaction at the foundation consisting of the normal and tangential components. For the dam to be in static equilibrium, the location of this reaction is such that the summation of forces and moments are equal to zero. The distribution of the normal component is assumed as linear, with a knowledge that the elastic and plastic properties of the foundation material and concrete affect the actual distribution.

2.3.11.1 THE PROBLEM OF DETERMINING THE actual distribution is complicated by the tangential reaction, internal stress relations, and other theoretical considerations. Moreover, variations of foundation materials with depth, cracks, and fissures that interrupt the tensile and shearing resistance of the foundation also make the problem more complex.

2.3.11.2 FOR OVERFLOW SECTIONS, the base width is generally determined by projecting the spillway slope to the foundation line, and all concrete downstream from this line is disregarded. If a vertical longitudinal joint is not provided at this point, the mass of concrete downstream from the theoretical toe must be investigated for internal stresses.

2.3.11.3 THE UNIT UPLIFT PRESSURE SHOULD BE ADDED to the computed unit foundation reaction to determine the maximum unit foundation pressure at any point.

2.3.11.4 INTERNAL STRESSES AND FOUNDATION PRESSURES should be computed with and without uplift to determine the maximum condition.

CHAPTER 3: COFFER DAMS

3.1 INTRODUCTION. This discussion is intended to provide guidance for the design of these structures. Geotechnical considerations, analysis and design procedures, construction considerations, and instrumentation are discussed. Special emphasis is placed on all aspects of cellular coffer dams, such as planning, hydraulic considerations, and layout.

3.1.1 DEFINITIONS. A list of symbols with their definitions are indicated elsewhere.

3.1.2 TYPES AND CAPABILITIES.

3.1.2.1 USES. Sheet pile cellular structures are used in a variety of ways, one of the principal uses being for cofferdams.

3.1.2.1.1 COFFERDAMS. When an excavation is in a large area overlain by water, such as a river or lake, cellular cofferdams are widely used to form a water barrier, thus providing a dry work area. Cellular structures are economical for this type of construction since stability is achieved relatively inexpensively by using the soil cell fill for mass. Ring or membrane tensile stresses are used in the interlocking steel sheet piling to effect a soil container. The same sheet piling may be pulled and reused unless it has been damaged from driving into boulders or dense soil deposits. Driving damage is not usually a major problem since it is rarely necessary to drive the piling to great depths in soil.

3.1.2.1.2 RETAINING WALLS AND OTHER STRUCTURES. Sheet pile cellular structures are also used for retaining walls; fixed crest dams and weirs; lock, guide, guard, and approach walls; and substructures for concrete gravity superstructures. Each of these structures can be built in the wet, thus eliminating the need for dewatering. When used as substructures, the cells can be relied upon to support moderate loads from concrete superstructures. Varying designs have been used to support the concrete loads, either on the fill or on the piling. When danger of rupture from large impact exists, the cells should be filled with tremie concrete. In the case of concrete guard walls for navigation locks, bearing piles have been driven within the cells to provide added lateral support for the load with the cell fill. Precautions must be taken to prevent loss of the fill which could result in instability of the pile-supported structure. Bearing piles driven within the cells should never be used to support structures subjected to lateral loads.

3.1.2.2 TYPES. There are three general types of cellular structures, each depending on the weight and strength of the fill for its stability. For typical arrangement of the three types of cells, see Figure 1-1.

3.1.2.2.1 CIRCULAR CELLS. This type consists of a series of complete circular cells connected by shorter arcs. These arcs generally intercept the cells at a point making an angle of 30 or 45 degrees with the longitudinal axis of the cofferdam. The primary advantages of circular cells are that each cell is independent of the adjacent cells, it can be filled as soon as it is constructed, and it is easier to form by means of templates.

3.1.2.2.2 DIAPHRAGM CELLS. These cells are comprised of a series of circular arcs connected by 120-degree intersection pieces or crosswalls (diaphragms). The radius of the arc is often made equal to the cell width so that there is equal tension in the arc and the diaphragm. The diaphragm cell will distort excessively unless the various units are filled essentially simultaneously with not over 5 feet of differential soil height in adjacent cells. Diaphragm cells are not independently stable and failure of one cell could lead to failure of the entire cofferdam.

3.1.2.2.3 CLOVERLEAF CELLS. This type of cell consists of four arc walls, within each of the four quadrants, formed by two straight diaphragm walls normal to each other, and intersecting at the center of the cell. Adjacent cells are connected by short arc walls and are proportioned so that the intersection of arcs and diaphragms forms three angles of 120 degrees. The cloverleaf is used when a large cell width is required for stability against a high head of water. This type has the advantage of stability over the individual cells, but has the disadvantage of being difficult to form by means of templates. An additional drawback is the requirement that the separate compartments be filled so that differential soil height does not exceed 5 feet.

3.1.3 DESIGN PHILOSOPHY.

3.1.3.1 CELLULAR COFFERDAMS, in most instances, serve as a high head or moderately high head dam for extended periods of time, protecting personnel, equipment, and completed work and maintaining the navigation pool. Planning, design, and construction of these structures must be accomplished by the same procedures and with the same high level of engineering competency as those required for permanent features of the work. Adequate foundation investigation and laboratory testing must be performed to determine soil and foundation parameters affecting the integrity of the cofferdam. Hydraulic and hydrologic design studies must be conducted to determine the most economical layout.

3.1.3.2 THE ANALYTICAL DESIGN of cellular cofferdams requires close coordination between the structural engineer and the geotechnical engineer. Close coordination is necessary, not only for the soil and foundation investigations noted above, but also to ensure that design strengths are applied correctly.

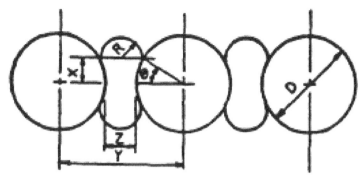

a. Plan circular cell

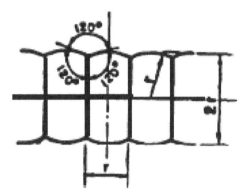

b. Plan arc and diaphragm cell

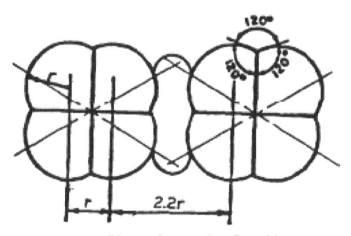

c. Plan clover leaf cell

Figure 1-1

Typical arrangement of circular, diaphragm,

and cloverleaf cells

and that assumptions used in the design, such as the saturation level within the cell fill, are realistic. Though cofferdams are often referred to as temporary structures, their importance, as explained above, requires that they be designed for the same factors of safety as those required for permanent structures.

3.1.3.3 TO ENSURE COMPLIANCE WITH ALL DESIGN requirements and conformity with safe construction practices, the cellular cofferdam construction should be subjected to intensive inspection by both construction and design personnel. Periodic and timely visits by design personnel to the construction site are required to ensure that: site conditions throughout the construction period are in conformance with design assumptions, contract plans, and specifications; project personnel are given assistance in adapting the plans and specifications to actual site conditions as they are revealed during construction; and any engineering problems not fully assessed in the original design are observed and evaluated, and appropriate action is taken. Coordination between construction and design should be sufficient to enable design personnel to respond in a timely manner when changed field conditions require modifications of design. (4) Not all features of a construction cofferdam will be designed by the Government. In particular, the design of the dewatering system, generally, will be the responsibility of the contractor so that the contractor can utilize his particular expertise and equipment. However, the dewatering system must be designed to be consistent with the assumptions made in the cofferdam design, including the elevation of the saturation level within the cell fill and the rate of dewatering. To achieve this, the requirements for the dewatering system must be explicitly stated in the contract specifications, and the contractor's design must be carefully reviewed by the cofferdam designer to ensure that the intent and provisions of the specifications are met.

3.2 PLANNING, LAYOUT, AND ELEMENTS OF COFFERDAMS

3.2.1 AREAS OF CONSIDERATION. For a construction cofferdam to be functional, it must provide a work area free from frequent flooding and of sufficient size to allow for necessary construction activities. These two objectives are dependent on several factors and are interrelated as described below.

3.2.1.1 HEIGHT OF PROTECTION. The top of the cofferdam should be established so that a dry working area can be economically maintained. To establish an economical top elevation for cofferdam and flooding frequency, stage occurrence and duration data covering the practical range of cofferdam heights must be evaluated, taking into account the required life of the cofferdam. Factors which affect the practical range of cofferdam heights include: effects on channel width to accommodate streamflow and navigation where required; increased flow velocity during high river stages and the resultant scour; effects on completed adjacent structures to which the cofferdam joins (the "tie-in"), i.e., these structures must be designed to resist pools to top of cofferdam; and practical limitations on the size of cell due to interlock stresses and sliding stability. By comparing these factors with the effects of lost time and dewatering and cleanup costs resulting from flooding, an economical top elevation of cofferdam can be established.

3.2.1.2 AREA OF ENCLOSURE. The area enclosed by the cofferdam should be minimized for reasons of economy but should be consistent with construction requirements. The area often will be limited by the need to maintain a minimum channel width and control scour and to minimize those portions of completed structures affected by the tie-in. The minimum area provided must be sufficient to accommodate berms, access roads, an internal drainage system, and a reasonable working area. Minimum functional area requirements should be established in coordination with construction personnel.

3.2.1.3 STAGING. When constructing a cofferdam in a river, the flow must continue to be passed and navigation maintained. Therefore, the construction must be accomplished in stages, passing the water temporarily through the completed work, and making provisions for a navigable channel. The number of stages should be limited because of the costs and time delays associated with the removal of the cells in a completed stage and the construction of the cells for the following stage. However, the number of stages must be consistent with the need to minimize streamflow velocities and their associated effects on scour, streambank erosion, upstream flooding, and navigation. When developing the layout for a multistage cofferdam, special attention should be given to maximizing the number of items common to each stage of the cofferdam. With proper planning some cells may be used for two subsequent stages. In those cells that will be common to more than one stage, the connecting tees or wyes that are to be utilized in a future stage must be located with care.

3.2..1.4 HYDRAULIC MODEL STUDIES. Hydraulic model studies are often necessary to develop the optimum cofferdam layout, particularly for a multistage cofferdam. From these studies, currents which might adversely affect navigation, the potential for scour, and various remedies can be determined.

3.2.2 ELEMENTS OF COFFERDAMS.

3.2.2.1 SCOUR PROTECTION. Flowing water can seriously damage a cofferdam cell by undermining and the subsequent loss of cell fill. Still further, scour caused by flowing water can lead to damage by increased underseepage and increased interlock stresses. The potential for this type of damage is dependent upon the velocity of the water, the eddies produced, and the erodibility of the foundation material. Damage can be prevented by protecting the foundation outside of the cell with riprap or by driving the piling to a sufficient depth beneath the anticipated scour. Deflectors designed to streamline flow are effective in minimizing scour along the face of the cofferdam. These deflectors consist of a curved sheet pile wall, with appropriate bracing, extending into the river from the outer upstream and downstream corners of the cofferdam. Figure 2-1 shows a schematic deflector layout. As noted previously, hydraulic model studies are useful in predicting the potential for scour and in developing the most efficient deflector geometry.

3.2.2.2 BERMS. A soil berm may be constructed inside the cells to provide additional sliding and overturning resistance. The berm will also serve to lengthen the seepage path and decrease the upward seepage gradients on the interior of the cells. However, a berm will require a larger cofferdam enclosure and an increase in the overall length of the cofferdam, and will increase construction and maintenance costs. Also, an inside berm inhibits inspection of the inside piling for driving damage and makes cell drainage maintenance more difficult. It is generally advisable, therefore, to increase the diameter of the cells instead of constructing a berm to achieve stability since the amount of piling per lineal foot of cofferdam is, essentially, independent of the diameter of the cells. Any increase in the diameter of the cells must be within the limitations of the maximum allowable interlock stress. In order for a berm to function as designed, the berm must be constantly maintained and protected against erosion and the degree of saturation must be consistent with design assumptions.

3.2.2.3 FLOODING FACILITIES. Flooding of a cofferdam by overtopping can cause serious damage to the cofferdam, perhaps even failure. An overflow can wash fill material from the cells and erode berm material. Before overtopping occurs, the cofferdam should therefore be filled with water in a controlled manner by providing floodgates or sluiceways. The floodgates or sluiceways can also be used to facilitate removal of the cofferdam by flooding. Floodgates are constructed in one or more of the connecting arcs by cutting the piling at the appropriate elevation and capping the arc with concrete to provide a nonerodible surface. Control is maintained by installing timber needle beams that can be removed when flooding is desired. Figure 2-2 shows a typical floodgate arrangement. Sluiceways consist of a steel pipe placed through a hole cut in the piling of a connecting arc. Flow is controlled by means of

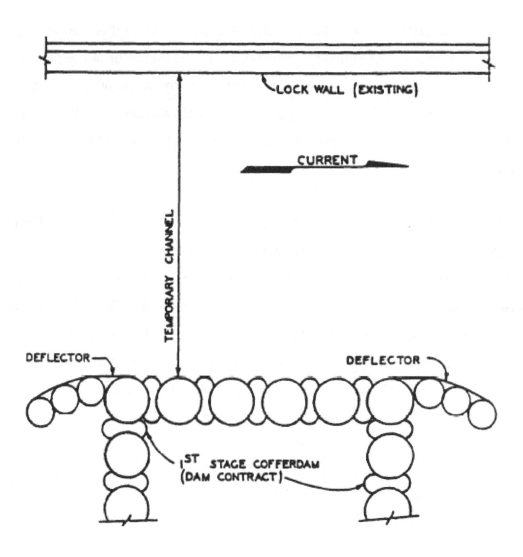

Figure 2-1

Schematic deflector layout

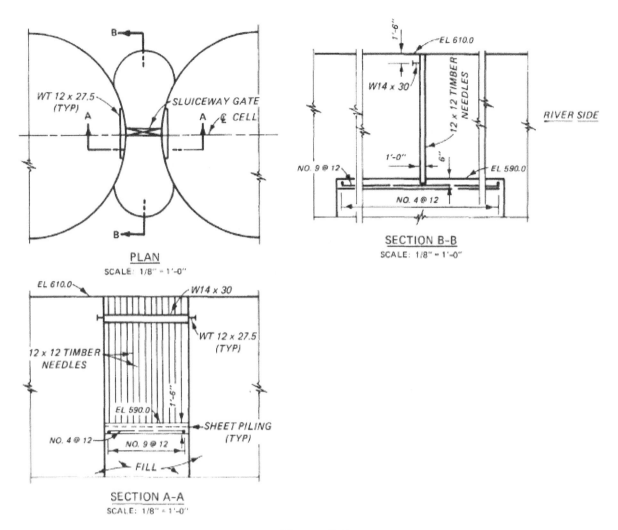

Figure 2-2

Typical floodgate arrangement

a slidegate or valve operated from the top of the cell. The size, number, and invert elevations of the flooding facilities are determined by comparing the volume to be filled with the probable rate of rise of the river. These elements must be sized so that it is possible to flood the cofferdam before it is overtopped. For either system, the adjacent berm must be protected against the flows by means of a concrete flume, a splash pad, or heavy stone.

3.2.2.4 TIE-INS. Cofferdams often must be connected to land and to completed portions of the structure.

3.2.2.4.1 TIE-IN TO LAND. Where the cofferdam joins a steep sloping shoreline, the first cell is usually located at a point where the top of the cell intersects the sloping bank. A single wall of steel sheet piling connected to the cell and extending landward to form a cutoff wall is often required to increase the seepage path and reduce the velocity of the water. The length of the cutoff wall will depend upon the permeability of the overburden. The wall should be driven to rock or to a depth in overburden as required by the permeability of the overburden. The depths of overburden into which the cells and cutoff wall are driven should be limited to 30 feet in order to prevent driving the piling out of interlock. Otherwise, it will be necessary to excavate a portion of the overburden prior to driving the piling. Where the cofferdam abuts a wide floodplain which is lower than the top of the cofferdam cells, protection from floodwaters along the land side can be obtained by constructing an earth dike with a steel sheet pile cutoff wall. The dike may join the upstream and downstream arms of the cofferdam or extend from the end of the cofferdam into the bank, depending upon the type of overburden, location of rock, and extent of the floodplain.

3.2.2.4.2 TIE-IN TO EXISTING STRUCTURES. Tie-ins to a vertical face of a structure can be accomplished by embedding a section of sheet piling in the structure to which a tee pile in the cell can be connected. Another method of tie-in to a vertical face consists of wedging a shaped-to-fit timber beam between the cell and the vertical face. As the cofferdam enclosure is dewatered, the hydrostatic pressure outside the cofferdam seats the beam, thus creating a seal. Tie-ins to a sloping face are somewhat more complicated, and it is necessary to develop details to fit each individual configuration. The most common schemes consist of timber bulkheads or timber cribs tailored to fit the sloping face. See Figure 2-3 for typical tie-in details.

3.2.2.5 CELL LAYOUT AND GEOMETRY. The cofferdam layout, generally, should utilize only one cell size which satisfies all design requirements. In some areas it might be possible to meet all stability requirements with smaller cells; however, the additional costs resulting from the construction and use of more than one size template will usually exceed the additional cost of an increase in the cell diameter. For individual cell and connecting arc geometry, the arrangements and criteria contained in the Steel Sheet Piling Handbook published by the U. S. Steel Corporation are recommended. These suggested arrangements should, however, be modified to require an odd number of piles between the connecting wyes or tees as shown in Figure 2-4.

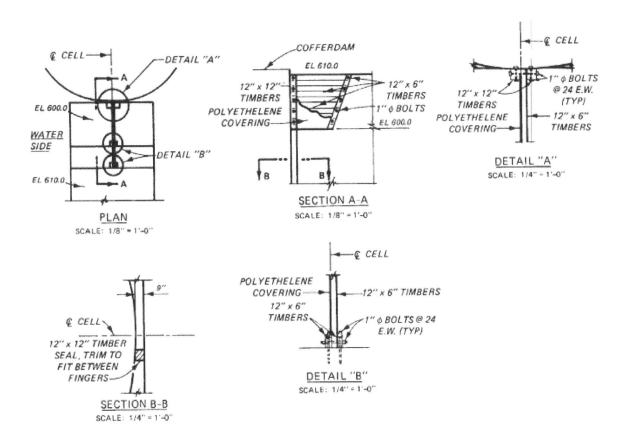

Figure 2-3
Typical tie-in details

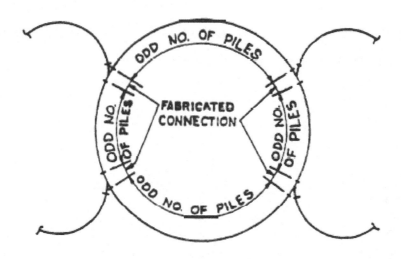

Figure 2-4

Arrangements of connecting wyes and tees

This will allow the use of only one type of fabricated wye or tee rather than two types if an even number of piles are used between connections. Although two additional piles might be required for each cell, this cost would be offset by the ease of checking shop drawings and simplifying construction, i.e., the tees or wyes could not be placed and driven in the wrong location. In developing details for other configurations, special attention should be given to the location of tees or wyes and the number of piles between connections. f. Protection and Safety Features. Other features which must be considered in the planning and layout of a cofferdam include: a rock or concrete cap on the cell to protect the cell fill from erosion and to provide a suitable surface for construction equipment; personnel safety facilities including sufficient stairways and an alarm system; and navigation warnings, including painting of cells, reflective panels, and navigation lights.

3.3 GEOTECHNICAL CONSIDERATIONS

3.3.1 SUBSURFACE INVESTIGATIONS

3.3.1.1 INTRODUCTION.

3.3.1.1.1 THE PLANNING, design, and construction of cofferdams should be approached as though the cofferdam is the primary structure of the project, the end result rather than the means to an end. The same degree of care, particularity, and competence should be exercised with the cofferdam as with the main structure. This necessarily involves detailed investigations because the foundation conditions, perhaps more than any other factor, impact on the cost and degree of difficulty in construction and eventual integrity of the cofferdam. Though impractical, if not impossible, to accurately determine all of the subsurface details, the major details should be determined to avoid needless delays and claims as well as possible failure resulting from inadequate subsurface investigations.

3.3.1.1.2 THE INVESTIGATIVE PROGRAM should be such that the cofferdam investigations form an integral part of the overall program for the main structure. By integrating the investigative programs for the various structures, the use of resources and information is maximized. Typically, there are three main investigative stages in the development of a project: the survey investigation which is a combination reconnaissance and feasibility stage made prior to Congressional authorization to determine the most favorable site, engineering feasibility, and costs; the definite project or specifications investigation which is made after Congressional authorization to provide required geologic and foundation data for preparation of contract plans and specifications; and the construction investigation which is made as the work progresses to fill in details.

3.3.2 PRELIMINARY INVESTIGATIONS.

3.3.2.1 OFFICE STUDIES. Office studies of the general area of the cofferdam location should be initiated prior to any field work. These preliminary studies should include a review of all geotechnical data compiled during the survey investigation stage for the project, including reports, maps, and aerial photographs. The investigation should include a study of the topography, physiography, geologic history, stratigraphy, geologic structure, petrography, and ground-water conditions. This information should include: bedrock type, occurrence, and general structural relationship; leakage and foundation problems possible if soluble rocks are present; possible results of glaciation (buried valleys, pervious divides, lacustrine deposits); presence or absence of faults and associated earthquake problems; extent of weathering; depth and character of overburden materials; general ground-water conditions; and availability of sources of construction materials. It is essential that the regional and local geology be known and understood prior to developing and implementing a plan of subsurface investigation.

3.3.2.2 FIELD STUDIES. As with the office studies, field reconnaissance and, perhaps, a limited number of subsurface borings and geophysical studies are conducted for the survey investigation. The results of those initial field studies should be incorporated into investigations which are designed to reveal specific information on the cofferdam foundation conditions. The resulting information should include if possible: nature and thickness of the overburden; maps of rock outcrops denoting type and condition of rock, discontinuities, presence or absence of geologic structure; and preliminary groundwater conditions.

3.3.3 DEVELOPMENT OF A BORING PLAN.

3.3.3.1 PRELIMINARY. After a careful evaluation of all available site data, a limited number of borings should be laid out along the center line of the proposed cofferdam location.

3.3.3.1.1 THIS INITIAL EXPLORATION can ordinarily be accomplished with splitspoon standard penetration sampling of the overburden and NX-size diamond coring of the bedrock, supplemented by a number of borings drilled with nonsampling equipment such as roller rock bits. Standard penetration resistances should be obtained at least at 5-foot-depth intervals or at material changes, whichever is the lesser. The NX-size (or comparable wireline equipment) is the smallest size coring equipment that should be used, and only then if acceptable recovery is obtained. The nonsampled borings will provide information on the overburden thickness, the presence or absence of boulders, and the top of rock configuration. A number of the borings should be selected to remain open and function as piezometers to provide ground-water data. If available, downhole geophysical equipment should be used to obtain additional data from each hole. The type of probes used will be necessarily dependent on the foundation material. For example, a

© J. Paul Guyer   2017   102

gamma probe would be one of the most useful tools in logging interbeds of sand and clay or limestone and shale while a caliper probe might prove invaluable in cavernous limestone. Other important geophysical instruments that might be utilized for the preliminary investigative stage are the portable seismograph and the electrical resistivity instrument. The portable seismograph may be used to obtain information on the bedrock surface that will be invaluable in planning the detailed exploration. The electrical resistivity apparatus may be used to determine approximate depths of weathering, the extent of buried gravel deposits, and the ground-water table. In using these instruments, the investigator must keep in mind that data derived from such tools are general in nature and intended to be used as supplemental data. Care must be exercised to prevent erroneous assumptions or interpretations on unsupported or unconfirmed geophysical data.

3.3.3.1.2 PRELIMINARY INVESTIGATIONS are intended to provide the general information necessary to supplement the office studies and provide the basis needed to plan a comprehensive final investigative program. Data obtained from the preliminary program should answer the general questions as to classification of materials including index properties, consistency or relative density, overburden thickness, and ground-water conditions.

3.3.3.2 FINAL.

3.3.3.2.1 AFTER THE PRELIMINARY investigative program has disclosed the general characteristics of the subsurface materials, a more specific program must then be designed. Economic and time limitations often control the amount of effort expended on subsurface investigations, and although there will never be enough time or money to uncover all defects and their locations, the program must be adequate to define the essential character of the subsurface materials. The program results should enable the investigators to determine the nature of the overburden and the bedrock.

3.3.3.2.2 AS WITH ANY DAM, the final number, spacing, and depth of borings for foundation exploration of a cellular cofferdam are determined by several factors, principal among which is the complexity of the geologic conditions. The holes should extend to top of rock if practicable, or at least to a depth where stresses from the structure are small. Again as a general rule, the borings should extend to a depth at least equal to the designed height of the cofferdam. In applying these general rules, care must be exercised to avoid formulating a plan of borings on a predetermined pattern to predetermined depths, possibly losing available information to be gained from the flexibility afforded by a knowledgeable use of the geology. Additional borings should be located, oriented, and drilled to depths to fully and carefully explore previously disclosed trouble areas, such as layers of weak compressible clay, fault zones or zones of highly dissolved rock, or irregular rock surfaces.

3.3.3.2.3 THE PROGRAM SHOULD BE detailed enough to adequately cover sources of common problems in cellular cofferdam construction. Typical obstacles such as boulders in the overburden may cause difficulty in driving the cell sheets and lead to interpretations of a false top of rock. The program should also fully cover foundation features which have resulted in past cellular cofferdam failures, i.e., foundation failures precipitated by faults, slip planes, and high uplift pressures.

3.3.3.2.4 THE FINAL PLAN of investigation should include continuous undisturbed sampling of the overburden to provide the necessary samples for laboratory testing. The type, number, and depth of undisturbed sample borings should be determined after an evaluation of information derived from the disturbed sample borings. Bedrock cores should be taken to adequately define the top of rock as well as the presence or absence of discontinuities in the rock. Large diameter cores for testing may not be necessary if such testing has been performed for the main structure and if there is no change in the geology.

3.3.3.2.5 THE LOCATION, ORIENTATION, and depth of core borings should be adjusted to recover as much information as possible on the more probable problem defects in the particular rock. For example, the major problem in sandstone is generally the jointing, especially if subjected to folding, whereas the major problem in limestone is generally associated with solution cavities. Regardless of the degree of care exercised in the core drilling operations, all core may not be recovered. Particular emphasis should be accorded this "lost core," and for design purposes, this loss should be attributed to soft or weak materials unless there is incontrovertible evidence to the contrary. The possibility of such potential sliding planes should always be considered regardless of rock type, because seemingly competent rock may contain weak clay seams and adversely oriented clay-filled joints along which sliding can take place. Borings for top of rock determination should go into the rock sufficiently to determine depth of discontinuities. This depth should be adjusted to fit conditions as determined by other studies, e.g., the depth should be increased if geologic interpretations from core borings and outcrops indicate an average depth of 25 feet of cavernous rock. In addition, if there is evidence of an abrupt change in the top of rock elevation, such as an erosional scarp or severe and widespread solution activity in the limestone, a number of roller rock bit borings should be drilled on close centers to better define the condition.

3.3.3.2.6 A RELIABLE ESTIMATE of water inflow as well as an accurate determination of the elevation and fluctuation of the ground-water table are primary concerns in the design and construction of any hydraulic structure. One method of obtaining information is the field pumping test which may be performed to determine the permeability of the foundation materials.

3.3.3.2.7 THE INVESTIGATION PLAN should be flexible so that information may be evaluated as soon as possible and adjustments made as needed. The program should begin as soon as possible and should carry through the design stage until adequate information is available for preparation of contract plans and specifications. The program must be refined through analysis of the geologic details to provide specific and reliable information on the character of the overburden; the depth to and configuration of the top of rock; the depth and character of bedrock weathering; the structures or discontinuities, such as faults, shear zones, folds, joints, solution channels, bedding, and schistosity; the physical properties of the foundation materials; the elevation and fluctuation limits of the ground water; and potential foundation problems and their treatment, such as leakage and stability.

3.3.3.2.8 THE FINAL INVESTIGATIONS should supply the necessary information to complete the interpretation or "picture" of subsurface conditions, dispel any reasonable doubts or fears on the practicability of the design, and provide adequate information for reliable estimates on foundation-related bid items for the contract.

3.3.4 PRESENTATION OF DATA.

3.3.4.1 REPORT. Following a complete and thorough evaluation of all geotechnical data, a report should be prepared for inclusion in the design memorandum. Scheduling on a project is usually such that design exploration and actual design are done concurrently. Consequently, the data should be discussed with the design engineers as the data are evaluated. In most cases, the cellular cofferdam will be included in the feature design memorandum for the primary structure. In any case, the report should include a brief summary of the topography, of the regional and site geology including the seismic history, and of the subsurface investigations and tests that were performed. The summary of the site physiography and geology should emphasize those conditions of engineering significance, i.e., those most pertinent to the engineering structure, in this case, the cofferdam. Such conditions that should be covered are the character and thickness of the overburden with particular note of any potential trouble materials; the estimated top of rock; the type, stratigraphic sequence, and geologic structure of the rock; the nature and depth of rock weathering; and site ground-water conditions. The detailed account of the geotechnical investigations should include the number, type, and location of the explorations as well as an explanation for the particular explorations. A brief description of the various pieces of equipment used should also be provided. The account should contain a summary of the type of tests, both field and laboratory, that were performed and the results that were obtained. And finally, any search for sources of construction materials, whether specifically for the cofferdam or not, should be summarized. The report should include a detailed account of the data that were obtained, conclusions as to the subsurface conditions and their impact on the cellular cofferdam, and recommendations, particularly as to foundation treatment and construction materials. The preponderance of the information contained in this report, minus interpretations, should be presented to contractors for bidding purposes in accordance with the Unified Soil Classification System.

3.3.4.2 DRAWINGS. Drawings are a necessary part of the report and should include a plan of exploration, boring logs, top of rock map, and geologic interpretations of subsurface conditions at the cofferdam site. The sections must provide the location of the borings; the character and thickness of the overburden with particular note of any potential trouble materials and an interpretation of their extent and configuration; the estimated top of rock line, both weathered and unweathered; the character of the weathered rock zone; overburden and bedrock classification; structural features such as faults, joints, and bedding planes; and ground-water conditions. The boring logs should show the designation and location, the surface elevation, and the overburden/bedrock contact, and should describe the material in terms of the Unified Soil Classification System. The boring logs should also show the depths of material change, blow counts in the overburden or areas of rapid drill penetration, defects, core loss, drill water increase or loss, water level data with dates obtained, pressure test data, the date hole was completed, percentage of core recovery, and size and type of hole. The results of any geophysical survey should be presented to support or supplement other exploratory data. The proposed cellular cofferdam location should be included on the drawings to more accurately depict the founding of the structure in relation to the subsurface conditions and to facilitate review.

3.3.5 INVESTIGATIONS DURING AND FOLLOWING CONSTRUCTION.

3.3.5.1 CONSTRUCTION AND POSTCONSTRUCTION DATA ACQUISITION. The subsurface investigations must continue throughout the construction and postconstruction period. Information on foundation conditions should be obtained and recorded whenever and wherever possible during construction and operation of the cofferdam. This information should include volume and thicknesses of any deleterious material such as a weak compressible clay bed that might necessarily be excavated and the depth or elevation of the excavation, increased sheet pile resistance and the reason for the increase such as lenses or zones of cobbles or boulders, depth of sheet pile refusals, and water inflows as evidenced by the volume of pumping required to maintain a dry working area. This information should be continuously compared with data developed for design and for preparation of plans and specifications. The information obtained during and following construction of the cofferdam may prove invaluable in the event that problems with the performance of the structure develop, resulting in remedial action and/or a contract claim.

3.3.5.2 CONSTRUCTION FOUNDATION REPORT. After completion of the cofferdam construction, an as-built foundation report on construction of the cofferdam must be prepared. Although this report in most cases will be included with the foundation report for the entire project, its initiation and completion should not be delayed. The report must contain all data pertinent to the foundation, including but not limited to a comparison of the foundation conditions anticipated and those actually encountered; a complete description of any materials necessarily excavated and the methods utilized in the excavation; a description, evaluation, and tabulation of the sheet pile driving including method, type, date, and depth; a description of the methods used and any problems encountered in the foundation treatment and any deviations from the design treatment (including reasons for such change); a tabulation and evaluation of the water pumping required as well as a comparison with the anticipated inflow; a detailed discussion of any solutions to problems encountered; and the results and evaluation of, and recommendations based on the instrumentation.

3.3.6 FIELD AND LABORATORY TESTING. Field and laboratory testing are used to estimate the engineering properties needed for the rational design of both the foundation and the structure. The foundation design requires an estimate of both the strength and seepage qualities of the foundation, The engineering properties of the cell fill can usually be estimated with sufficient accuracy from laboratory index tests.

3.3.7 FIELD TESTING. During the initial phase of exploration, field tests are generally made to obtain a rough estimate of the strength of the foundation. Later stages may require similar testing to refine or extend the subsurface profile, or more sophisticated testing may be required where better estimates are needed. Field or in situ testing of rock strength is usually expensive and difficult; consequently, most such testing is reserved for projects that are large and/or have complicated or difficult foundation problems. Preliminary estimates are commonly made for soils using the methods listed in Table 3-1. Two of these methods are summarized below.

Method	Remarks
Penetration resistance from standard penetration test	In clays, test provides data helpful in a relative sense; i.e., in comparing different deposits. Generally not helpful where number of blows per foot, N , is low
	In sand, N-values less than about 15 indicate low relative densities
Natural water content of disturbed or general type samples	Useful when considered with soil classification and previous experience is available
Hand examination of disturbed samples	Useful where experienced personnel are available who are skilled in estimating soil shear strengths
Position of natural water contents relative to liquid limit (LL) and plastic limit (PL)	Useful where previous experience is available
	If natural water content is close to PL, foundation shear strength should be high
	Natural water contents near LL indicate sensitive soils with low shear strengths
Torvane or pocket penetrometer tests on intact portions of general samples	Easily performed and inexpensive, but results may be excessively low; useful for preliminary strength estimates
Vane shear	
Quasi-static cone penetration	See FHWA-TS-78-209, 1977 (item 26)

Table 3-1

Methods of Preliminary Appraisal of Foundation Strengths

3.3.7.1 VANE SHEAR TESTS. The vane shear test is an in situ test and is often valuable for soft clay foundations where considerable disturbance may occur during sampling. A disturbance, especially when using conventional sampling methods, usually reduces the undrained strength of the sampled soil to a value that often would result in an uneconomical design. Because this test is performed in situ, sample disturbance is minimized. The test usually overestimates the soil's undrained strength and must be reduced by an applicable correction factor. Bjerrum (item 9) recommended a correction that is a function of the clay's plasticity index and varies as shown in Figure 3-1. The testing procedure should be followed closely because the results can be very sensitive to the testing details. Because of the uncertainty of the results from this test, an independent method of estimating the foundation shear strength should be included in any testing program. Often unconsolidated-undrained triaxial testing of good quality undisturbed samples is a good independent check. A bracket for estimating the foundation shear strength can sometimes be established by taking corrected field vane tests as an upper bound and good quality undisturbed samples tested in undrained compression as a lower bound for shear strength.

3.3.7.2 STANDARD PENETRATION TEST. This test is one of the most widely used methods for soil exploration in the United States. It is a means of measuring

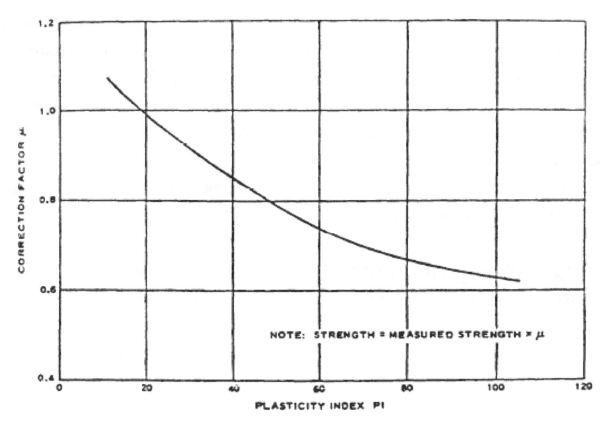

Figure 3-1

Vane shear correction chart

the penetration resistance to the split-spoon sampler as well as obtaining a disturbed-type sample. Correlations between the penetration resistance and the consistency of cohesive soils and the relative density of granular soils have been published and are often used to estimate soil strength. These correlations are very rough and, except for the smallest of structures, should be liberally supplemented with other, better quality, strength testing. The designer should be aware of the severe limitations of using the results of this test. For example, the values obtained when testing soft clays, coarse gravels, or micaceous soils are often of little or no value. 3-8. Field Seepage Testing. The permeability of pervious foundation soils can usually be estimated with sufficient accuracy by using existing correlations with the foundation's grain-size distribution, Figure 3-2. Field pumping tests are a much more accurate means for determining the permeability of the foundation soils, especially for stratified deposits. These tests, however, are expensive and are usually justified only for unusual site

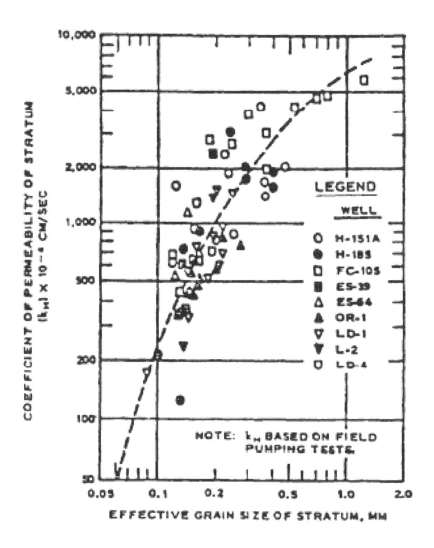

Figure 3-2

Effective grain size of stratum versus in situ coefficient of permeability.

(Based on data collected in the Mississippi River Valley and Arkansas

River Valley.)

conditions. Clay foundations are usually considered impervious for estimates of seepage

quantities. However, the effects of discontinuities and thin beds of granular materials are

important and should not be neglected. A number of field tests are available to measure

rock mass permeability including pumping tests, tracer tests, and injection or pressure

tests. The most frequently used field test is the borehole pressure test which is relatively

simple and inexpensive. Among the three types of pressure tests, water, pressure drop, and air, water is the most common because it is simple to perform, is not overly time-consuming, and can be performed above or below the ground-water table.

3.3.8 LABORATORY TESTING.

3.3.8.1 A LABORATORY TESTING program should be designed to supplement and refine the information obtained from the subsurface investigation and field tests. The amount and type of testing depends on the type and variability of the foundation and borrow areas, the size of the structure, the consequences of failure, and the experience of the designer with local conditions.

3.3.8.2 THE LABORATORY TESTING PROGRAM is typically performed in phases that follow the subsurface investigation program. Initially, index tests are performed on samples obtained from the exploration program. These results are then used as a basis for the selection of samples and the design of a laboratory testing program.

3.3.9 INDEX TESTS.

3.3.9.1 INDEX TESTS are used to classify soil in accordance with the Unified Soil Classification System (Table 3-2), to develop accurate foundation soil profiles, and as an aid in correlating the results of engineering property tests to areas of similar soil conditions. Both disturbed and undisturbed soil samples should be subjected to index-type tests. Index tests should be initiated, if possible, during the course of field investigations. All samples furnished to the laboratory should be visually classified and natural water content determinations made; however, no water content tests need be run on clean sands or gravels. Mechanical analyses (gradations) of a large number of samples are not usually required for identification purposes, Atterberg limits tests should be performed on representative fine-grained samples selected after evaluation of the boring profile. For selected borings, Atterberg limits should be determined at frequent intervals on the same samples for which natural water contents are determined.

3.3.9.2 NORMALLY, ATTERBERG LIMITS, determinations, and mechanical analyses are performed on a sufficient number of representative samples from preliminary borings to establish the general variation of these properties within the foundation, borrow, or existing fill soils. A typical boring log is shown in Figure 3-3.

3.3.9.3 ALL ROCK CORES SHOULD BE LOGGED in the field by a geologist, preferably as the cores come from the hole. A number of laboratory classification and index tests for rock are available. These tests include water content, unit weight, porosity, and unconfined compression, all of which should be performed on representative samples

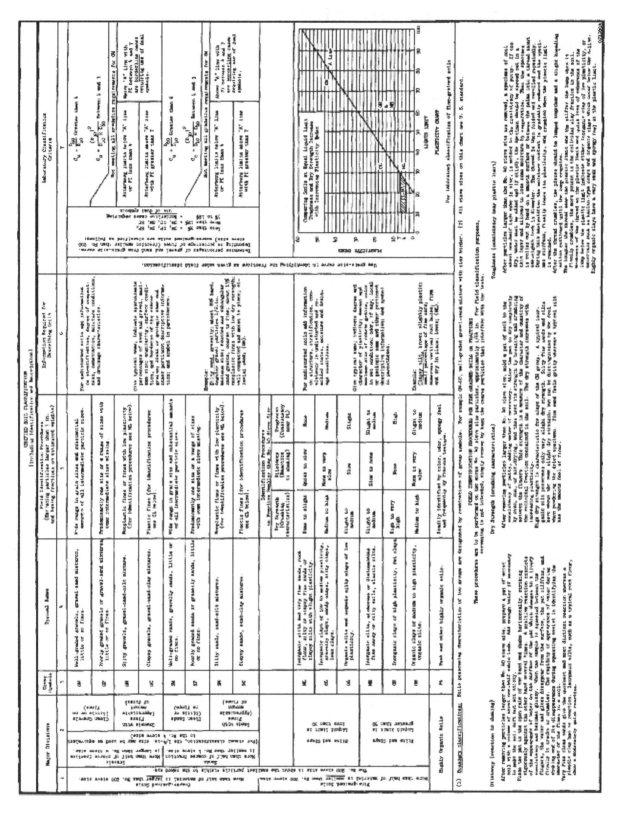

Table 3-2

Unified Soil Classification

3.3.10 ENGINEERING PROPERTY TESTS. A good estimate of the strength and seepage characteristics of the foundation is necessary for an adequate foundation design. The estimate of the foundation strength is usually the most critical design parameter. Seepage characteristics are usually estimated based on the gradation of the foundation soils and an evaluation of geologic properties, especially discontinuities. The properties of the cell fill are usually estimated based on gradation analyses and the anticipated method of placement of the fill.

3.3.11 PERMEABILITY OF SOILS.

3.3.11.1 FINE-GRAINED SOILS. There is generally no need for laboratory permeability tests on fine-grained fill material or clay foundation deposits. In underseepage analyses, simplifying assumptions must be made relative to thickness and soil types. Furthermore, stratification, root channels, and other discontinuities in fine-grained materials can significantly affect seepage conditions.

3.3.11.2 COARSE-GRAINED SOILS. The problem of foundation underseepage requires reasonable estimates of permeability of coarse-grained pervious deposits. However, because of the difficulty and expense in obtaining undisturbed samples of sand and gravel, laboratory permeability tests are rarely performed on foundation deposits. Instead, correlations developed between grain size and coefficient of permeability, such as that shown in Figure 3-2 are generally utilized. This correlation explains the need for performing gradation tests on pervious materials where underseepage problems are indicated.

3.3.12 PERMEABILITY OF ROCK. The determination of rock mass permeability quite often depends on secondary porosity produced through fracturing and solution rather than on primary porosity of the rock. Consequently, geologic interpretations and evaluations are extremely important in determining the discontinuities that serve as ready passageways for ground-water flows.

3.3.13 SHEAR STRENGTH--GENERAL.

3.3.13.1 THERE ARE THREE PRIMARY TYPES of shear strength tests for soils, each representing a certain loading condition. The Q-test represents unconsolidated-undrained conditions; the R-test, consolidated-undrained conditions; and the S-test, consolidated-drained conditions. The unconsolidated undrained strength generally governs the design of foundations on fine-grained deposits. R-tests are generally not needed for most cellular structure designs. S-tests are used where long-term stability of a fine-grained foundation is to be checked or if the soil to be tested is a granular material.

3.3.13.2 Q- AND R-TESTS ARE PERFORMED in triaxial testing devices while S-tests are performed using direct shear and triaxial testing devices. The unconfined compression (UC) test is a special case of the Q-test in that it also represents unconsolidated-undrained conditions but is run with no confining pressure. Also, rough estimates of unconsolidated-undrained strength of clay can be obtained through the use of simple hand devices such as the pocket penetrometer or Torvane. However, these devices should be correlated with the results of Q- and UC-tests.

3.3.13.3 THE DISCUSSION RELATES the applicability of each test to the different general soil types. The applicability of the results of the different shear tests to field loading conditions and the different cases of stability are important.

3.3.13.4 THERE ARE TWO BASIC TYPES of shear strength tests utilized to obtain values of cohesion and angles of internal friction to determine strength parameters of the foundation rock: the triaxial and the direct shear. The data to determine rock strength in an undrained state under three-dimensional loading are obtained from the triaxial test. This test is performed on intact cylindrical rock samples not less than NX core size, i.e., approximately 2-1/8 inches in diameter. The direct shear test, an undrained type, is performed on core samples ranging from 2 to 6 inches in diameter. In this test, the samples are oriented such that the normal load is applied perpendicular to the feature being tested. These normal loads should be comparable to those loads anticipated in the field. For moisture-sensitive rocks such as indurated clays and compaction shales, soil property test procedures described in EM 1110-2- 1906 should be used.

3.3.14 SHEAR STRENGTH--SAND. Since consolidation of sand occurs simultaneously with loading, the appropriate shear strength of sands for use in design is the consolidated-drained, S-strength. However, the shear strength of sand in the foundation or cell, regardless of the method of placement, is not normally a critical or controlling factor in design. Therefore, excessive laboratory testing to determine the shear strength of sand is usually not warranted. Satisfactory approximations for most sand can also be made from correlations with standard penetration resistances and relative densities. Such correlations can be found in most standard engineering texts on soil mechanics (Figure 3-4). Seepage forces, discussed in detail in Chapter 4, can reduce the shear resistance, especially at the toe of the structure, to undesirable levels.

3.3.15 SHEAR STRENGTH--CLAY AND SILT.

3.3.15.1 THE UNDRAINED SHEAR STRENGTH parameters should be determined for all fine-grained materials in the foundation. In areas of soft, fine-grained foundations, it is imperative that an adequate shear testing program be accomplished to establish the variation in unconsolidated-undrained shear strength with depth within the foundation (usually expressed as the ratio of undrained shear strength s_u to effective vertical stress as shown in Figure 3-5. A sufficient number of Q-tests, supplemented by UC tests, where appropriate, should be performed throughout the critical foundation stratum or strata. Data obtained from any field vane shear strength tests may also be helpful in establishing this variation.

3.3.15.2 R-TESTS CAN BE HELPFUL in estimating the variation in undrained shear strength with depth, and in determining the increase in undrained shear strength with increased effective consolidation stress. This may be necessary in estimating the gain in shear strength with time after loading. c. The results of S-tests are used in evaluating the long-term stability of the foundation and in judging the stability of structures where pore pressure data, such as those obtained from piezometers, are available.

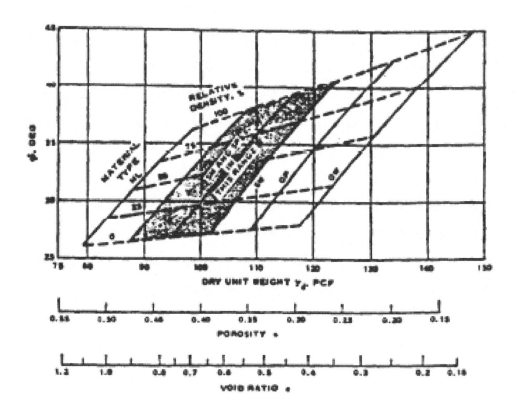

Figure 3-4

Angle of internal friction versus density for coarse-grained soils

3.3.16 PROCEDURES. In performing these tests, one should be sure that field conditions are duplicated as closely as possible. Confining pressures for triaxial tests and normal loads for direct shear tests should be chosen such that the anticipated field pressures are bracketed by the laboratory pressures based on depth and location of sample and anticipated field loadings. All samples should be sheared at a rate of loading slow enough that there will be no significant time-rate effect. The specimen size should also be chosen such that scale effects are minimized. Standard size of samples for triaxial testing of soils is 1.4 inches in diameter by 3 inches in height. However, if the sample is fissured or contains an appropriate amount of large particles such as shells, gravel, etc., then a larger size sample (2.8 inches in diameter by 6 inches in height) can be utilized in order to obtain valid results.

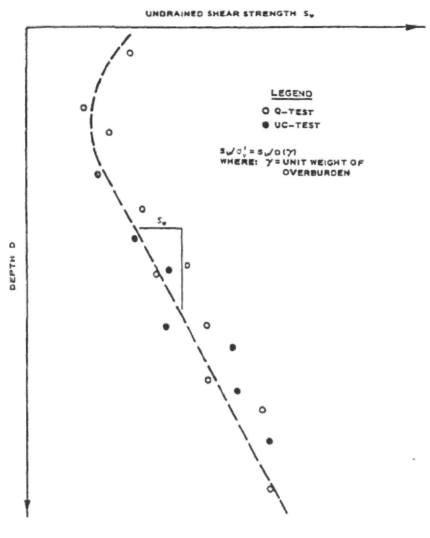

Figure 3-5

Typical plot showing variation of unconsolidated, undrained shear strength with depth

3.4 FOUNDATION TREATMENT

3.4.1 PROBLEM FOUNDATIONS AND TREATMENT.

3.4.1.1 FOUNDATION TREATMENT is sometimes considered for foundations with insufficient bearing capacity or problem seepage conditions. Problem seepage conditions can be the result of excessive seepage quantities or high seepage forces.

3.4.1.2 THE FOLLOWING FOUNDATION treatment methods can be used to improve a deficient foundation.

3.4.1.2.1 REMOVAL OF OBJECTIONABLE material. Removal may be before or after the piles are driven to form the cell.

3.4.1.2.2 IN SITU COMPACTION. Several methods are available and include vibroflotation, compaction piles, surcharge loads, and dynamic surface loads.

3.4.1.2.3 DEEP PENETRATION OF SHEET PILING. For design purposes, a trial penetration of two thirds of the cell height is usually considered when the cell is sited on a pervious foundation. An adjustment of this length should be based on a careful analysis of the seepage forces at the toe of the structure.

3.4.1.2.4 BERMS AND BLANKETS. Impervious blankets may be located on the outside of the cells to reduce seepage quantities and pressures. Interior berms reduce the likelihood of boiling at the toe of the structure.

3.4.1.2.5 CONSOLIDATION. The strength of foundation material, especially fine-grained material, may be increased by consolidation. Surface preloading of the foundation and the use of sand drains are two of the methods used to accelerate consolidation of the foundation.

3.4.2 GROUTING.

3.4.2.1 CORRECTIONAL METHODS. As for all such structures, foundation treatment should be carefully considered for cellular cofferdams. In many cases, removal of the unfavorable foundation material may be impracticable, if not impossible, and other methods of treatment must be selected. Grouting is one such method which should be considered, especially in instances where the piling of a cellular cofferdam will be driven to rock. During the evaluation of the data developed during the subsurface investigations, special note should be made of any unfavorable foundation condition that would justify at least some consideration of grouting. Such unfavorable foundation conditions might be noted as a result of evidences of solution activity such as soluble rock or drill rods dropping during drilling, open joints or bedding planes, joints or bedding planes filled with easily erodible material, faults, loss of drill fluid circulation, or unusual ground-water conditions, Generally, the problems related to such unfavorable foundation conditions can be grouped into two categories: problems related to the strength of the foundation material and problems related to the permeability of the foundation material.

3.4.2.2 PROBLEMS RELATED TO STRENGTH. Among the problems related to strength that should be anticipated are: insufficient bearing capacity, insufficient resistance to sliding failure, and general structural weaknesses due to underground caverns or solution channels, or due to voids that develop during or following construction. Problem 3 is closely related to Problems 1 and 2 and should be considered jointly. In developing parameters for allowable bearing capacity, deficiencies noted in Problem 3 must be carefully considered. All too often, rock strength parameters are used in stability analyses that are based on rock sample strengths rather than mass rock strengths. The various discontinuities that reduce the foundation rock strengths may result in consequential reductions in the ultimate bearing capacity. As mentioned above, the bedrock may contain bedding plane cavities and solution channels that can extend to considerable depth (low crossbed shear strength). In recognizing the presence of such discontinuities, the possibility must also be recognized that an unfavorable combination of these discontinuities could exist under the cellular cofferdam, thus adversely affecting the sliding stability of the structure. The presence of these weak planes must be carefully considered when doing a sliding stability analysis.

3.4.2.3 PROBLEMS RELATED TO PERMEABILITY. Among the problems related to permeability that should be anticipated are: reduction in the strength of the foundation materials due to high seepage forces, high uplift forces at the base of the structure, and inability to economically maintain the coffered area in an unwatered state. In many cases, the piling of a cellular cofferdam will be driven to rock. The presumption should be that some seepage will occur not only at the piling/bedrock contact, but also through openings in the bedrock. This seepage may result in piping of materials through the bedrock openings below the cofferdam, greatly reducing the strength of the foundation. These openings along bedding planes can also result in high uplift pressures. Quite often, the vertical permeability of the rock above the open bedding plane is only a small fraction of the permeability along the plane. If such a situation exists, it is possible that the high uplift pressures will jack the foundation. The size and continuity of solution channels acting as water passageways may have a serious economic impact on the dewatering of the work area within the cofferdam. Unfortunately, there is no way to accurately estimate the dewatering problems and costs that might result from such solution channels in the foundation.

3.4.2.4 SELECTION OF TREATMENT. Treatment of the cofferdam foundation by grouting may be used to lessen, if not eliminate, defects in the foundation, resulting in a strengthened foundation with reduced seepage. Grouting should be selected as a method of foundation treatment only after a careful and thorough evaluation of all pertinent factors. Primary factors that must be necessarily considered before selection of grouting as the method of treatment are the engineering design requirements, the subsurface conditions, and the economic aspects. Although cost is just one factor to consider, in many circumstances, cost may be the controlling factor. The cost of grouting must be weighed against such other costs as that of pumping, delays, claims, and/or failure. It may be that there is no benefit in reducing minor leakage by costly grouting.

3.4.2.4.1 GENERAL. Information obtained and evaluated during the subsurface investigations for design of the structure should be adequate to plan the grouting program. If the grouting program is properly designed and conducted, it becomes an integral part of the ongoing subsurface investigations. A comprehensive program must necessarily take into account the type of structure, the purpose of the structure, and the intent of the grout program. As an example, foundation grouting for a cellular cofferdam is not intended to be permanent nor 100 percent effective. The program should be designed to provide the desired results as economically as possible. The program should be flexible enough to be revised during construction and performed only where there is a known need.

3.4.2.4.2 TO STRENGTHEN. Grouting has been used on occasion to strengthen the foundation by area or consolidation grouting under the cells to increase the load-bearing capacity of the rock. This may be a viable option if the grouting is intended to increase the already acceptable factor of safety. However, if it appears that the factor of safety falls appreciably below the allowable factor of safety, total reliance should not be placed on grouting. The effectiveness of such grouting is impossible to predict or to evaluate. Certainly complete grouting is impossible because of the irregularity of the openings as well as the amount and character of any filling material.

3.4.2.4.3 TO REDUCE SEEPAGE. The principal purpose of grouting for cellular cofferdams has been in conjunction with seepage control and drainage. Curtain grouting is one method used to reduce uplift pressures and leakage under the cofferdam and thus reduce total dewatering costs. Although a single line curtain will suffice in most cases, the rock conditions may be such that it will be necessary to install a multiline curtain or a curtain with multiline segments. The exact location of the grout curtain will be influenced by a number of factors including the type of structure, the foundation conditions peculiar to the site, and the time the curtain is installed. For most cellular cofferdams, the grout curtain is located on or near the axis of the structure. However, if the curtain is not installed until the cofferdam has been constructed, it may be impracticable to drill holes through the cell fill. In this case the grout line should be moved off and just outside the cells. When installing the grout curtain, the flow of grout must be carefully controlled to prevent the grout from flowing too far, resulting in grout waste. To prevent such waste, it may be necessary to limit the quantity of grout injected, or to add a "stopper" line grouted at low pressure. The orientation and inclination of the holes should be adjusted to intercept the principal water passageways. Occasionally, however, conditions may render this impracticable and it may be that vertical holes on closer centers are more feasible.

3.4.3 DEVELOPMENT OF PROGRAM. Following an evaluation of the foundation conditions and the selection of grouting as a method of foundation treatment, the evaluations, conclusions, and recommendations should be included in the report of the subsurface investigations. Using data developed from the investigations, the pertinent reference manuals, and especially past experience, plans and specifications should be prepared for the grouting program. After having reviewed all available and pertinent data and having decided on the particular grouting program to be implemented, a number of basic factors must be decided: the area selected for grouting, the selection of the grout, the selection of the type of grouting, and the need for special instructions, provisions, or restrictions.

3.4.3.1 SELECTION OF LOCATION. The area indicated for grouting should be a zone large enough to include any anticipated treatment. This is especially important in installing a grout curtain for a cellular cofferdam. This should be coincidental with provisions to provide for grouting anytime within the contract period without additional mobilization and demobilization costs to the Government. The drawings rightfully should show a grout curtain to be installed beneath the cofferdam along an approximate alignment and to definite limits. However, because of the numerous unknowns inherent in a grouting program, the plans and specifications should provide that the area of grouting extend some distance beyond the limits shown.

3.4.3.2 SELECTION OF GROUT. The selection of the grout should be made only after a careful evaluation of the foundation conditions or materials being tested. The type of grout used in reducing or stopping high velocity flows would be different from that used for slow seeps, or the grout used to fill large cavities might be different from that used to fill small voids. A factor to be considered in sealing high velocity flow would be the time of set; the large quantities and costs would necessarily be considered in filling large cavities; while in filling small voids, the size of the void and the particle size of the grout are necessary considerations.

3.4.3.3 SELECTION OF TYPE OF GROUTING. Grouting may be done before, during, and/or after installation of the cofferdam or other construction activities in any given area. In the installation of a grout curtain, all or portions of the curtain may be constructed from the original ground surface and/or from floating plant in the river. If done from floating plant, in general, stop-grouting methods should be used because it is not practical to stage drill and grout from floating plant. Drilling and grouting from floating plant by the stop-grouting method should be considerably less costly than stage grouting, the holes being drilled and grouted to the bottom of the curtain in one setup.

3.4.3.4 SPECIAL INSTRUCTIONS. In drilling from floating plant, it should be expressly understood that the depth of water penetrated will not be credited to the drilling footage for payment. If drilling and grouting are performed from the cofferdam, only drilling that is required below the original ground surface should be paid for. To effectively grout water-bearing openings associated with cavernous rock, the following general procedure should be followed: the grout holes should be drilled through the overburden and the casing should be seated a minimum of 1 foot in rock; the hole should be drilled at least 5 feet into rock, if the top of rock is lower than anticipated; if stop-grouting methods are used, grouting of the rock should be performed through a packer set just below the bottom of the casing; should a special feature be encountered in the hole, the packer setting may be varied to isolate and treat this feature. Grouting of the overburden, if necessary, can then be done immediately following the rock grouting. The specifications should provide that if, as the work progresses, supplemental grouting is required at any area within specified limits at any time, such additional grouting will be at the established contract unit prices for the items of work involved. Although pressure testing should be provided for in the specifications, the condition of the foundation may be such that all grout holes should be grouted, in which case, pressure testing would not be necessary. If at all possible, the initial dewatering of the cofferdam should be performed at the lowest possible river stage or other measures should be taken to ensure a stable cofferdam capable of being unwatered until the foundation and the adequacy of the foundation treatment can be checked.

3.5 SOURCES AND PROPERTIES OF CELL FILL

3.5.1 BORROW AREA. Borrow-related problems occur frequently in earth-workrelated construction, and sometimes result in costly design changes and contract modifications. Special diligence during the exploration and characterization of borrow fill will be beneficial during both the design and construction of the project.

3.5.2 LOCATION. Borrow areas are generally located as close to the project site as possible to reduce hauling costs. The final selection of the borrow site, however, is governed by several additional considerations.

3.5.2.1 CELL FILL PROPERTIES. When the most desirable cell fill is not locally available, the cost of processing or designing the structure around marginal cell fill should be compared with the increase in cost due to longer haul distances.

3.5.2.2 LAND USE. Although cell fill is often dredged from river channels, it is sometimes desirable to locate the borrow areas outside of the river. When this occurs, special consideration and planning should be initiated to provide proper reclamation of the area.

3.5.2.3 ENVIRONMENTAL ASPECTS. Environmental considerations may restrict the use of certain potential borrow sites. An early review of the probable borrow sites for any detrimental environmental consequences should be considered. These consequences are sometimes mitigated by placing restrictions on the use of the borrow area and by special reclamation of the site. For example, wildlife habitats or recreational areas can sometimes be created at these sites with a small additional cost.

3.5.3 SELECTION OF CELL FILL.

3.5.3.1 ALMOST ALL MODERN CELLULAR sheet pile structures are designed based on the assumption that a free-draining granular fill will be available near the construction site. Soils with less than about 5 percent of the particles by weight passing the No. 200 sieve and 15 percent passing the No. 100 sieve are usually termed free draining. Granular fills with many fines and even fine-grained fills have occasionally been used in the past; however, the poor performance of these fills usually favors use of better quality fill. b. The performance of the sheet pile structure is directly related to the drainage characteristics of the cell fill. Free-draining fill will have a lower seepage line within the fill than less pervious material. The lower seepage line improves the cell performance by:

3.5.3.1.1 REDUCING THE SHEET pile interlock force. (Reducing this force is especially beneficial for high cells or where marginal material is used. However, a reduction in the interlock force may reduce the stiffness of the structure, with slightly larger structural movements.)

3.5.3.1.2 INCREASING THE EFFECTIVE stress at the base of the cell, increasing lateral sliding resistance.

3.5.3.1.3 INCREASING the internal shear resistance.

3.6 SEEPAGE CONTROL

6.1 SEEPAGE THROUGH CELL.

3.6.1.1 THE LOCATION OF the free water surface in a cell is usually estimated using empirical relationships based on the type of cell fill. The recommendations in Figure 3-6 serve as a guide and starting point for estimating the location of the seepage line. These recommendations are conservative for most applications; however, each design should be evaluated for conditions that would tend to raise the seepage line. If both the quality of the cell fill and the assurance of proper inspection cannot be guaranteed during the design of the project, full saturation of the cell should be considered for design purposes. Some conditions that require evaluation are:

- Possible leakage from pipelines crossing the cells.
- Waves overtopping the outboard piles.
- Excessive leakage through the outboard piles.
- Poor drainage through the inboard piles.
- Lower permeability than expected of the cell fill.
- Hydraulic filling of cell fill.

3.6.1.2 THE QUANTITY OF SEEPAGE THROUGH the cell is a function of both the tightness and integrity of the outboard piles and the type of cell fill, the chief barrier being the outboard piling. The tightness of the outboard piling depends on the physical condition of the piling and the piling interlock force. An increase in seepage through the cell can generally be expected.

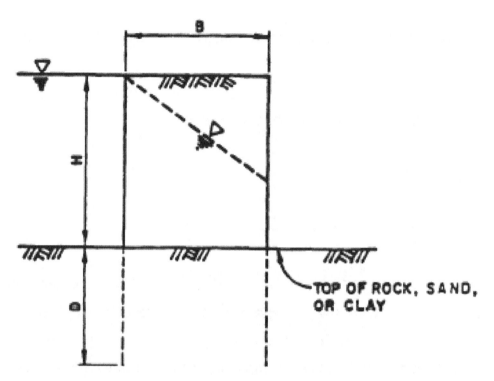

SLOPE OF FREE-WATER SURFACE IN CELLS DE-
PENDS ON PERMEABILITY OF CELL FILL. UNLESS
SPECIAL DRAINAGE IS PROVIDED AND SLOPE IS
CONTROLLED, ASSUME THE FOLLOWING:

- FREE-DRAINING COARSE-GRAINED FILL (GW,
  GP, SW, SP): SLOPE 1 VERTICAL TO 1
  HORIZONTAL

- SILTY COARSE-GRAINED FILL (GM, GC, SM,
  SC): SLOPE 1 VERTICAL TO 2 HORIZONTAL

- FINE-GRAINED FILL: SLOPE 1 VERTICAL
  TO 3 HORIZONTAL

Figure 3-6

Estimate of free water location in fill

3.6.1.2.1 SECOND-HAND PILING is used. New piling in good condition should be considered for major structures. For other structures, used piling may be considered when either seepage conditions are slight or pose little threat to the safety to the structure.

3.6.1.2.2 ROUGH DRIVING IS EXPERIENCED during construction. The foundation exploration program should investigate conditions that lead to rough driving. Contract specifications should restrict hard driving.

3.6.1.2.3 THE INTERLOCK FORCES are small. The increase in seepage due to this condition is usually small, and is usually not considered.

3.7 FOUNDATION UNDERSEEPAGE.

3.7.1 FOUNDATION UNDERSEEPAGE is generally not a problem for structures built on clay or good quality rock foundations. Problems almost always are confined to coarse-grained soil such as gravel and sand and sometimes silty materials. The most treacherous conditions occur where undetected pervious seams exist in the foundation.

3.7.2 COFFERDAMS ON SAND ARE often designed using a trial sheet pile penetration of two thirds of the height of the structure above-the dredgeline. A flow net is most often used to estimate the seepage forces. If the exit gradient at the toe of the structure is large, a loaded filter or a wide-base berm should be considered.

3.7.3 DEPENDING ON THE SITE conditions, up to 50 percent of the passive resistance, even with 2/3H penetration, at the toe can be lost due to seepage forces. This loss increases the possibility of excessive penetration of the inboard piles.

3.8 SEISMIC CONSIDERATIONS

3.8.1 STRUCTURE-FOUNDATION INTERACTION. The susceptibility of cellular structures to damage due to earthquake loadings depends on the complex interaction of the structure and the foundation. In addition to these loads, a reduction in strength of the foundation, cell fill, or backfill behind a cellular bulkhead can also simultaneously occur during an earthquake. Structures founded on saturated, cohesionless materials or cohesive soils that contain lenses of saturated, cohesionless soil can lose practically all of their foundation support when subjected to a vibratory loading, such as an earthquake.

Similarly, the cell fill or the backfill can also liquefy, increasing the lateral loading against the cell.

3.8.2 LIQUEFACTION POTENTIAL.

3.8.2.1 THE SIGNIFICANT FACTORS influencing the liquefaction potential of the foundation or fill include: soil type, relative density or void ratio, initial confining pressure, intensity of ground shaking, and duration of ground shaking. The vulnerability of liquefaction-susceptible foundations can be initially estimated using simplified methods and charts that incorporate the most important variables that contribute to liquefaction.

3.8.2.2 FIGURES 3-7 AND 3-8 DEFINE CONDITIONS where liquefaction is: very likely to occur, not very likely to occur, or a marginal condition exists where additional factors or further analysis should be considered. Charts of this nature are frequently updated and improved. For this reason, more recent material should be consulted for marginal or complex conditions.

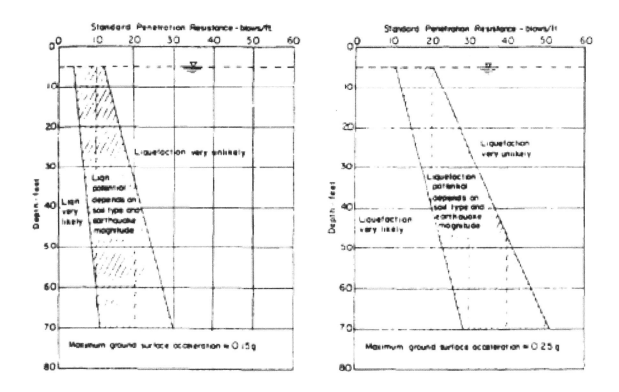

Figure 3-7

Liquefaction potential evaluation charts

for sands with water table at depth of about 5 feet

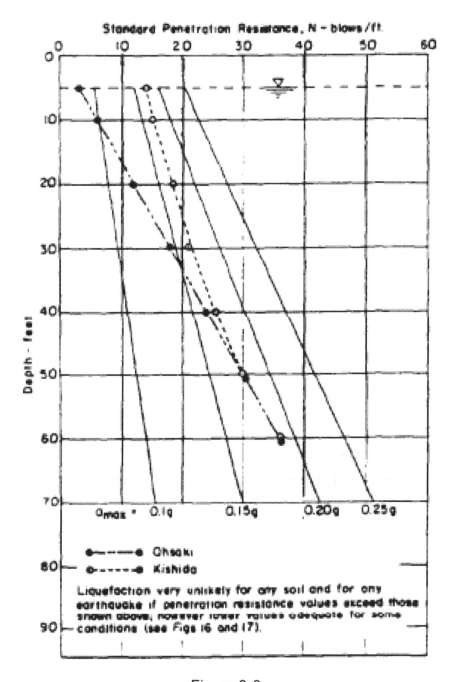

Figure 3-8

Penetration resistance values for which liquefaction

is unlikely to occur under any conditions

CHAPTER 4: ARCH DAM EARTHQUAKE ANALYSIS

4.1 INTRODUCTION. A dynamic method of analysis is required to properly assess the safety of existing concrete arch dams and to evaluate proposed designs for new dams that are located in regions with significant seismicity. Dynamic analysis is also performed to determine the adequacy of structural modifications proposed to improve the seismic performance of old dams. The prediction of the actual dynamic response of arch dams to earthquake loadings is a very complicated problem and depends on several factors including intensity and characteristics of the design earthquakes, interaction of the dam with the foundation rock and reservoir water, computer modeling, and the material properties used in the analysis. Detailed descriptions of the recommended dynamic analysis procedures are provided in the "Theoretical Manual for Analysis of Arch Dams" (Ghanaat 1993b). Guidance concerning the seismic studies needed to specify the design earthquake ground motions, methods of analysis, parameters influencing the dam response, and the presentation and evaluation of the analysis results are discussed in this chapter.

4.2 GEOLOGICAL-SEISMOLOGICAL INVESTIGATION. Estimation of appropriate seismic excitation parameters is an important aspect of the seismic design, analysis, and evaluation of new and existing dams. Concrete arch dams built in seismic regions may be subjected to ground shaking due to an earthquake at the dam site or, more likely, to ground motions induced by distant earthquakes. In addition, large dams may experience earthquakes triggered at the dam site immediately following the reservoir impoundment or during a rapid drawdown. However, such reservoir-induced earthquakes are usually no greater than those to be expected without the reservoir, and they do not augment the seismicity of the region. The estimation of future earthquake ground motions at a dam site requires geological, seismological, geophysical, and geotechnical investigations. The primary purposes of these studies are to establish the tectonic and geologic setting at and in the vicinity of the dam site, to identify active faults and seismic sources, to collect and analyze the historic and instrumental seismic data, and to study the foundation conditions at the dam site that form the basis for estimating the ground motions. However, the lack of necessary data or difficulty in obtaining them, as well as numerous uncertainties

© J. Paul Guyer   2017   135

associated with the source mechanism and the seismic wave propagation, often complicate the estimation process of ground motions. Therefore, at the present time seismic parameters for dam projects are approximated by empirical relations and through simplified procedures that decouple or neglect the effects of less understood phenomena. The primary factors that must be considered in determination of the seismic parameters for dam projects are discussed in the following paragraphs.

4.2.1 REGIONAL GEOLOGIC SETTING. A study of regional geology is required to understand the overall geologic setting and seismic history of a dam site. The study area, as a minimum, should cover a 100 km radius around the site. But in some cases it may be extended to as far as 300 km in order to include all significant geologic features such as major faults and to account for area-specific attenuation of earthquake ground motion with distance. A typical geologic study consists of: Description of the plate tectonic setting of the dam region together with an account of recent movements. Regional geologic history and physiographic features. Description of geologic formations, rock types, soil deposits. Compilation of active faults in the site region and assessment of the capability of faults to generate earthquakes. Characterization of each capable fault in terms of its maximum expected earthquake, recurrence intervals, total fault length, slip rate, slip history, and displacement per event, etc. Field work such as an exploratory trench or bulldozer cuts may also be required to evaluate the seismic history.

4.2.2 REGIONAL SEISMICITY. The seismic history of a region provides information on the occurrence of past earthquakes that help to identify seismicity patterns and, thus, give an indication of what might be expected in the future. Procedures for estimating the ground motion parameters at a particular site are primarily based on historic and instrumentally recorded earthquakes and other pertinent geologic considerations. It is important, therefore, to carefully examine such information for accuracy, completeness and consistency. When possible, the following investigations may be required: Identification of seismic sources significant to the site, usually within about a 200-km radius. Development of a catalog of the historical and instrumentally recorded earthquakes for the dam site region. The data, whenever possible, should include locations, magnitudes or epicentral intensity, date and time of occurrence, focal depth, and focal mechanism. Illustration of the compiled information by means of appropriate regional and local seismicity maps. Analysis of seismicity data to construct recurrence curves of the frequency of earthquakes for the dam region, to examine spatial patterns of epicenters for possible connection with the identified geologic structures, and to evaluate the catalog for completeness and accuracy. A review of the likelihood of reservoir induced seismicity (RIS) at the dam site, although this is not expected to influence the design earthquake parameters as previously mentioned.

4.2.3 LOCAL GEOLOGIC SETTING. Local geology should be studied to evaluate some of the site-specific characteristics of the ground motion at the dam site. Such data include rock types, surface structures, local faults, shears and joints, and the orientation and spacing of joint systems. In some cases, there may be geologic evidence of primary or sympathetic fault movement through the dam foundation. In those situations, a detailed geologic mapping, and geophysical and geotechnical exploration should be carried out to assess the potential, amount, and the type of such movements at the dam site.

4.3 DESIGN EARTHQUAKES. The geological and seismological investigations described in the previous paragraph provide the basis for estimating the earthquake ground motions to be used in the design and analysis of arch dams. The level of such earthquake ground motions depend on the seismic activity in the dam site vicinity, source-to-site distance, length of potential fault ruptures, source mechanism, surface geology of the dam site, and so on. Two approaches are available for estimating the ground motion parameters: deterministic and probabilistic. Both approaches require specification of seismic sources, assessing maximum magnitudes for each of the sources, and selecting ground motion attenuation relationships. The probabilistic analysis requires the additional specification of the frequency of earthquake recurrence for each of the sources in order to evaluate the likelihood of exceeding various level of ground motion at the site. The earthquake ground motions for which arch dams should be designed or analyzed include OBEs and MDEs. The ground motions defined for each of these earthquakes are discussed in the following paragraphs.

4.3.1 OPERATING BASIS EARTHQUAKE (OBE). The OBE is defined as the ground motion with a 50 percent probability of being exceeded in 100 years. In design and safety evaluation of arch dams, an OBE event should be considered as an unusual loading condition. The dam, its appurtenant structures, and equipment should remain fully operational with minor or no damage when subjected to earthquake ground motions not exceeding the OBE.

4.3.2 MAXIMUM DESIGN EARTHQUAKE (MDE). The MDE is the maximum level of ground motion for which the arch dam should be analyzed. The MDE is usually equated to the MCE which, by definition, is the largest reasonably possible earthquake that could occur along a recognized fault or within a particular seismic source zone. In cases where the dam failure poses no danger to life or would not have severe economic consequences, an MDE less than the MCE may be used for economic reasons. An MDE event should be considered as an extreme loading condition for which significant damage is acceptable, but without a catastrophic failure causing loss of life or severe economic loss.

4.3.3 RESERVOIR-INDUCED EARTHQUAKE (RIE). The reservoir-induced earthquake is the maximum level of ground motion that may be triggered at the dam site during filling, rapid drawdown, or immediately following the reservoir impoundment. Statistical analysis of the presumed RIE cases have indicated a relation between the occurrence of RIE and the maximum water depth, reservoir volume, stress regime, and local geology. The likelihood of an RIE is normally considered for dams higher than about 250 feet and reservoirs with capacity larger than about 105 acre-feet, but the possibility of an RIE occurring at new smaller dams located in tectonically sensitive areas should not be ruled out. The possibility of RIE's should therefore be considered when designing new high dams, even if the region shows low historical seismicity. The determination of whether the RIE should be considered as a dynamic unusual or a dynamic extreme loading condition should be based on the probability of occurrence but recognizing that the RIE is no greater than the expected earthquake if the reservoir had not been built.

4.4 EARTHQUAKE GROUND MOTIONS. The earthquake ground motions are characterized in terms of peak ground acceleration, velocity, or displacement values, and seismic response spectra or acceleration time histories. For the evaluation of arch dams, the response spectrum and/or time-history representation of earthquake ground motions should be used. The ground motion parameters for the OBE are determined based on the probabilistic method. For the MCE, however, they are normally estimated by deterministic analysis, but a probabilistic analysis should also be considered so that the likelihood of a given intensity of ground motion during the design life of the dam structure can be determined. The earthquake ground motions required as input for the seismic analysis of arch dams are described in the following subparagraphs.

4.4.1 DESIGN RESPONSE SPECTRA. The ground motion used for the seismic analysis of arch dams generally is defined in the form of smooth response spectra and the associated acceleration time histories. In most cases site specific response spectra are required, except when the seismic hazard is very low; in which case a generic spectral shape such as that provided in most building codes may suffice. When site-specific response spectra are required, the effects of magnitude, distance, and local geological conditions on the amplitude and frequency content of the ground motions should be considered. In general, the shape of the response spectrum for an OBE event is different from that for the MCE, due to differences in the magnitude and the earthquake sources as shown in Figure 7-1. Thus, two separate sets of smooth response spectra may be required, one for the OBE and another for the MCE. The smooth response spectra for each design earthquake should be developed for both horizontal and vertical components of the ground motion. The design spectra are typically developed for 5 percent damping. Estimates for other damping values can be obtained using available relationships (Newmark and Hall 1982). The vertical response spectra can be estimated using the simplified published relationships between the vertical and horizontal spectra which will be described in a future engineer manual. The relationship used should recognize the significant influence of the source-to-site distance and of the particular period range (£ 0.2 sec) on the vertical response spectra.

4.4.2 ACCELERATION TIME HISTORIES. When acceleration time histories of ground motions are used as seismic input for the dynamic analysis of arch dams, they should be established with the design response spectra and should have appropriate strong motion duration and number of peaks. The duration of strong motion is commonly measured by the bracketed duration. This is the duration of shaking between the first and last accelerations of the accelerogram exceeding 0.05 g.

4.4.2.1 ACCELERATION TIME HISTORIES are either selected from recorded ground motions appropriate to the site, or they are synthetically developed or modified from one or more ground motions. In the first approach, several records are usually required to ensure that the response spectra of all records as a whole do not fall below the smooth design spectra. This procedure has the advantage that the dam is analyzed for natural motions, several dynamic analyses should be performed. In addition, the response spectrum of individual records may have peaks that substantially exceed the design response spectra.

4.4.2.2 ALTERNATIVELY, ACCELERATION TIME HISTORIES are developed either by artificially generating an accelerogram or by modifying a recorded accelerogram so that the response spectrum of the resulting accelerogram closely matches the design response spectra. The latter technique is preferred, because it starts with a natural accelerogram and thus preserves the duration and phasing of the original record and produces time histories that look natural. An example of this procedure which shows a good match with the smooth design response spectrum is demonstrated in Figure 7-2.

4.4.2.3 FOR LARGE THIN ARCH DAMS with fundamental periods near 0.5 to 1 sec located at close distances to the earthquake source, it is desirable to include a strong intermediate-to-long period pulse (0.5 to 5 sec) to account for the "fling" characteristic of near-source ground motion.

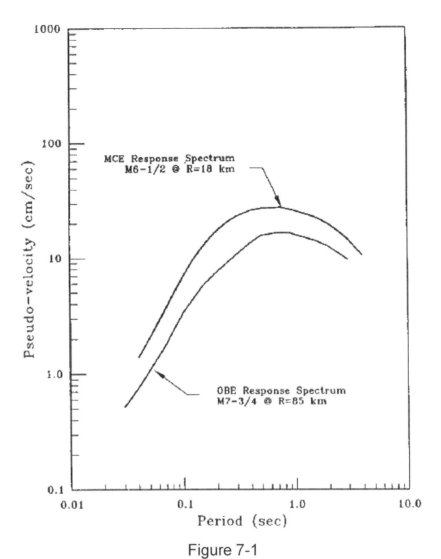

Figure 7-1

Smooth design response-spectrum examples for OBE

and MCE events for 5 percent damping

4.5 FINITE ELEMENT MODELING FACTORS AFFECTING DYNAMIC RESPONSE.
Dynamic analysis of arch dams for earthquake loading should be based on a 3-D idealization of the dam-water-foundation system which accounts for the significant interaction effects of the foundation rock and the impounded water. To compute the linear response of the dam, the concrete arch and the foundation rock are modeled by standard finite elements, whereas the interaction effects of the impounded water can be represented with any of three different level of refinement. In addition, the dynamic response of arch dams is affected by the damping and by the intensity and spatial variation of the seismic input. These factors and the finite element modeling of various components of an arch dam are discussed in the following sections.

4.5.1 ARCH DAM. The finite element model of an arch dam for dynamic analysis is essentially identical to that developed for the static analysis. In a linear-elastic analysis, the arch dam is modeled as a monolithic structure with no allowance for the probable contraction joint opening during earthquake excitation. Thin and moderately thin arch dams are adequately modeled by a single layer of shell elements, whereas thick gravity-arch dams should be represented by two or more layers of solid elements through the dam thickness. The size of the mesh should be selected following the general guidelines for static analysis and shown in Figure 6-1. In addition, the dynamic response of the appurtenant structures attached to the dam may be significant and also should be considered. For example, the power intakes attached to the dam may include free-standing cantilevers that could vibrate during the earthquake shaking. The power intakes in this case should be included as part of the dam model to ensure that the dam stresses induced by the vibration of these components are not excessive.

4.5.2 DAM-FOUNDATION ROCK INTERACTION. Arch dams are designed to resist the major part of the water pressures and other loads by transmitting them through arch action to the canyon walls. Consequently, the effects of foundation rock on the earthquake response of arch dams are expected to be significant and must be considered in the dynamic analysis. However, a complete solution of the dam foundation interaction effects is very complicated and such procedures have not yet been fully developed. There are two major factors contributing to this complex interaction problem. First is the lack of a 3-D model of the unbounded foundation rock region to account for energy loss due to the radiation of vibration waves. The other and even more important contributing factor is related to the prescription of spatial variation of the seismic input at the dam-foundation interface, resulting from wave propagation of seismic waves through the foundation rock and from scattering by the canyon topography. Faced with these difficulties, an overly simplified model of the foundation rock (Clough 1980) is currently used in practice. This widely used simplified model ignores inertial and damping effects and considers only the flexibility of the foundation rock. The foundation model for the dynamic analysis is therefore similar to that described for the static analysis. As shown in Figures 6-1a and 6-3, an appropriate volume of the foundation rock should be idealized by the finite element discretization of the rock region. Each foundation element is represented by a solid element having eight or more nodes and characterized by its dynamic deformation modulus and Poisson's ratio.

4.5.2.1 SHAPE OF FOUNDATION MODEL. Using the finite element procedure, a foundation model can be developed to match the natural topography of the foundation rock region. However, such a refined model is usually not required in practice. Instead, a prismatic model employed in the GDAP program and described in Chapter 6 may be used. This foundation model depicted in Figure 6-1a is constructed on semicircular planes cut into the canyon walls normal to the dam-foundation contact surface; in moving from the base to the dam crest, each semicircle is rotated about a diameter always oriented in the upstream-downstream direction.

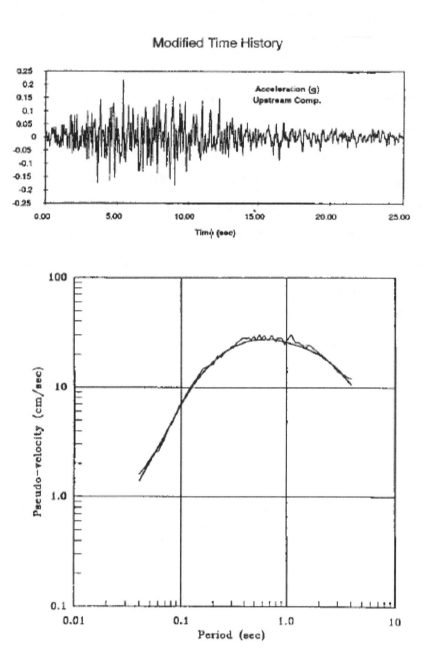

Figure 7-2

Comparison of response spectrum of modified time history and smooth-design response spectrum for 5 percent damping

4.5.2.2 SIZE OF FOUNDATION MODEL. The size of the foundation model considered in GDAP is controlled by the radius (Rf) of the semicircular planes described in the previous paragraph. In the static analysis discussed previously, Rf was selected so that the static displacements and stresses induced in the dam were not changed by further increase of the foundation size. In the dynamic analysis, the natural frequencies and mode shapes of vibration control the dam response to earthquakes. Therefore, the size of a foundation model should be selected so that the static displacements and stresses, as well as the natural frequencies and mode shapes, are accurately computed. The natural frequencies of the dam-foundation system decrease as the size of the flexible foundation rock increases (Clough et al. 1985 and Fok and Chopra 1985), but for the massless foundation, the changes are negligible when the foundation size Rf is greater than one dam height, except for the foundation rocks with very low modulus of elasticity. For most practical purposes, a massless foundation model with Rf equal to one dam height is adequate. However, when the modulus ratio of the rock to concrete is less than one-half, a model with Rf equal to two times dam height should be used.

4.5.3 DAM-WATER INTERACTION. Interaction between the dam and impounded water is an important factor affecting the dynamic response of arch dams during earthquake ground shaking. In the simplest form, this interaction can be represented by an "added mass" attached to the dam first formulated by Westergaard (1933). A more accurate representation of the added mass is obtained using a finite element formulation which accounts for the complicated geometry of the arch dam and the reservoir (Kuo 1982 (Aug)). Both approaches, however, ignore compressibility of water and the energy loss due to radiation of pressure waves in the upstream direction and due to reflection and refraction at the reservoir bottom. These factors have been included in a recent and more refined formulation (Fok and Chopra 1985 (July)), but computation of the resulting frequency-dependent hydrodynamic pressure terms requires extensive efforts and requires consideration of a range of reservoir-bottom reflection coefficients.

4.5.3.1 GENERALIZED WESTERGAARD ADDED MASS. Westergaard (1933) demonstrated that the effects of hydrodynamic pressures acting on the vertical face of a rigid gravity dam could be represented by an added mass attached to the dam, if the compressibility of water is neglected. A general form of this incompressible added-mass concept has been applied to the analysis of arch dams (Kuo 1982 (Aug)). This generalized formulation, also described by Ghanaat (1993b), is based on the same parabolic pressure distribution in the vertical direction used by Westergaard, but it recognizes the fact that the hydrodynamic pressures acting on the curved surface of an arch dam are due to the total accelerations normal to the dam face. Although the resulting added mass calculated in this manner is often used in the analysis of arch dams, it does not properly consider the hydrodynamic effects. In fact, there is no rational basis for the assumed parabolic pressure distribution used for the arch dams, because limitations imposed in the original Westergaard formulation are violated. The original Westergaard formulation assumed a rigid dam with a vertical upstream face and an infinite reservoir. However, the procedure is very simple and provides a reasonable estimate of the hydrodynamic effects for preliminary or feasibility analysis. The generalized added-mass formulation has been implemented in the GDAP program and is available as an option. The program automatically calculates the added mass for each nodal point on the upstream face of the dam; the resulting added mass of water is then added to the mass of concrete to account for the hydrodynamic forces acting on the dam.

4.5.3.2 INCOMPRESSIBLE FINITE ELEMENT ADDED MASS. In more refined analyses of new and existing arch dams, the effects of reservoir-water interaction due to seismic loading is represented by an equivalent added mass of water obtained from the hydrodynamic pressures acting on the face of the dam. The procedure is based on a finite element solution of the pressure wave equation subjected to appropriate boundary conditions (Kuo 1982 (Aug) and Ghanaat (1993b)). The nodal point pressures of the incompressible water elements are the unknowns. The bottom and sides of the reservoir, as well as a vertical plane at the upstream end, are assumed to be rigid. In addition, the hydrodynamic pressures at the water-free surface are set to zero; thus the effects of surface waves are neglected, but these have little effect on the seismic response. In general, a finite element model of the reservoir water can be developed to match the natural canyon topography, but a prismatic reservoir model available in the GDAP program is quite adequate in most practical situations. The GDAP reservoir model is represented by a cylindrical surface generated by translating the dam-water interface nodes in the upstream direction as shown in Figure 7-3a. The resulting water nodes generated in this manner match those on the dam face and are usually arranged in successive planes parallel to the dam axis, with the distance between the planes increasing with distance from the dam. Experience shows that the reservoir water should include at least three layers of elements that extend upstream a distance at least three times the water depth.

4.5.3.3 COMPRESSIBLE WATER WITH ABSORPTIVE RESERVOIR BOTTOM. The added mass representation of hydrodynamic effects ignores both water compressibility effects and the energy absorption mechanism at the reservoir bottom. These factors have been included in a recent formulation of the dam-water interaction mechanism which is fully described by Fok and Chopra (1985 (July)). It introduces frequency-dependent hydrodynamic terms in the equations of motion that can be interpreted as an added mass, an added damping, and an added force. The added damping term arises from the refraction of hydrodynamic pressure waves into the absorptive reservoir bottom and also from the propagation of pressure waves in the upstream direction. The energy loss at the reservoir bottom is approximated by the wave reflection coefficient a, which is defined as the ratio of reflected-to-incident wave amplitude of a pressure wave striking the reservoir bottom. The values of a can be varied from a = 1.0, for a rigid, nonabsorptive boundary similar to that used in the GDAP model, to a = 0.0, indicating total absorption.

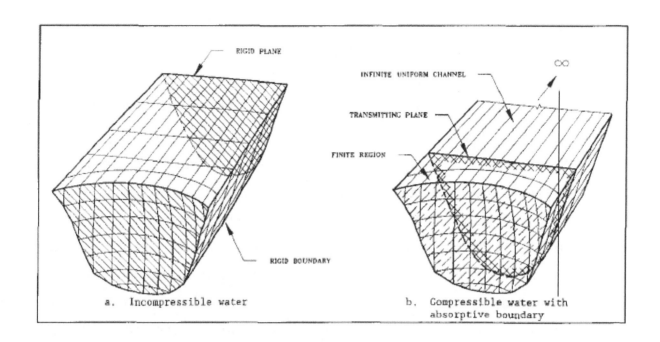

RIGID PLANE

INFINITE UNIFORM CHANNEL

TRANSMITTING PLANE

FINITE REGION

∞

RIGID BOUNDARY

a.   Incompressible water

b.   Compressible water with
     absorptive boundary

Figure 7·3

Finite element models of fluid domain with and without water compressibility

4.5.3.3.1 THE RESPONSE ANALYSIS OF AN ARCH DAM including the effects of damwater interaction, water compressibility, and reservoir-bottom absorption can be performed using the EACD-3D program (Fok, Hall, and Chopra 1986 (July)). The finite element idealizations of the dam and foundation rock employed in this program are essentially equivalent to those employed by the GDAP program; the fluid region near the dam is modeled by liquid finite elements similar to those in GDAP, but, unlike the GDAP, these elements are of compressible water and are connected to a uniform channel extending to infinity to permit pressure waves to radiate away from the dam (Figure 7-3b).

4.5.3.3.2 ANOTHER MAJOR DIFFERENCE of the EACD-3D model is that the reservoir boundary is absorptive and thus dissipation of hydrodynamic pressure waves in the reservoir bottom materials is permitted. However, this method requires considerable computational effort and is too complicated for most practical applications. An even more important consideration is the lack of guidance or measured data for determining an appropriate a factor for use in the analysis. Consequently, such analyses must be repeated for a range of a factors in order to establish a lower and upper bound estimate of the dam response. It is also important to note that the significance of water compressibility depends on the dynamic characteristics of the dam and the impounded water. Similar to gravity dams (Chopra 1968), the effects of water compressibility for an arch dam can be neglected if the ratio of the natural frequency of the reservoir water to the natural frequency of the arch dam-foundation system without water is greater than 2.

4.5.4 DAMPING. Damping has a significant effect on the response of an arch dam to earthquake and other dynamic loads. The energy loss arises from several sources including the concrete arch structure, foundation rock, and the reservoir water. Dissipation of energy in the concrete arch structure is due to internal friction within the concrete material and at construction joints. In the foundation rock this energy loss is facilitated by propagation of elastic waves away from the dam (radiation damping) and by hysteretic losses due to sliding on cracks and fissures within the rock volume. An additional source of damping, as discussed in paragraph 7-5c(3), is associated with the energy loss due to refraction of hydrodynamic pressure waves into the reservoir bottom materials and propagation of pressure waves in the upstream direction.

4.5.4.1 THE CURRENT STANDARD EARTHQUAKE ANALYSIS of arch dams is based on a massless foundation rock model and employs incompressible added mass for representing the hydrodynamic effects. In this type of analysis, only the material damping associated with the concrete structure is explicitly considered. The overall damping constant for the entire model in such linear elastic analyses is normally specified based on the amplitude of the displacements, the opening of the vertical contraction joints, and the amount of cracking that may occur in the concrete arch. Considering that the measured damping values for concrete dams subjected to earthquake loading are scarce and that the effects of contraction joints, lift surfaces, and cracks cannot be precisely determined, the damping value for a moderate shaking such as an OBE event should be limited to 5 percent.

4.5.4.2 HOWEVER, UNDER THE MCE EARTHQUAKE ground motions, damping constants of 7 or 10 percent may be used depending on the level of strains developed in the concrete and the amount of nonlinear joint opening and/or cracking that occurs. In more severe MCE conditions, especially for large dams, additional damping can be incorporated in the analysis by employing a dam-water interaction model which includes water compressibility and permits for the dissipation of energy at the reservoir boundary.

4.6. METHOD OF ANALYSIS. The current earthquake response analysis of arch dams is based on linear-elastic dynamic analysis using the finite element procedures. It is assumed that the concrete dam and the interaction mechanisms with the foundation rock and the impounded water exhibit linear-elastic behavior. Using this method, the arch dam and the foundation rock are treated as 3-D systems idealized by the finite element discretization discussed in previous paragraphs and in Chapter 6. Under the incompressible added-mass assumption for the impounded water, the response analysis is performed using the response-spectrum modal-superposition or the time-history method. For the case of compressible water, however, the response of the dam to dynamic loads must be evaluated using a frequency-domain procedure, in order to deal with the frequency-dependent hydrodynamic terms. These methods of analyses are discussed in the following paragraphs.

4.6.1 RESPONSE-SPECTRUM ANALYSIS. The response-spectrum method of analysis uses a response-spectrum representation of the seismic input motions to compute the maximum response of an arch dam to earthquake loads. This approximate method provides an efficient procedure for the preliminary analyses of new and existing arch dams. It may also be used for the final analyses, if the calculated maximum stress values are sufficiently less than the allowable stresses of the concrete. Using this procedure, the maximum response of the arch dam is obtained by combining the maximum responses for each mode of vibration computed separately.

4.6.1.1 A COMPLETE DESCRIPTION OF THE METHOD is given in the theoretical manual by Ghanaat (1993b). First, the natural frequencies and mode shapes of undamped free vibration for the combined dam-water-foundation system are evaluated; the free vibration equations of motion are assembled considering the mass of the dam-water system and the stiffness of the combined dam and foundation rock models. The maximum response in each mode of vibration is then obtained from the specified response spectrum for each component of the ground motion, using the modal damping and the natural period of vibration for each particular mode. The same damping constant is used in all modes as represented by the response- spectrum curves. Since each mode reaches its maximum response at a different time, the total maximum response quantities for the dam, such as the nodal displacements and the element stresses, are approximated by combining the modal responses using the square root of the sum of the squares (SRSS) or complete quadratic combination (CQC) procedure. Finally, the resulting total maximum responses evaluated independently for each component of the earthquake ground motion are further combined by the SRSS method for the three earthquake input components, two horizontal and one vertical.

4.6.1.2 FOR A LINEAR-ELASTIC RESPONSE, only a few lower modes of vibration are needed to express the essential dynamic behavior of the dam structure. The appropriate number of vibration modes required in a particular analysis depends on the dynamic characteristics of the dam structure and on the nature of earthquake ground motion. But, in all cases, a sufficient number of modes should be included so that at least 90 percent of the "exact" dynamic response is achieved. Since the "exact" response values are not known, a trial-anderror procedure may be adapted, or it may be demonstrated that the participating effective modal masses are at least 90 percent of the total mass of the structure.

4.6.2 TIME-HISTORY ANALYSIS. Time-history analysis should be performed when the maximum stress values computed by the response-spectrum method are approaching or exceeding the tensile strength of the concrete. In these situations, linear-elastic time-history analyses are performed to estimate the maximum stresses more accurately as well as to account for the time-dependent nature of the dynamic response. Time-history analyses provide not only the maximum stress values, but also the simultaneous, spatial extent and number of excursions beyond any specified stress value. Thus, they can indicate if the calculated stresses beyond the allowable values are isolated incidents or if they occur repeatedly and over a significant area.

4.6.2.1 THE SEISMIC INPUT IN TIME-HISTORY analyses is represented by the acceleration time histories of the earthquake ground motion. Three acceleration records corresponding to three components of the specified earthquake are required; they should be applied at the fixed boundaries of the foundation model in the channel, across the channel, and in the vertical directions. The acceleration time histories are established following the procedures described.

4.6.2.2 THE STRUCTURAL MODELS OF THE DAM, foundation rock, and the impounded water for a time-history analysis are identical to those developed for response-spectrum analysis. However, the solution to the equations of motion is obtained by a step-by-step numerical integration procedure. Two methods of solution are available: direct integration and mode superposition (Ghanaat, technical report in preparation). In the direct method, step-by-step integration is applied to the original equations of motion with no transformation being carried out to uncouple them. Hence, this method requires that the damping matrix to be represented is in explicit form. In practice, this is accomplished using Raleigh damping (Clough and Penzien 1975), which is of the form

$$c = a_0 \, m + a_1 \, k$$

where coefficients a_0 and a_1 are obtained from two given damping ratios associated with two frequencies of vibration. The direct integration method is most effective when the response is required for a relatively short duration. Otherwise, the mode superposition method in which the step-by-step integration is applied to the uncoupled equations of motion will be more efficient. In the mode superposition method, first the undamped vibration mode shapes and frequencies are calculated, and the equations of motion are transformed to those coordinates. Then the response history for each mode is evaluated separately at each time-step, and the calculated modal response histories are combined to obtain the total response of the dam structure. It should be noted that the damping in this case is expressed by the modal damping ratios and need not be specified in explicit form.

4.7 EVALUATION AND PRESENTATION OF RESULTS. The earthquake performance of arch dams is currently evaluated using the numerical results obtained from a linear-dynamic analysis. The results of linear analysis provide a satisfactory estimate of the dynamic response to low- or moderate-intensity OBE earthquake motions for which the resulting deformations of the dam are within the linear-elastic range. In this case, the performance evaluation is based on simple stress checks in which the calculated elastic stresses are compared with the specified strength of the concrete. Under the MCE ground motions, it is possible that the calculated stresses would exceed the allowable values and that significant damage could occur. In such extreme cases, the dam should retain the impounded water without rupture, but the actual level of damage can be estimated only by a nonlinear analysis that takes account of the basic nonlinear behavior mechanisms such as the joint opening, tensile cracking, and the foundation failure. However, a complete nonlinear analysis is not currently possible, and linear analysis continues to be the primary tool for assessing the seismic performance of arch dams subjected to damaging earthquakes. Evaluation of the seismic performance for the MCE is more complicated, it requires some judgement and elaborate interpretations of the results before a reasonable estimate of the expected level of damage can be made or the possibility of collapse can be assessed.

4.7.1 EVALUATION OF RESPONSE-SPECTRUM ANALYSIS. The first step in response spectrum analysis is the calculation of vibration mode shapes and frequencies. The mode shapes and frequencies provide insight into the basic dynamic response behavior of an arch dam. They provide some advance indication of the sensitivity of the dynamic response to earthquake ground motions having various frequency contents. Figure 7-4 demonstrates a convenient way for presenting the mode shapes. In this figure the vibration modes are depicted as the plot of deflected shapes along the arch sections at various elevations. After the calculation of mode shapes and frequencies, the maximum dynamic response of the dam structure is computed. These usually include the maximum nodal displacements and element stresses. In particular, the element stresses are the primary response quantity used for the evaluation of earthquake performance of the dam.

4.7.1.1 DYNAMIC RESPONSE. The basic results of a response-spectrum analysis include the extreme values of the nodal displacements and element stresses due to the earthquake loading. As discussed earlier, these extreme response values are obtained by combining the maximum responseś developed in each mode of vibration using the SRSS or CQC combination rule. In addition, they are further combined by the SRSS method to include the effects of all three components of the earthquake ground motion. Thus, the resulting dynamic response values obtained in this manner have no sign and should be interpreted as being either positive or negative. For example the response-spectrum stress values are assumed to be either tension or compression.

4.7.1.2 TOTAL RESPONSE. The evaluation of earthquake performance of an arch dam using the response-spectrum method of analysis involves comparison of the total stresses due to both static and earthquake loads with the expected strength of the concrete. To obtain the total stress values, the response spectrum estimate of the dynamic stresses (s_d) should be combined with the effects of static loads (s_{st}). The static stresses in a dam prior to the earthquake are computed using the procedures described. The static loads to be considered include the self-weight, hydrostatic pressures, and the temperature changes that are expected during the normal operating condition as discussed. Since response-spectrum stresses have no sign, combination of static and dynamic stresses should consider dynamic stresses to be either positive or negative, leading to the maximum values of total tensile or compressive stresses:

$$\sigma_{max} = \sigma_{st} \pm \sigma_d$$

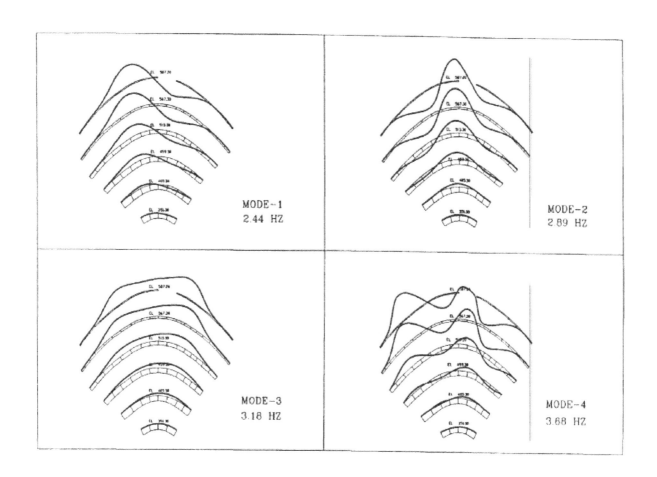

Figure 7-4

Four lowest vibration modes of Portugues Arch Dam

4.7.1.2.1 THIS COMBINATION OF STATIC AND DYNAMIC stresses is appropriate if Sst and Sd are oriented similarly. This is true for arch or cantilever stresses at any point on the dam surface, but generally is not true for the principal stresses. In fact, it is not possible to calculate the principal stresses from a response-spectrum analysis, because the maximum arch and cantilever stresses do not occur at the same time; therefore, they cannot be used in the principal stress formulas.

4.7.1.2.2 THE COMPUTED TOTAL ARCH AND CANTILEVER stresses for the upstream and downstream faces of the dam should be displayed in the form of stress contours as shown in Figure 7-5. These represent the envelopes of maximum total arch and cantilever stresses on the faces of the dam, but because they are not concurrent they cannot be combined to obtain envelopes of principal stresses, as was mentioned previously.

4.7.2 RESULTS OF TIME-HISTORY ANALYSIS. Time-history analysis computes time-dependent dynamic response of the dam model for the entire duration of the earthquake excitation. The results of such analyses provide not only the maximum response values, but also include time-dependent information that must be examined and interpreted systematically. Although evaluation of the dynamic response alone may sometimes be required, the final evaluation should be based on the total response which also includes the effects of static loads.

4.7.2.1 MODE SHAPES AND NODAL DISPLACEMENTS. Vibration mode shapes and frequencies are required when the mode-superposition method of time-history analysis is employed. But it is also a good practice to compute them for the direct method. The computed vibration modes may be presented as shown in Figure 7-4 and discussed previously. The magnitude of nodal displacements and deflected shape of an arch dam provide a visual means for the evaluation of earthquake performance. As a minimum, displacement time histories for several critical nodal points should be displayed and evaluated. Figure 7-6 shows an example of such displacement histories for a nodal point on the dam crest.

4.7.2.2 ENVELOPES OF MAXIMUM AND MINIMUM ARCH AND CANTILEVER STRESSES. Examination of the stress results for a time-history analysis should start with presentation of the maximum and minimum arch and cantilever stresses. These stresses should be displayed in the form of contour plots for the upstream and downstream faces of the dam. The contour plots of the maximum arch and cantilever stresses represent the largest computed tensile (positive) stresses at all locations in the dam during the earthquake ground shaking (Figure 7-5). Similarly, the contour plots of the minimum stresses represent the largest compressive (negative) arch and cantilever stresses in the dam. The maximum and the minimum stresses at different points are generally reached at different instants of time. Contour plots of the maximum arch and cantilever stresses provide a convenient means for identifying the overstressed

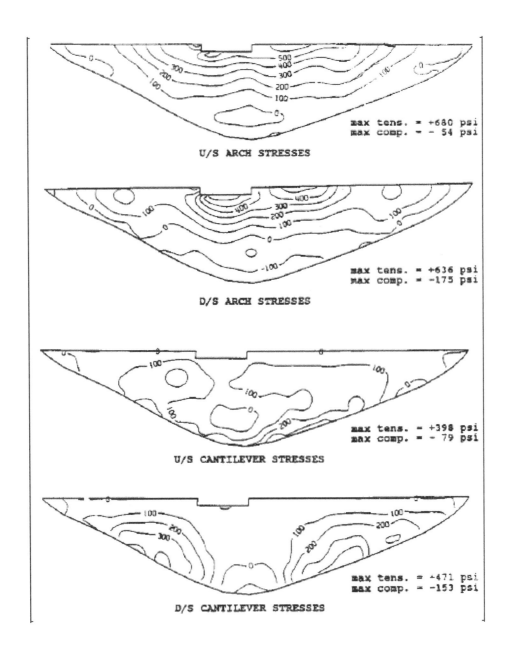

Figure 7-5

Envelope of maximum arch and cantilever stress (in psi)

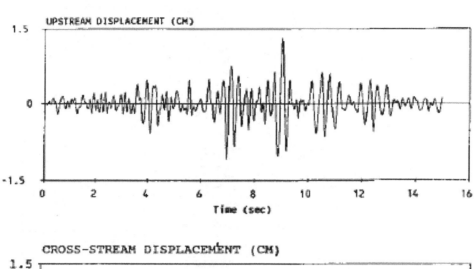

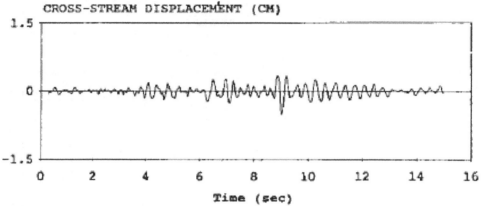

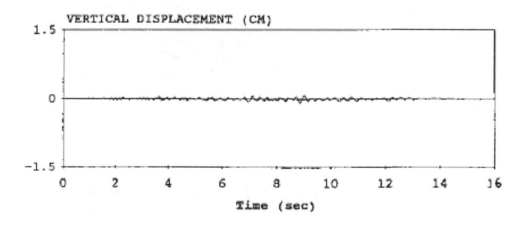

Figure 7-6

Displacement time history of a crest node in upstream,

cross-stream, and vertical direction

areas where the maximum stresses approach or exceed tensile strength of the concrete.

Based on this information, the extent and severity of tensile stresses are determined, and

if necessary, further evaluation which accounts for the time-dependent nature of the dynamic response should be made as described in the following sections. Contour plots of the minimum stresses show the extreme compressive stresses that the dam would experience during the earthquake loading. The compressive stresses should be examined to ensure that they meet the specified safety factors for the dynamic loading.

4.7.2.3 CONCURRENT STRESSES. The envelopes of maximum and minimum stresses discussed demonstrate the largest tensile and compressive stresses that are developed at different instants of time. They serve to identify the overstressed regions and the times at which the critical stresses occur. This information is then used to produce the concurrent (or simultaneous) state of stresses corresponding to the time steps at which the critical stresses in the overstressed regions reach their maxima. The concurrent arch and cantilever stresses in the form of contour plots (Figure 7-7) can be viewed as snap shots of the worst stress conditions.

4.7.2.4 ENVELOPES OF MAXIMUM AND MINIMUM PRINCIPAL STRESSES. The time histories of principal stresses at any point on the faces of the dam are easily computed from the histories of arch, cantilever, and shear stresses at that point. When the effects of static loads are considered, the static and dynamic arch, cantilever, and shear stresses must be combined for each instant of time prior to the calculation of the total principal stresses for the same times. The resulting time histories of principal stresses are used to obtain the maxima and minima at all points on both faces of the dam which are then presented as vector plots as shown in Figures 7-8 and 7-9. (5) Time History of Critical Stresses. When the maximum and concurrent stresses show that the computed stresses exceed the allowable value, the time histories of critical stresses should be presented for a more detailed evaluation (Figure 7-10). In this evaluation the time histories for the largest maximum arch and cantilever stresses should be examined to determine the number of cycles that the maximum stresses exceed the allowable value. This would indicate whether the excursion beyond the allowable value is an isolated case or is repeated many times during the ground motion. The total duration that the allowable value (or cracking stress) is exceeded by these excursions should also be estimated to demonstrate whether the maximum stress cycles are merely spikes or they are of longer duration and, thus, more damaging. The number of times that the allowable stress can safely be exceeded has not yet been established. In practice, however, up to five stress cycles have been permitted based on judgement but have not been substantiated by experimental data. The stress histories at each critical location should be examined for two opposite points on the upstream and downstream faces of the dam as in Figure 7-10. For example, a pair of cantilever stress histories can demonstrate if stresses on both faces are tension, or if one is tension and the other is compression. The implication of cantilever stresses being tension on both faces is that the tensile cracking may penetrate through the dam section, whereas in the case of arch stresses, this indicates a complete separation of the contraction joint at the location of maximum tensile stresses.

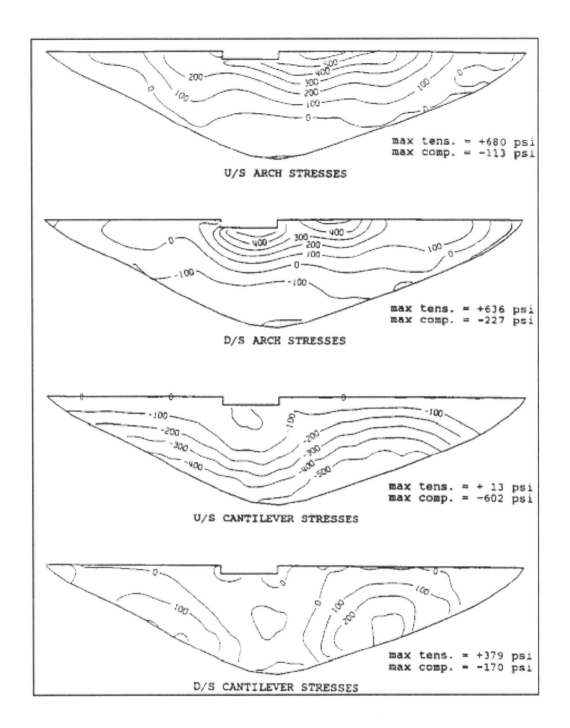

Figure 7-7

Concurrent arch and cantilever stresses (in psi) at time-step corresponding to maximum

arch stress

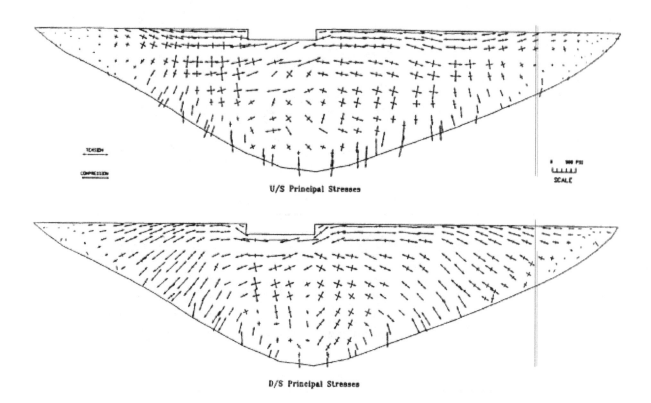

U/S Principal Stresses

D/S Principal Stresses

Figure 7-8. Envelope of maximum principal stresses with their corresponding
perpendicular pair

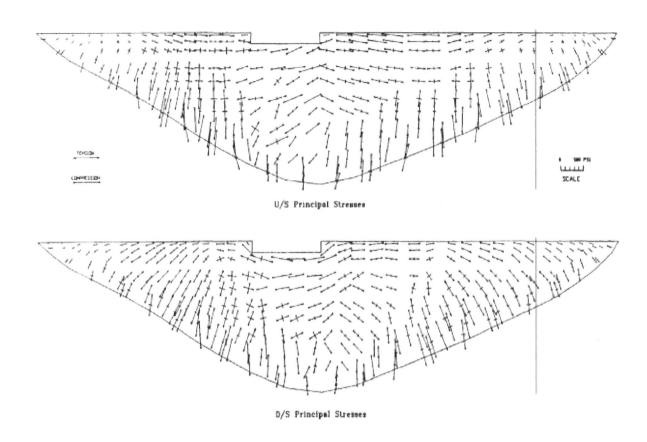

U/S Principal Stresses

D/S Principal Stresses

Figure 7-9

Envelope of maximum-minimum principal stresses with their corresponding perpendicular

pair

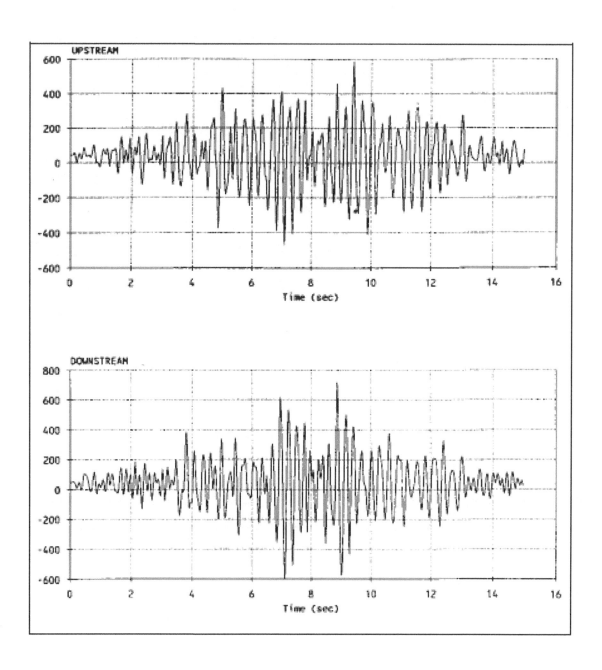

Figure 7-10

Time histories of arch stresses (in psi) at two opposite points on upstream and downstream faces of dam

CHAPTER 5: ARCH DAM CONCRETE PROPERTIES

5.1 INTRODUCTION. Unlike gravity dams that use the weight of the concrete for stability, arch dams utilize the strength of the concrete to resist the hydrostatic loads. Therefore, the concrete used in arch dams must meet very specific strength requirements. In addition to meeting strength criteria, concrete used in arch dams must meet the usual requirements for durability, permeability, and workability. Like all mass concrete structures, arch dams must keep the heat of hydration to a minimum by reducing the cement content, using low-heat cement, and using pozzolans. This chapter discusses the material investigations and mixture proportioning requirements necessary to assure the concrete used in arch dams will meet each of these special requirements. This chapter also discusses the testing for structural and thermal properties that relates to the design and analysis of arch dams, and it provides recommended values which may be used prior to obtaining test results.

5.2 MATERIAL INVESTIGATIONS. The material discussed in the next few paragraphs is intended to provide guidance in the investigations that should be performed for arch dams.

5.2.1 CEMENT. Under normal conditions the cementitious materials used in an arch dam will simply be a Type II portland cement (with heat of hydration limited to 70 cal/gm) in combination with a pozzolan. However, Type II cement may not be available in all project areas. The lack of Type II cement does not imply that massive concrete structures, such as arch dams, are not constructable. It will only be necessary to investigate how the available materials and local conditions can be utilized. For example, the heat of hydration for a Type I cement can be reduced by modifying the cement grinding process to provide a reduced fineness. Most cement manufacturers should be willing to do this since it reduces their cost in grinding the cement. However, there would not necessarily be a cost savings to the Government, since separate silos would be required to store the specially ground cement. In evaluating the cement sources, it is preferable to test each of the available cements at various fineness to determine the heat generation characteristics of each. This information is useful in performing parametric thermal studies.

5.2.2 POZZOLANS. Pozzolans are siliceous or siliceous and aluminous materials which in themselves possess little or no cementitious value; however, pozzolans will chemically react, in finely divided form and in the presence of moisture, with calcium hydroxide at ordinary temperatures to form compounds possessing cementitious properties. There are three classifications for pozzolans: Class N, Class F, and Class C.

5.2.2.1 CLASS N POZZOLANS are naturally occurring pozzolans that must be mined and ground before they can be used. Many natural pozzolans must also be calcined at high temperatures to activate the clay constituent. As a result, Class N pozzolans are not as economical as Classes F and C, if these other classes are readily available.

5.2.2.2 CLASSES F AND C are fly ashes which result from burning powdered coal in boiler plants, such as electric generating facilities. The ash is collected to prevent it from entering the atmosphere. Being a byproduct of another industry, fly ash is usually much cheaper than cement. However, some of the savings in material may be offset by the additional material handling and storage costs. Fly ash particles are spherical and are about the same fineness as cement. The spherical shape helps reduce the water requirement in the concrete mix.

5.2.2.3 IT IS IMPORTANT THAT THE POZZOLAN source produce consistent material properties, such as constant fineness and constant carbon content. Otherwise, uniformity of concrete will be affected. Therefore, an acceptable pozzolan source must be capable of supplying the total project needs.

5.2.2.4 IN MASS CONCRETE, pozzolans are usually used to replace a portion of the cement, not to increase the cementitious material content. This will reduce the amount of portland cement in the mixture proportions. Not only will this cement reduction lower the heat generated within the mass, but the use of pozzolans will improve workability, long-term strength, and resistance to attack by sulphates and other destructive agents. Pozzolans can also reduce bleeding and permeability and control alkali-aggregate reaction.

5.2.2.5 HOWEVER, WHEN POZZOLANS are used as a cement replacement, the time rate of strength gain will be adversely affected; the more pozzolan replacement, the lower the early strength. As a result, an optimization study should be performed to determine the appropriate amount of pozzolan to be used. The optimization study must consider both the long-term and the early-age strengths. The early-age strength is important because of the need for form stripping, setting of subsequent forms, and lift joint preparation. The practical limits on the percentage of pozzolan that should be used in mass concrete range from 15 to 50 percent. During the planning phases and prior to performing the optimization study, a value of 25 to 35 percent can be assumed.

5.2.3 AGGREGATES. Aggregates used in mass concrete will usually consist of natural sand, gravel, and crushed rock. Natural sands and gravels are the most common and are used whenever they are of satisfactory quality and can be obtained economically in sufficient quantity. Crushed rock is widely used for the larger coarse aggregates and occasionally for smaller aggregates including sand when suitable materials from natural deposits are not economically available. However, production of workable concrete from sharp, angular, crushed fragments usually requires more vibration and cement than that of concrete made with well-rounded sand and gravel.

5.2.3.1 SUITABLE AGGREGATE SHOULD be composed of clean, uncoated, properly shaped particles of strong, durable materials. When incorporated into concrete, it should satisfactorily resist chemical or physical changes such as cracking, swelling, softening, leaching, or chemical alteration. Aggregates should not contain contaminating substances which might contribute to deteriorating or unsightly appearance of the concrete.

5.2.3.2 THE DECISION TO DEVELOP an on-site aggregate quarry versus hauling from an existing commercial quarry should be based on an economic feasibility study of each. The study should determine if the commercial source(s) can produce aggregates of the size and in the quantity needed. If an on-site quarry is selected, then the testing of the on-site source should include a determination of the effort required to produce the aggregate. This information should be included in the contract documents for the prospective bidders.

5.2.4 WATER. All readily available water sources at the project site should be investigated during the design phase for suitability for mixing and curing water. The purest available water should be used. When a water source is of questionable quality, it should be tested in accordance with approved standards. When testing the water, the designer may want to consider including ages greater than the 7 and 28 days required in the typical specification. This is especially true when dealing with a design age of 180 or 360 days, because the detrimental effects of the water may not become apparent until the later ages. Since there are usually differences between inplace concrete using an on-site water source and lab mix designs using ordinary tap water, the designer may want to consider having the lab perform the mix designs using the anticipated on-site water.

5.2.5 ADMIXTURES. Admixtures normally used in arch dam construction include air-entraining, water-reducing, retarding, and water-reducing/ retarding admixtures. During cold weather, accelerating admixtures are sometimes used. Since each of these admixtures is readily available throughout the United States, no special investigations are required.

5.3 MIX DESIGNS. The mixture proportions to be used in the main body of the dam should be determined by a laboratory utilizing materials that are representative of those to be used on the project. The design mix should be the most economical one that will produce a concrete with the lowest practical slump that can be efficiently consolidated, the largest maximum size aggregate that will minimize the required cementitious materials, adequate early-age and later-age strength, and adequate long-term durability. In addition, the mix design must be consistent with the design requirements discussed in the other chapters of this manual. A mix design study should be performed to include various mixture proportions that would account for changes in material properties that might reasonably be expected to occur during the construction of the project. For example, if special requirements are needed for the cement (such as a reduced fineness), then a mix with the cement normally available should be developed to account for the possibility that the special cement may not always be obtainable. This would provide valuable information during construction that could avoid prolonged delays.

5.3.1 COMPRESSIVE STRENGTH. The required compressive strength of the concrete is determined during the static and dynamic structural analyses. The mixture proportions for the concrete should be selected so that the average compressive strength (fcr) exceeds the required compressive strength (f'c) by a specified amount. The amount that fcr should exceed f'c depends upon the classification of the concrete (structural or nonstructural) and the availability of test records from the concrete production facility. Concrete for an arch dam meets the requirements for both structural concrete and nonstructural (mass) concrete. That is, arch dams rely on the strength of the concrete in lieu of its mass, which would classify it as a structural concrete. However, the concrete is unreinforced mass concrete, which would classify it as a nonstructural concrete. For determining the required compressive strength (fcr) for use in the mix designs, the preferred method would be the method for nonstructural concrete. Assuming that no test records are available from the concrete plant, this would require the mix design to be based on the following relationship:

$$f_{cr} = f'_c + 600 \text{ psi} \qquad\qquad (9-1)$$

This equation will be valid during the design phase, but it can change during construction of the dam as test data from the concrete plant becomes available.

5.3.2 WATER-CEMENT RATIO. Table 9-1 shows the maximum water-cement ratios recommended for concrete used in most arch dams. These water-cement ratios are based on minimum durability requirements. The strength requirements in the preceding paragraph may dictate an even lower value. In thin structures (thicknesses less than approximately 30 feet) it may not be practical to change mixes within the body of the structure, so the lowest water-cement ratio should be used throughout the dam. In thick dams (thicknesses greater than approximately 50 feet), it may be practical to use an interior class of concrete with a water-cement ratio as high as 0.80. However, the concrete in the upstream and downstream faces should each extend into the dam a minimum of 15 to 20 feet before transitioning to the interior mixture. In addition, the interior concrete mix should meet the same strength requirements of the surface mixes.

Location in Dam	Severe or Moderate Climate	Mild Climate
Upstream face above minimum pool	0.47	0.57
Interior (for thick dams only)	0.80	0.80
Downstream face and upstream face below minimum pool	0.52	0.57

TABLE 9-1

Maximum Permissible Water-Cement Ratio

5.3.3 MAXIMUM SIZE AGGREGATE. Typical specifications recommend 6 inches as the nominal maximum size aggregate (MSA) for use in massive sections of dams. However, a 6-inch MSA may not always produce the most economical mixture proportion. Figure 9-1 shows that for a 90-day compressive strength of 5,000 psi, a 3-inch MSA would require less cement per cubic yard than a 6-inch MSA. Therefore, the selection of a MSA should be based on the size that will minimize the cement requirement. Another consideration in the selection of the MSA is the availability of the larger sizes and the cost of handling additional sizes. However, if the various sizes are available, the savings in cement and the savings in temperature control measures needed to control heat generation should offset the cost of handling the additional aggregate sizes.

5.3.4 DESIGN AGE. The design age for mass concrete is usually set between 90 days and 1 year. This is done to limit the cement necessary to obtain the desired strength. However, there are early-age strength requirements that must also be considered: form removal, resetting of subsequent forms, lift line preparation, construction loading, and impoundment of reservoir. There are also some difficulties that must be considered when selecting a very long design age. These include: the time required to develop the mixture proportions and perform the necessary property testing and the quality assurance evaluation of the mixture proportions during construction.

5.3.4.1 DURING THE DESIGN PHASE, selection of a 1-year design strength would normally require a minimum of 2 years to complete the mix design studies and then perform the necessary testing for structural properties. This assumes that all the material investigations discussed earlier have been completed and that the laboratory has an adequate supply of representative material to do all required testing. In many cases, there may not be sufficient time to perform these functions in sequence, thus they must be done concurrently with adjustments made at the conclusion of the testing.

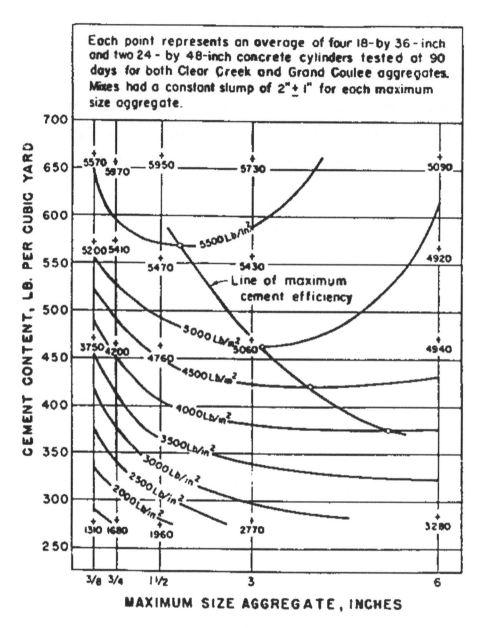

Each point represents an average of four 18-by 36-inch and two 24 - by 48-inch concrete cylinders tested at 90 days for both Clear Creek and Grand Coulee aggregates. Mixes had a constant slump of 2"± 1" for each maximum size aggregate.

Figure 9-1

Variation of cement content with maximum size aggregate for various compressive strengths. Chart shows that compressive strength varies inversely with maximum size aggregate for minimum cement content

5.3.4.2 DURING THE CONSTRUCTION PHASE of the project, the problems with extended design ages become even more serious. The quality assurance program requires that the contracting officer be responsible for assuring that the strength requirements of the project are met. With a Government-furnished mix design, the Contracting Officer must perform strength testing to assure the adequacy of the mixture proportions and make adjustments in the mix proportions when necessary. The problem with identifying variability of concrete batches with extended design ages of up to 1 year are obvious. As a result, the laboratory should develop a relationship between accelerated tests and standard cured specimens. EM 1110-2-2000 requires that, during construction, two specimens be tested in accordance with the accelerated test procedures, one specimen be tested at an information age, and two specimens be tested at the design age. The single-information-age specimen should coincide with form stripping or form resetting schedules. For mass concrete construction where the design age usually exceeds 90 days, it is recommended that an additional information-age specimen be tested at 28 days.

5.4 TESTING DURING DESIGN. During the design phase of the project, test information is needed to adequately define the expected properties of the concrete. For the purposes of this manual, the type of tests required are divided into two categories: structural properties testing and thermal properties testing. The number of tests and age at which they should be performed will vary depending on the type of analyses to be performed. However, Table 9-2 should be used in developing the overall testing program. If several mixes are to be investigated by the laboratory for possible use, then the primary mix should be tested in accordance with Table 9-2 and sufficient isolated tests performed on the secondary mixes to allow for comparisons to the primary mix results. ACI STP 169B (1978) and Neville (1981) provide additional information following tests and their significance.

5.4.1 STRUCTURAL PROPERTIES TESTING.

5.4.1.1 COMPRESSIVE STRENGTH. Compressive strength testing at various ages will be available from the mix design studies. However, additional companion compressive tests at various ages may be required for correlation to other properties, such as tensile strength, shear strength, modulus of elasticity, Poisson's ratio, and dynamic compressive strength. Most of the compression testing will be in accordance with a standard which is a uniaxial test. However, if the stress analysis is to consider a biaxial stress state, then additional biaxial testing may need to be performed.

Test	Age of Specimens (days)						
	1	3	7	28	90	180	1 yr.
Compression test	*	*	*	*	*	*	*
Modulus of rupture	*	*	*	*	*	*	*
Splitting tensile test				*			*
Shear test				*			*
Modulus of elasticity	*	*	*	*	*	*	*
Poisson's ratio	*	*	*	*	*	*	*
Dynamic tests							*
Creep	*	*	*	*	*	*	*
Strain capacity		*	*	*			*
Coefficient of thermal expansion	*	*	*	*			*
Specific heat	*	*	*	*			*
Thermal diffusivity	*	*	*	*			*

TABLE 9-2

Recommended Testing Program

5.4.1.2 TENSILE STRENGTH. The limiting factor in the design and analysis of arch dams will usually be the tensile strength of the concrete. Currently there are three accepted methods of obtaining the concrete tensile strength: direct tension test; the splitting tension test; and the modulus of rupture test. The direct tension test can give the truest indication of the tensile strength but is highly susceptible to problems in handling, sample preparation, surface cracking due to drying, and testing technique; therefore, it can often give erratic results. The splitting tension test also provides a good indication of the true tensile strength of the concrete and it has the advantage of being the easiest tension test to perform. In addition, the splitting tensile test compensates for any surface cracking and gives consistent and repeatable test results. However, when using the splitting tension test results as a criteria for determining the acceptability of a design, the designer should be aware that these results represent nonlinear performance that is normally being compared to a tensile stress computed from a linear analysis. The modulus of rupture test gives a value that is calculated based on assumed linear elastic behavior of the concrete. It gives consistent results and has the advantage of also being more consistent with the assumption of linear elastic behavior used in the design. A more detailed discussion of the importance of these tests in the evaluation of concrete dams is presented the ACI Journal (Raphael 1984). In testing for tensile strength for arch dam projects, the testing should include a combination of splitting tension tests and modulus of rupture tests.

5.4.1.3 SHEAR STRENGTH. The shear strength of concrete results from a combination of internal friction (which varies with the normal compressive stress) and cohesive strength (zero normal load shear strength). Companion series of shear strength tests should be conducted at several different normal stress values covering the range of normal stresses to be expected in the dam. These values should be used to obtain a curve of shear strength versus normal stress. Shear strength is determined in accordance with approved standards.

5.4.1.4 MODULUS OF ELASTICITY. When load is applied to concrete it will deform. The amount of deformation will depend upon the magnitude of the load, the rate of loading, and the total time of loading. In the analysis of arch dams, three types of deformations must be considered: instantaneous modulus of elasticity; sustained modulus of elasticity; and dynamic modulus of elasticity. Dynamic modulus of elasticity will be discussed in a separate section.

5.4.1.4.1 THE INSTANTANEOUS MODULUS OF ELASTICITY is the static modulus of elasticity. The modulus of elasticity in tension is usually assumed to be equal to that in compression. Therefore, no separate modulus testing in tension is required. Typical values for instantaneous (static) modulus of elasticity will range from 3.5 × 106 psi to 5.5 × 106 psi at 28 days and from 4.3 × 106 psi to 6.8 × 106 psi at 1 year. (b) The sustained modulus of elasticity includes the effects of creep, and can be obtained directly from creep tests. This is done by dividing the sustained load on the test specimen by the total deformation. The age of the specimen at the time of loading and the total time of loading will affect the result. It is recommended that the age of a specimen at the time of loading be at least 1 year and that the total time under load also be 1 year. The sustained modulus under these conditions will typically be approximately twothirds that of the instantaneous modulus of elasticity.

5.4.1.5 DYNAMIC PROPERTIES. Testing for concrete dynamic properties should include compressive strength, modulus of rupture or splitting tensile strength, and modulus of elasticity. The dynamic testing can be performed at any age for information but is only required at the specified design age. The rate of loading used in the testing should reflect the actual rate with which the dam will be stressed from zero to the maximum value. This rate should be available from a preliminary dynamic analysis. If the rate of loading is not available, then several rates should be used covering a range that can be reasonably expected. For example, a range of rates that would cause failure at 20 to 150 milliseconds could be used.

5.4.1.6 POISSON'S RATIO. Poisson's ratio is defined in American Society for Testing and Materials (ASTM) E6 (ASTM 1992) as "the absolute value of the ratio of transverse strain to the corresponding axial strain below the proportional limit of the material." In simplified terms, it is the ratio of lateral strain to axial strain. Poisson's ratio for mass concrete will typically range from 0.15 to 0.20 for static loads, and from 0.24 to 0.25 for dynamic loads.

5.4.1.7 CREEP. Creep is time-dependent deformation due to sustained load. Creep can also be thought of as a relaxation of stress under a constant strain. In addition to using the creep test to determine the sustained modulus of elasticity (discussed previously), creep is extremely important in the thermal studies. However, unlike the sustained modulus of elasticity, the thermal studies need early age creep information.

5.4.1.8 STRAIN CAPACITY. Analyses that are based on tensile strain capacity rather than tensile strength will require some information on strain capacity. Examples of these types of analyses include the closure temperature and NISA as discussed in Chapter 8. Strain capacity can be measured in accordance with approved standards or can be estimated from the results of the modulus of elasticity, modulus of rupture, and specific creep tests.

5.4.2 THERMAL PROPERTIES. Understanding the thermal properties of concrete is vital in planning mass concrete construction. The basic properties involved include coefficient of thermal expansion, specific heat, thermal conductivity, and thermal diffusivity.

5.4.2.1 COEFFICIENT OF THERMAL EXPANSION. Coefficient of thermal expansion is the change in linear dimension per unit length divided by the temperature change. The coefficient of thermal expansion is influenced by both the cement paste and the aggregate. Since these materials have dissimilar thermal expansion coefficients, the coefficient for the concrete is highly dependent on the mix proportions, and since aggregate occupies a larger portion of the mix in mass concrete, the thermal expansion coefficient for mass concrete is more influenced by the aggregate. Coefficient of thermal expansion is expressed in terms of inch per inch per degree Fahrenheit (in./in./ F). In many cases, the length units are dropped and the quantities are expressed in terms of the strain value per F. This abbreviated form is completely acceptable. Typical values for mass concrete range from 3.0 to 7.5×10^{-6} /F. Testing for coefficient of thermal expansion should be in accordance with approved standards. However, the test should be modified to account for the temperature ranges to which the concrete will be subjected, including the early-age temperatures.

5.4.2.2 SPECIFIC HEAT. Specific heat is the heat capacity per unit temperature. It is primarily influenced by moisture content and concrete temperature. Specific heat is typically expressed in terms of Btu/pound· degree Fahrenheit (Btu/lb-F). Specific heat for mass concrete typically ranges from 0.20 to 0.25 Btu/lb-F. Testing for specific heat should be in accordance with approved standards.

5.4.2.3 THERMAL CONDUCTIVITY. Thermal conductivity is a measure of the ability of the material to conduct heat. It is the rate at which heat is transmitted through a material of unit area and thickness when there is a unit difference in temperature between the two faces. For mass concrete, thermal conductivity is primarily influenced by aggregate type and water content, with aggregate having the larger influence. Within the normal ambient temperatures, conductivity is usually constant. Conductivity is typically expressed in terms of Btu-inch per hour-square foot-degree Fahrenheit (Btu-in./ hr-ft2-F). Thermal conductivity for mass concrete typically ranges from 13 to 24 Btu-in./hr-ft2-F. It can be determined in accordance with approved standards or it can be calculated by the following equation:

$$k = \sigma c \rho \qquad\qquad (9\text{-}2)$$

where

k = thermal conductivity
σ = thermal diffusivity
c = specific heat
ρ = density of concrete

5.4.2.4 THERMAL DIFFUSIVITY. Thermal diffusivity is a measure of the rate at which temperature changes can take place within the mass. As with thermal conductivity, thermal diffusivity is primarily influenced by aggregate type and water content, with aggregate having the larger influence. Within the normal ambient temperatures, diffusivity is usually constant. Diffusivity is typically expressed in terms of square feet/hour (ft^2/hr). For mass concrete, it typically ranges from 0.02 to 0.06 ft^2/hr .

5.4.2.5 ADIABATIC TEMPERATURE RISE. The adiabatic temperature rise should be determined for each mix anticipated for use in the project. The adiabatic temperature rise is determined using appropriate standards.

5.5 PROPERTIES TO BE ASSUMED PRIOR TO TESTING. During the early stages of design analysis it is not practical to perform in-depth testing. Therefore, the values shown in Tables 9-3, 9-4, and 9-5 can be used as a guide during the early design stages and as a comparison to assure that the test results fall within reasonable limits.

Compressive strength (f'_c)	$\geq$ 4,000 psi
Tensile strength (f'_t)	10% of f'_c

Shear strength (v)
    Cohesion
    Coef. of internal friction
	10% of f'_c
	1.0 (ϕ = 45°)

Instantaneous modulus of elasticity (E_i)	4.5×10^6 psi
Sustained modulus of elasticity (E_s)	3.0×10^6 psi
Poisson's ratio (μ_s)	0.20
Unit weight of concrete (ρ_c)	150 pcf

Strain capacity (ε_c)
    rapid load
    slow load
	100×10^{-6} in/in
	150×10^{-6} in/in

TABLE 9-3

Static Values (Structural)

Compressive strength (f'_{cd})	130% f'_c
Tensile strength (f'_{td})	130% f'_t
Modulus of elasticity (E_D)	5.5×10^6 psi
Poisson's ratio (μ_d)	0.25

TABLE 9-4

Dynamic Values (Structural)

Coefficient of thermal expansion (e)	5.0×10^{-6} per °F
Specific heat (c)	0.22 Btu/lb-°F
Thermal conductivity (k)	16 Btu-in./hr-ft^2-°F
Thermal diffusivity (σ)	0.04 ft^2/hr

TABLE 9-5

Thermal Values

CHAPTER 6: ARCH DAM CONSTRUCTION

6.1 INTRODUCTION. The design and analysis of any dam assumes that the construction will provide a suitable foundation (with minimal damage) and uniform quality of concrete. The design also depends on the construction to satisfy many of the design assumptions, including such items as the design closure temperature. The topics in this chapter will deal with the construction aspects of arch dams as they relate to the preparation of planning, design, and construction documents. Contract administration aspects are beyond the scope of this discussion.

6.2 DIVERSION.

6.2.1 GENERAL. With arch dams, as with all concrete dams, flooding of the working area is not as serious an event as it is with embankment dams, since flooding should not cause serious damage to the completed portions of the dam and will not cause the works to be abandoned. For this reason, the diversion scheme can be designed for floods of relatively high frequencies corresponding to a 25-, 10-, or 5-year event. In addition, if sluices and low blocks are incorporated into the diversion scheme, the design flood frequency is reduced and more protection against flooding is added as the completed concrete elevation increases.

6.2.2 THE DIVERSION SCHEME. The diversion scheme will usually be a compromise between the cost of diversion and the amount of risk involved. The proper diversion plan should minimize serious potential flood damage to the construction site while also minimizing expense. The factors that the designer should consider in the study to determine the best diversion scheme include: streamflow characteristics; size and frequency of diversion flood; available regulation by existing upstream dams; available methods of diversion; and environmental concerns. Since streamflow characteristics, existing dams, and environmental concerns are project-specific items, they will not be addressed in this discussion.

6.2.2.1 SIZE AND FREQUENCY OF DIVERSION FLOOD. In selecting the flood to be used in any diversion scheme, it is not economically feasible to plan on diverting the largest flood that has ever occurred or may be expected to occur at the site. Consideration should be given to how long the work will be under construction (to determine the number of flood seasons which will be encountered), the cost of possible damage to work completed or still under construction if it is flooded, the delay to completion of the work, and the safety of workers and the downstream inhabitants caused by the diversion works. For concrete dams, the cost of damage to the project would be limited to loss of items such as formwork and stationary equipment. Delays to the construction would be primarily the cleanup of the completed work, possible replacement of damaged equipment, and possible resupply of construction materials (Figure 13-1). It is doubtful that there would be any part of the completed work that would need to be removed and reworked. The risk to workers and downstream residents should also be minimal since the normal loads on the dam are usually much higher than the diversion loads, and overtopping should not cause serious undermining of the dam. And, as the concrete construction increases in elevation, the size of storm that the project can contain is substantially increased. Based on these considerations, it is not uncommon for the diversion scheme to be designed to initially handle a flood with a frequency corresponding to a 5- or 10-year event.

Figure 13-1

View from right abutment of partially complete Monticello Dam in California showing water
flowing over low blocks

6.2.2.2 PROTECTION OF THE DIVERSION WORKS. The diversion scheme should be
designed to either allow floating logs, ice and other debris to pass through the diversion
works without jamming inside them and reducing their capacity, or prevent these items
from entering the diversion works (International Commission on Large Dams 1986).
Systems such as trash structures or log booms can be installed upstream of the
construction site to provide protection by collecting floating debris.

6.2.2.3 OTHER CONSIDERATIONS. If partial filling of the reservoir before completing the
construction of the dam is being considered, then the diversion scheme must take into
account the partial loss of flood storage. In addition, partial filling will require that some

method (such as gates, valves, etc.) be included in the diversion scheme so that the reservoir level can be controlled.

6.2.3 METHODS OF DIVERSION. Typical diversion schemes for arch dams include tunnels, flumes or conduits, sluices, low blocks, or a combination of any of these. Each of these will require some sort of cofferdam across the river upstream of the dam to dewater the construction site. A cofferdam may also be required downstream. The determination of which method or methods of diversion should be considered will rest on the site conditions.

6.2.3.1 TUNNELS AND CULVERTS. Tunnels are the most common method of diversion used in very narrow valleys (Figure 13-2). The main advantage to using a diversion tunnel is that it eliminates some of the need for the staged construction required by other diversion methods. They also do not interfere with the foundation excavation or dam construction. The main disadvantage of a tunnel is that it is expensive, especially when a lining is required. As a result, when a tunnel is being evaluated as part of the diversion scheme, an unlined tunnel should be considered whenever water velocities and the rock strengths will allow. At sites where a deep crevice exists at the stream level and restitution concrete is already required to prevent excessive excavation, a culvert may be more economical than a tunnel (Figure 13-3). All tunnels and culverts will require a bulkhead scheme at their intakes to allow for closure and will need to be plugged with concrete once the dam is completed. The concrete plug will need to have a postcooling and grouting system (Figure 13-4).

6.2.3.2 CHANNELS AND FLUMES. Channels and flumes are used in conjunction with sluices and low blocks to pass the stream flows during the early construction phases. Channels and flumes should be considered in all but very narrow sites (Figures 13-5 and 13-6). Flumes work well where flows can be carried around the construction area or across low blocks. Channels will require that the construction be staged so that the dam can be constructed along one abutment while the river is being diverted along the base of the other abutment (Figure 13-6). Once the construction reaches sufficient height, the water is diverted through sluices in the completed monoliths (Figure 13-7). A similar situation applies for flumes. However, flumes can be used to pass the river through the construction site over completed portions of the dam. If the flume has adequate clearance, it is possible to perform limited work under the flume.

6.2.3.3 SLUICES AND LOW BLOCKS. Sluices and low blocks are used in conjunction with channels and flumes. Sluices and low blocks are used in the intermediate and later stages of construction when channels and flumes are no longer useful. Sluices become very economical if they can be incorporated into the appurtenant structures, such as part of the outlet works or power penstock (Figure 13-7). It is acceptable to design this type of diversion scheme so that the permanent outlet works or temporary low-level outlets handle smaller floods, with larger floods overtopping the low blocks (Figure 13-8). The low blocks are blocks which are purposely lagged behind the other blocks. As with the tunnel option, sluices not incorporated into the appurtenant structures will need to be closed off and concrete backfilled at the end of the construction (Figure 13-9).

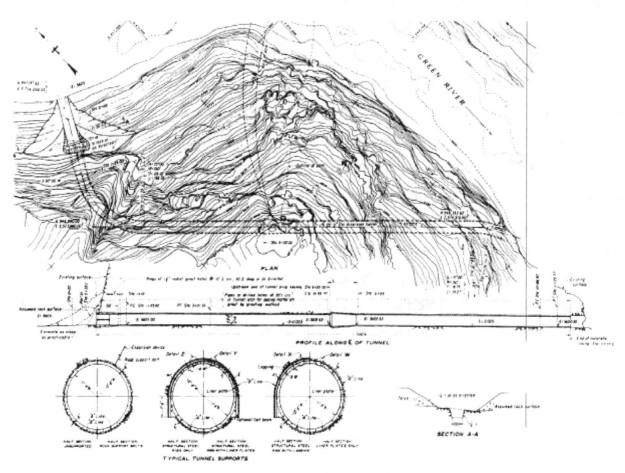

Figure 13- 2

Diversion tunnel for Flaming Gorge Dam

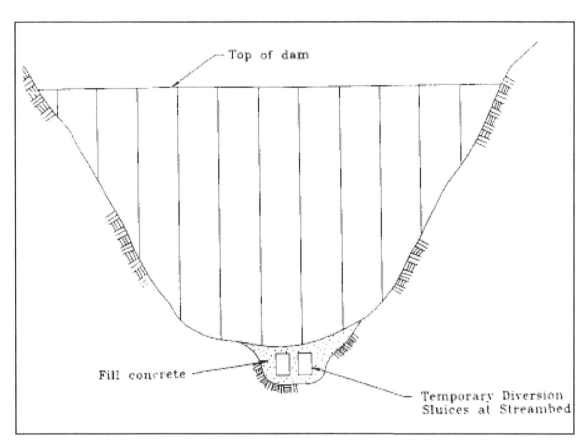

Figure 13-3

Diversion sluices through concrete pad at the base of a dam

6.3 FOUNDATION EXCAVATION. Excavation of the foundation for an arch dam will require that weak areas be removed to provide a firm foundation capable of withstanding the loads applied to it. Sharp breaks and irregularities in the excavation profile should be avoided (U.S. Committee on Large Dams 1988). Excavation should be performed in such manner that a relatively smooth foundation contact is obtained without excessive damage due to blasting. However, it is usually not necessary to over-excavate to produce a symmetrical site. A symmetrical site can be obtained by the use of restitutional concrete such as dental concrete, pads, thrust blocks, etc. Foundations that are subject to weathering and that will be left exposed for a period of time prior to concrete placement should be protected by leaving a sacrificial layer of the foundation material in place or by protecting the final foundation grade with shotcrete that would be removed prior to concrete placement.

6.4 CONSOLIDATION GROUTING AND GROUT CURTAIN. The foundation grouting program should be the minimum required to consolidate the foundation and repair any damage done to the foundation during the excavation and to provide for seepage control. The final grouting plan must be adapted to suit field conditions at each site.

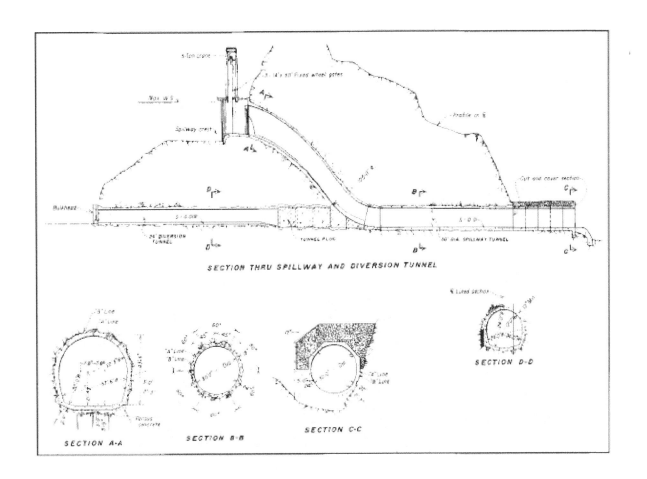

Figure 13-4

Typical diversion tunnel plug

Figure 13-5

Complete diversion flume at Canyon Ferry dam site in use for first stage diversion

6.4.1 CONSOLIDATION GROUTING. Consolidation grouting consists of all grouting required to fill voids, fracture zones, and cracks at and slightly below the excavated foundation. Consolidation grouting is performed before any other grouting is done and is usually accomplished from the excavated surface using low pressures. In cases of very steep abutments, the grouting can be performed from the top of concrete placements to prevent "slabbing" of the rock. Holes should be drilled normal to the excavated surface, unless it is desired to intersect known faults, shears, fractures, joints, and cracks. Depths vary from 20 to 50 feet depending on the local conditions. A split spacing process should be used to assure that all groutable voids, fracture zones, and cracks have been filled.

6.4.2 GROUT CURTAIN. The purpose of a grout curtain is to control seepage. It is installed using higher pressures and grouting to a greater depth than the consolidation grouting. The depth of the grout curtain will depend on the foundation characteristics but will typically vary from 30 to 70 percent of the hydrostatic head. To permit the higher pressures, grout curtains are usually installed from the foundation gallery in the dam or from the top of concrete placements. If lower pressures in the upper portion can be used, the grout curtain can also be placed at foundation grade prior to concrete placement and supplemented by a short segment of grouting from the gallery to connect the dam and the previously installed curtain. Grout curtains can also be installed from adits or tunnels that extend into the abutments. The grout curtain should be positioned along the footprint of the dam in a region of zero or minimum tensile stress. It should extend into the foundation so that the base of the grout curtain is located at the vertical projection of the heel of the dam. To assist in the drilling operations, pipes are embedded in the concrete. Once the concrete placement has reached an elevation that the grouting pressure will not damage or lift the concrete, the grout holes are drilled through these pipes and into the foundation. Grout curtain operations should be performed before concrete placement starts or after the monolith joints have been grouted to prevent damage to the embedded grout stops and to prevent leakage into the monolith joints.

Figure 13-6

Channel used in the Phase 1 Diversion for Smith Mountain Dam

Figure 13·7

Sluice used in the Phase II diversion for Smith Mountain dam

Figure 13-8

Aerial view of partially completed New Bullards Bar Dam in California showing water

flowing over the low blocks

6.5 CONCRETE OPERATIONS. The concrete operations discussed in this section are the special concrete construction requirements for arch dams which are not covered in other chapters of this and other manuals.

6.5.1 FORMWORK. Although modern arch dams are almost always curved in both plan and section, the forms used in constructing these dams are usually comprised of short, plane segments. The length of these segments correspond to the length of form panels, which should not exceed 8 feet. The detailed positioning of the formwork is usually supplied to the contractor as part of the contract documents. The computer program Arch Dam Construction Program (ADCOP) has been developed to generate the information needed to establish the location of the formwork along the upstream and downstream faces of each lift. ADCOP is a PC based computer program which requires arch dam geometry information similar to the ADSAS program and generates data which will control layout of formwork for each monolith in the body of an arch dam. ADCOP generates contraction joints which are spiral in nature.

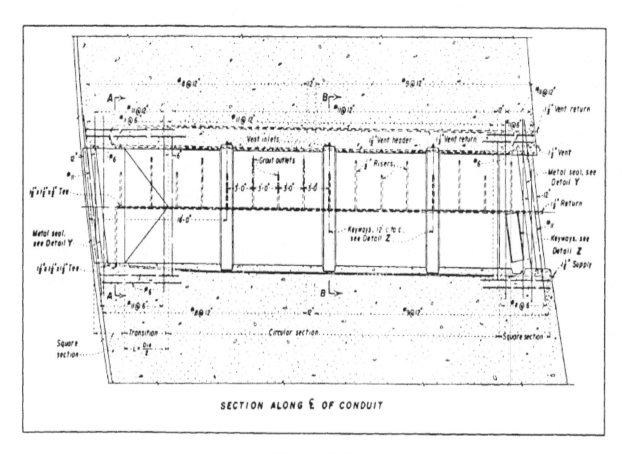

SECTION ALONG ℄ OF CONDUIT

Figure 13-9

Diversion conduit through Morrow Point Dam

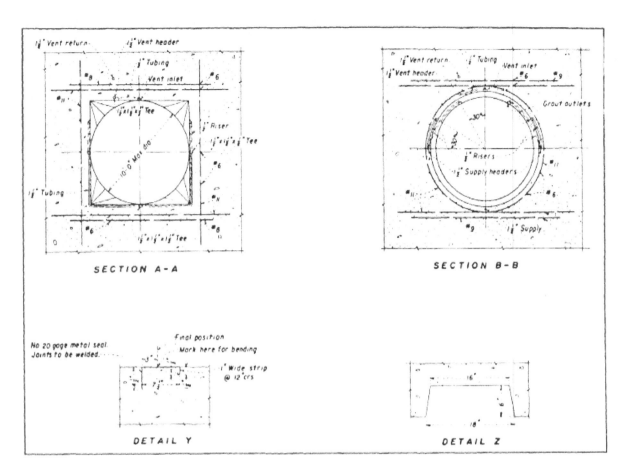

Figure 13-9 (continued)

Diversion conduit through Morrow Point Dam

6.5.2 HEIGHT DIFFERENTIAL. The maximum height differential between adjacent monoliths should be no more than 40 feet and no more than 70 to 100 feet between highest and lowest blocks in the dam. For dams with a pronounced overhang, limits should be placed on the placement of concrete with respect to ungrouted monoliths to avoid undesirable deflection and coincident tensile stresses.

6.5.3 CONCRETE QUALITY. The current trend in mass concrete construction is to minimize the use of cement by varying the concrete mixes used in different parts of the structure. This is usually done between the upstream and downstream faces and in areas of localized high stresses. For large dams, this practice can be very beneficial. For smaller dams, where the thicknesses are less than 30 feet, it may not be practical to vary the mix from the upstream to downstream faces. The benefits of adjusting the concrete mix for localized areas should be addressed, but the minimum requirements discussed in

Chapter 9 should not be relaxed. Also, the effects of the differing cement contents created by these adjustments must be considered in the construction temperature studies.

6.5.4 JOINT CLEANUP. A good bond between horizontal construction joints is essential if cantilever tensile load transfer is to be achieved and leakage to the downstream face is to be minimized. Bond strength between lift lines can be almost as strong as the in-place concrete if the joints are properly prepared and the freshly placed concrete is well consolidated. Vertical contraction (monolith) joints that are to be grouted should be constructed so that no bond exists between the blocks in order that effective joint opening can be accomplished during the final cooling phase without cracking the mass concrete.

6.5.5 EXPOSED JOINT DETAILS. Joints should be chamfered at the exposed faces of the dam to give a desirable appearance and to minimize spalling.

6.6 MONOLITH JOINTS.

6.6.1 SPACING OF MONOLITH JOINTS. The width of a monolith is the distance between monolith joints as measured along the axis of the dam. The determination of the monolith width will be part of the results of the closure temperature analysis discuss.. Once set, the monolith joint spacing should be constant throughout the dam if possible. In general, if the joints are to be grouted, monolith widths will be set at a value that allows for sufficient contraction to open the joints for grouting. However, there are limits to monolith widths. In recent construction, monolith widths have been commonly set at approximately 50 feet but with some structures having monoliths ranging from 30 to 80 feet. The USBR (1977) recommends that the ratio of the longer to shorter dimensions of a monolith be between 1 and 2. Therefore, a dam with a base thickness of 30 feet could have a maximum monolith width of 60 feet. However, the thickness of the base of a dam will vary significantly as the dam progresses from the crown cantilever up the abutment contacts. Near the upper regions of the abutment contacts, the base thickness of the dam will drop to a value close to the thickness of the crest. If a dam has a base thickness of 30 feet at the crown cantilever and a crest thickness of only 10 feet, then the limit on the monolith width at the crown cantilever will be 60 feet while the limit near the ends of the dam would be 20 feet. Since monolith spacing must be uniform throughout the dam, the maximum limit can be set at the higher value determined for the crown cantilever, with the understanding that some cracking could be expected along the abutment contacts near the ends of the dam. In any case, the joint spacing should never exceed 80 feet.

6.6.2 LIFT HEIGHTS. Lift heights are typically set at 5, 7.5, or 10 feet. This height should be uniform as the construction progresses from the base.

6.6.3 WATER STOPS AND GROUT STOPS. The terms water stop and grout stop describe the same material used for different purposes. A water stop is used to prevent seepage from migrating through a joint, typically from the upstream face of the dam to the downstream face. A grout stop is used to confine the grout within a specified area of a joint. General layout for the water stops/grout stops for a grouted dam is shown in Figure 13-10. Figure 13-20 also shows a typical layout at the upstream face for ungrouted cases where a joint drain is added. Proper installation and protection of the water stop/grout stop during construction is as important as the shape and material.

6.6.4 SHEAR KEYS. Shear keys are installed in monolith joints to provide shearing resistance between monoliths. During construction they also help maintain alignment of the blocks, and they may help individual blocks "bridge" weak zones within the foundation. Vertical shear keys, similar to those shown in Figure 13-11, will also increase the seepage path between the upstream and downstream faces. Other shapes of shear key, such as the dimple, waffle, etc., shown in Figures 13-12 and 13-13, have also been used. The main advantage in these other shaped keys is the use of standard forms and, in the case of the dimple shape, less interference with the grouting operations. However, these other shapes provide additional problems in concrete placement and are not as effective in extending the seepage path as are the vertical keys.

6.6.5 VERTICAL VERSUS SPIRAL JOINTS. Arch dams can be constructed with either vertical or spiral radial contraction joints. Vertical radial joints are created when all monolith joints, at each lift, are constructed on a radial line between the dam axis and the axis center on the line of centers (see Chapter 5 for definition of dam axis and axis center). Therefore, when looking at a plan view of a vertical joint, the joints created at each construction lift will fall on the same radial line. Spiral joints are created when the monolith joints, at each lift, are constructed on a radial line between the dam axis and with the intrados line of centers for that lift elevation. Therefore, when looking at a plan view of a spiral joint, each joint created at each construction lift will be slightly rotated about the dam axis. Spiral radial joints will also provide some keying of the monoliths, while vertical radial joints will not.

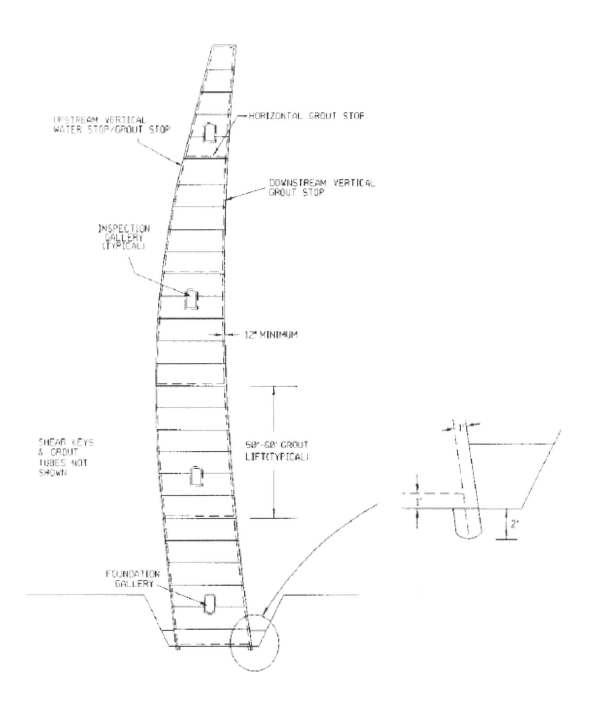

Figure 13-10

Water stop and grout stop details

Figure 13-11

Vertical keys at monolith joints of Monar Dam, Scotland

202

a)   b)

Figure 13-12

Dimple shear keys

6.6.6 MONOLITH JOINT GROUTING. The purpose of grouting monolith joints is to provide a monolithic structure at a specified "closure temperature." To accomplish this, grout is injected into the joint by a means of embedded pipes similar to those shown in Figures 13-14 and 13-15. To ensure complete grouting of the joint prior to grout set, and to prevent excessive pressure on the seals and the blocks, grout lifts are usually limited to between 50 and 60 feet.

6.6.6.1 JOINT OPENING FOR GROUTING. The amount that a joint will open during the final cooling period will depend upon the spacing of the monolith joints, the thermal properties of the concrete, and the temperature drop during the final cooling period. The typical range of joint openings needed for grouting is 1/16 to 3/32 inch. With arch dams, where it is desirable to have a monolithic structure that will transfer compressive loads across the monolith joints, it is better to have an opening that will allow as thick a grout as practical to be placed. A grout mixture with a water-cement ratio of 0.66 (1 to 1 by volume) will usually require an opening on the higher side of the range given (Water Resources Commission 1981). In the initial design process, a required joint opening of 3/32 inch should be assumed.

6.6.6.2 LAYOUT DETAILS. The supply loop at the base of the grout lift provides a means of delivering the grout to the riser pipes. Grout can be injected into one end of the looped pipe. The vertical risers extend from the supply loop at 6-foot intervals. Each riser contains outlets spaced 10 feet on centers, which are staggered to provide better grout coverage to the joint. The risers are discontinued near the vent loop (groove) provided at the top of the grout lift. Vent pipes are positioned at each end of the vent groove to permit air, water, and thin grout to escape in either direction. Normally, these systems are terminated at the downstream face. Under some conditions, they can be terminated in galleries. The ends of the pipes have nipples that can be removed after completion of the grouting and the remaining holes filled with dry pack. Typical grout outlet and vent details are shown in Figures 13-14 and 13-15.

Figure 13-13

Waffle keys at the vertical monolith joints of Gordon Dam, Australia

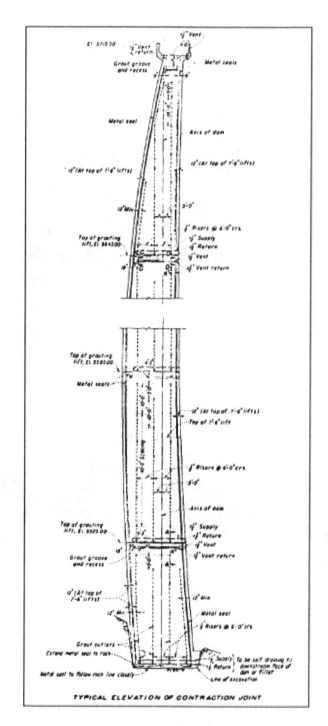

Figure 13-14

Contraction joint and grouting system for East Canyon Dam

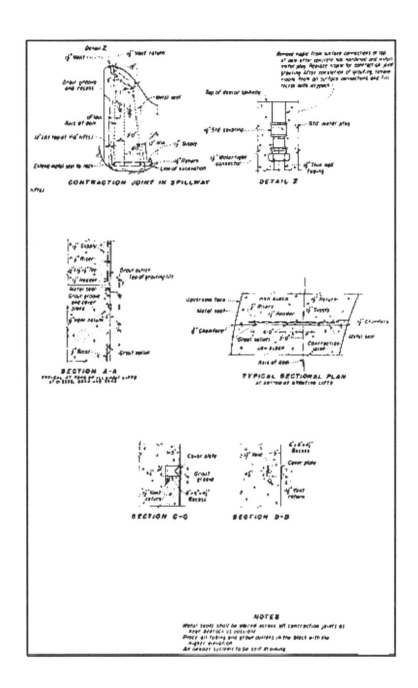

Figure 13-14 (continued)

Contraction joint and grouting system for East Canyon Dam

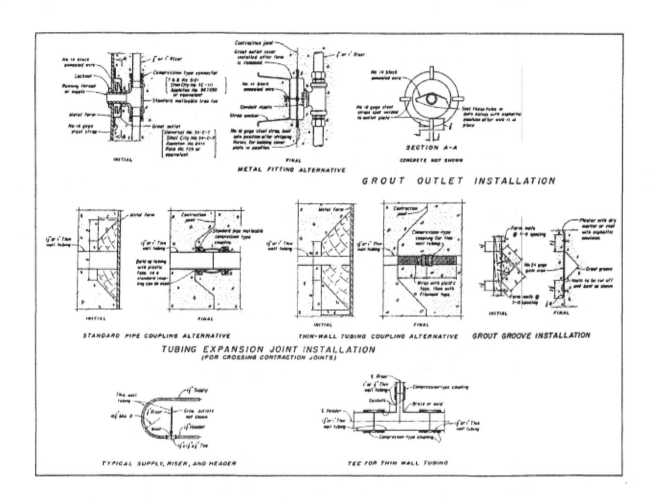

Figure 13-15

Example of grouting system details

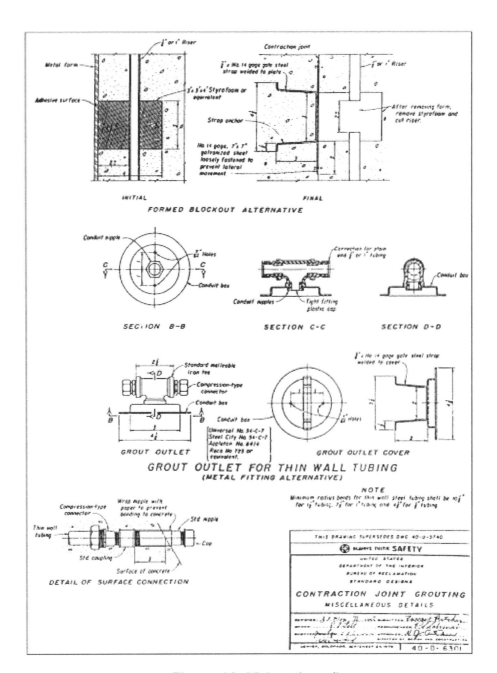

Figure 13-15 (continued)

Example of grouting system details

6.6.6.3 PREPARATION FOR GROUTING. Prior to grouting, the system is tested to assure that obstructions do not exist. The monolith joint is then cleaned with air and water under pressure. The joint is filled with water for a period of 24 hours. The water is drained from the joint to be grouted. Joints of two or more ungrouted joints on either side are filled with water, but not pressurized. Once the grout reaches the top of the grout lift, the lift above is filled with water to protect the upper grout stop. Immediately after a grouting operation is completed, the water is drained from the joints in the lift above. Water in the adjacent ungrouted joints should remain in place for at least 6 hours after the grouting operation is completed.

6.6.6.4 GROUT MIX. The grout should consist of the thickest mix that will enter the joint, fill all of the small voids, and travel to the vent. Grout mixes usually vary from water-cement ratios of 2 to 1 by volume (1.33 by weight) at the start of the grouting operation to thicker mixes (1 to 1 by volume or 0.66 by weight) as the operation progresses. If the joints are sufficiently open to accept a thicker grout, then mixes with ratios of 0.70 to 1 by volume (0.46 by weight) should be used to finish the job.

6.6.6.5 GROUTING OPERATION. Grout is injected in the supply loop at the bottom of the grout lift so that grout first comes in contact with the riser farthest from the supply portal. This will allow for the most favorable expulsion of air, water, and diluted grout as the grouting operation progresses. Once the grout appears at the return end of the supply loop, the return end is closed and the grout is forced up the risers and into the joint. The grouting must proceed at a rate fast enough that the grout will not set before the entire joint is filled with a thick grout. However, the rate of grouting must also be slow enough to allow the grout to settle into the joint. When thick grout reaches one end of the vent loop at the top of the grout lift, grouting operations are stopped for a short time (5 to 10 minutes) to allow the grout to settle in the joint. After three to five repetitions of a showing of thick grout, the valves are closed and the supply pressure increased to the allowable limit (usually 30 to 50 psi) to force grout into all small openings of the joint and to force the excess water into the pores of the concrete, leaving a grout film of lower water-cement ratio and higher density in the joint. The limiting pressure is set at a value that will avoid excessive deflection in the block and joint opening in the grouted portions below the joint.

The system is sealed off when no more grout can be forced into the joint as the pressure is maintained.

6.7 GALLERIES AND ADITS.

6.7.1 GENERAL. Adits are near horizontal passageways that extend from the surface into the dam or foundation. Galleries are the internal passageways within the dam and foundation and can be horizontal, vertical, or sloped. Chambers or vaults are created when galleries are enlarged to accommodate equipment. Galleries serve a variety of purposes. During construction, they can provide access to manifolds for the concrete postcooling and grouting operations. The foundation gallery also provides a work space for the installation of the grout and drainage curtains. During operation, galleries provide access for inspection and for collection of instrumentation data. They also provide a means to collect the drainage from the face and gutter drains and from the foundation drains. Galleries can also provide access to embedded equipment such as gates or valves. However, with all the benefits of galleries, there are also many problems. Galleries interfere with the construction operations and, therefore, increase the cost of construction. They provide areas of potential stress concentrations, and they may interfere with the proper performance of the dam. Therefore, galleries, as well as other openings in the dam, should be minimized as much as possible.

6.7.2 LOCATION AND SIZE. Typical galleries are 5 feet wide by 7.5 feet high. Figure 13-16 shows the most typical shapes of galleries currently being used. Foundation galleries are somewhat larger to allow for the drilling and grouting operations required for the grout and drainage curtains. Foundation galleries can be as large as 6 feet wide by 8.5 feet high. Personnel access galleries, which provide access only between various features within the dam, can be as small as 3 feet wide by 7 feet high. Spiral stairs should be 6 feet 3 inches in diameter to accommodate commercially available metal stairs.

6.7.3 LIMITATIONS OF DAM THICKNESS. Galleries should not be put in areas where the thickness of the dam is less than five times the width of the gallery.

6.7.4 REINFORCEMENT REQUIREMENTS. Reinforcement around galleries is not recommended unless the gallery itself will produce localized high tensile stresses or if it is positioned in an area where the surrounding concrete is already in tension due to the other external loads being applied to it. Even under these conditions, reinforcement is only required if the cracking produced by these tensions is expected to propagate to the reservoir. Reinforcement will also be required in the larger chambers formed to accommodate equipment.

6.7.5 LAYOUT DETAILS (Figure 13-17). Gallery and adit floors should be set at the top of a placement lift for ease of construction. Galleries should be at a slope comfortable for walking. Ramps can be used for slopes up to 10 degrees without special precautions and up to 15 degrees if nonslip surfaces are provided. Stairs can be used for slopes up to 40 degrees. Spiral stairs or vertical ladders can be used in areas where slopes exceed these limits. Landings should be provided approximately every 12 vertical feet when spiral stairs or ladders are used. Landings should also be provided in stairways if at all possible. Handrails should be provided in all galleries where the slope is greater than 10 degrees. There should be a minimum of 5 feet between the floor of the foundation gallery and the rock interface. There should also be a minimum 5-foot spacing between a gallery or adit and the monolith joints and external faces. The preferable location of the galleries is near the center of the monolith to minimize its impact on the section modulus of the cantilever. As a minimum, galleries should be located away from the upstream face at a distance that corresponds to 5 percent of the hydrostatic head.

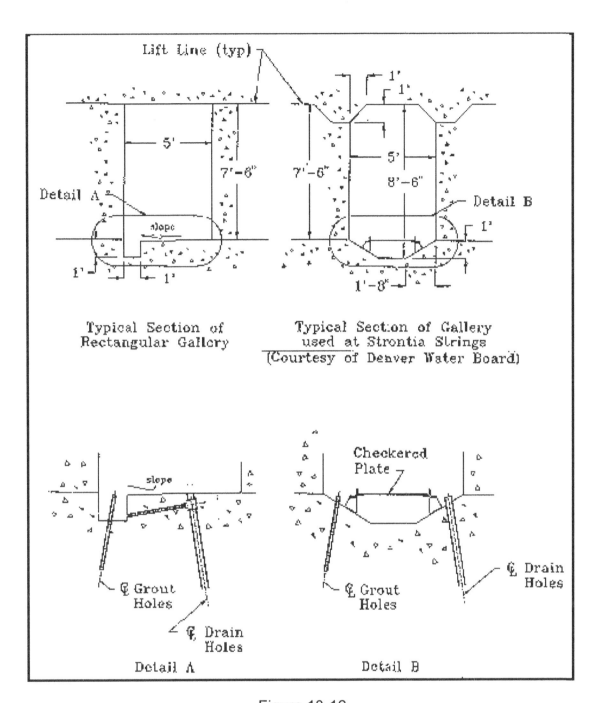

Figure 13-16

Typical gallery details

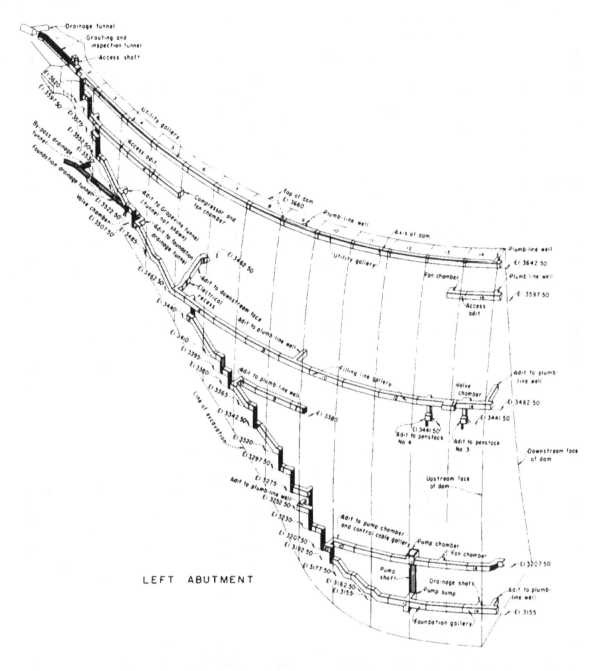

LEFT ABUTMENT

Figure 13-17

Left abutment gallery system for Yellowtail Dam

6.7.6 UTILITIES. Water and air lines should be embedded in the concrete to help in future maintenance operations. Lighting and ventilation should also be provided for the convenience and safety of personnel working in the galleries. Telephones should be located in gate rooms or chambers within the dam, as well as scattered locations throughout the galleries, for use in emergencies and for convenience of operation and maintenance personnel.

6.8 DRAINS. Drains fall into two categories: foundation drains and embedded drains. Embedded drains include face drains, gutter drains, and joint drains. All arch dams should include provisions for foundation drains and for face and gutter drains. Joint drains are not recommended if the monolith joints are to be grouted because they can interfere with the grouting process. Providing water/grout stops on each side of the drain help alleviate that problem, but the addition of the drain and the grout stop reduces the available contact area between the grout and mass concrete and thereby reduces the area for load transfer between adjacent monoliths.

6.8.1 FOUNDATION DRAINS. Foundation drains provide a way to intercept the seepage that passes through and around the grout curtain and thereby prevents excessive hydrostatic pressures from building up within the foundation and at the dam/foundation contact. The depth of the foundation drains will vary depending on the foundation conditions but typically ranges from 20 to 40 percent of the reservoir depth and from 35 to 75 percent of the depth of the grout curtain. Holes are usually 3 inches in diameter and are spaced on 10- foot centers. Holes should not be drilled until after all foundation grouting in the area has been completed. Foundation drains are typically drilled from the foundation gallery, but if no foundation gallery is provided, they can be drilled from the downstream face. Foundation drains can also be installed in adits or tunnels that extend into the abutments.

6.8.2 FACE DRAINS. Face drains are installed to intercept seepage along the lift lines or through the concrete. They help minimize hydrostatic pressure within dam as well as staining on the downstream face. Face drains extend from the crest of the dam to the foundation gallery. If there is no foundation gallery, then the drains are extended to the downstream face and connected to a drain pipe in the downstream fillet. They should be 5 or 6 inches in diameter and located 5 to 10 feet from upstream face. If the crest of the dam is thin, the diameter of the drains can be reduced and/or the distance from the upstream face can be reduced as they approach the crest. Face drains should be evenly spaced along the face at approximately 10 feet on centers (Figure 13-18).

6.8.3 GUTTER DRAINS. Gutter drains are drains that connect the gutters of the individual galleries to provide a means of transporting seepage collected in the upper galleries to the foundation gallery and eventually to the downstream face or to a sump. These drains are 8-inch-diameter pipes and extend from the drainage gutter in one gallery to the wall or drainage gutter in the next lower gallery. These drains are located in approximately every fourth monolith (Figure 13-19).

6.8.4 JOINT DRAINS. As noted earlier, joint drains are not normally installed in arch dams because of the joint grouting operations. However, if the monolith joints are not to be grouted, or if drains are required in grouted joints and can effectively be installed, then a 5- or 6-inch drain similar to the face drains should be used. The drain should extend from the crest of the dam to the foundation and should be connected to the foundation gallery (Figure 13-20).

6.9 APPURTENANT STRUCTURES. The appurtenant structures should be kept as simple as possible to minimize interference with the mass concrete construction. Outlet works should be limited to as few monoliths as possible. Conduits should be aligned horizontally through the dam and should be restricted to a single construction lift. Vertical sections of the conduits can be placed outside the main body of the dam.

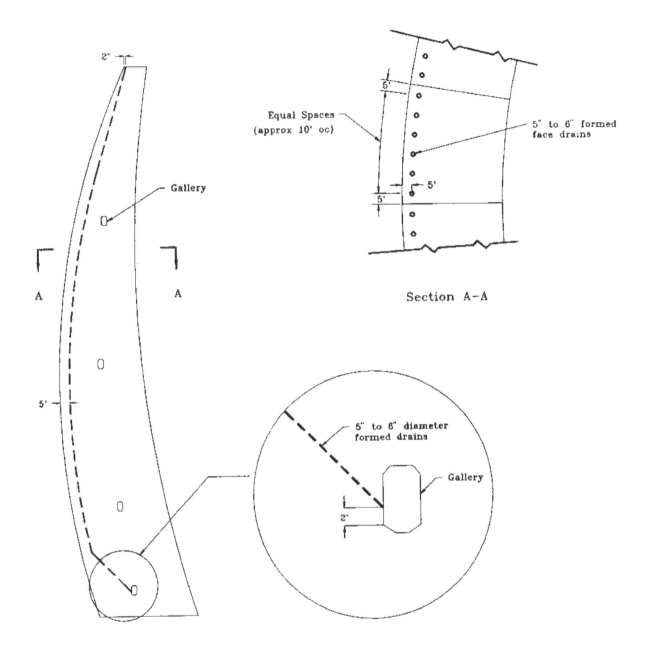

2'

Gallery

A

A

5'

Equal Spaces
(approx 10' oc)

5'

5'

5'

5" to 6" formed
face drains

Section A-A

5" to 6" diameter
formed drains

Gallery

2'

Figure 13-18

Details of face drains

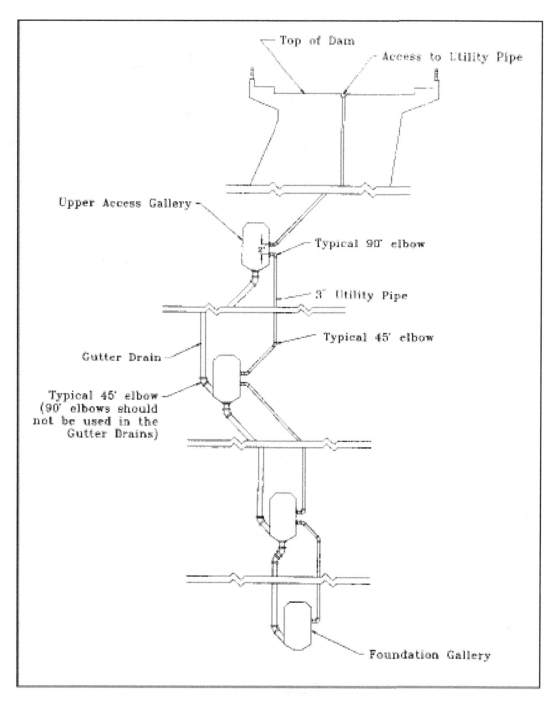

Figure 13-19

Details of gutter drains and utility piping

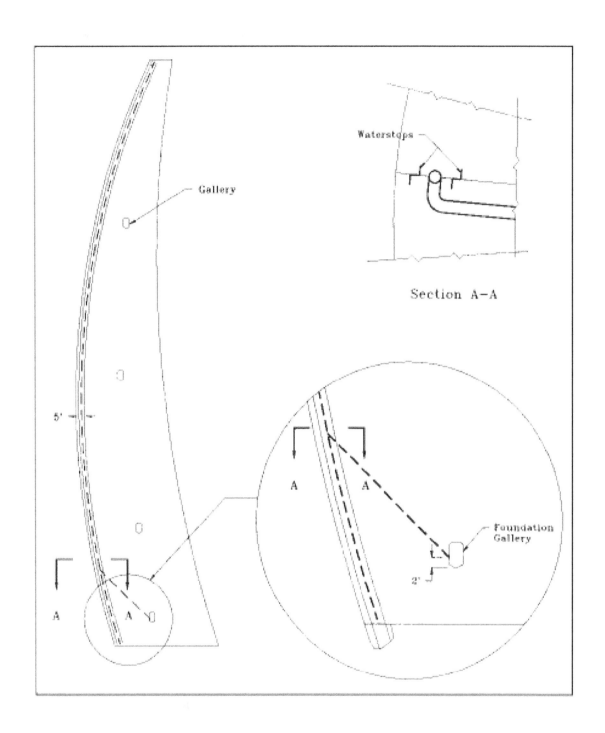

Figure 13-20

Details of joint drains

CHAPTER 7: FOUNDATION INVESTIGATIONS FOR ARCH DAMS

7.1 INTRODUCTION. Foundation investigations for arch dams generally must be accomplished in more exacting detail than for other dam types because of the critical relationship of the dam to its foundation and to its abutments. This chapter will describe the procedures which are commonly followed in accomplishing each phase of these investigations including the foundation analysis. It is very important that these investigations employ the latest state-of-the-art techniques in geological and rock mechanics investigations. This work is usually accomplished in relatively discrete increments or phases with each leading to the succeeding one and building upon the previous one. These phases are described as separate sections in the following text and are covered in chronological order as they are normally accomplished. Usually, a considerable amount of geological information and data are available in the form of published literature, maps, remotely sensed imagery, etc. which should be assembled and studied prior to initiation of field investigations. This information is very useful in forming the basis for a very preliminary appraisal of site adequacy and also serves as the basis for initiation of the succeeding phase of the investigation.

7.2 SITE SELECTION INVESTIGATIONS. This phase of foundation investigation is performed for the purpose of locating the safest and most economical site(s) on which to construct the arch dam. It also will serve to verify the suitability of the foundation to accommodate an arch dam.

7.2.1 IT IS IMPORTANT TO DETERMINE the rock types, rock quality, and suitable founding depths for the dam. This information will be required in estimating foundation treatment and excavation depths necessary for the construction of a dam at each of the potential sites being investigated. This information is utilized in the development of cost comparisons between the various sites being evaluated for site selection. Investigational techniques at this stage normally consist of geophysical surveys and limited core borings used together to prepare a subsurface interpretation along the alignment of each site under evaluation. Sufficient data must be obtained to preclude the likelihood of missing major foundation defects which could change the order of comparison of the sites evaluated. The type and spacing of core borings as well as the geophysical surveys must be designed by a competent engineering geologist taking into account the foundation rock types and conditions and anticipated structural configuration of the dam. This must be done in close coordination with the dam designer.

7.2.2 A MAJOR FACTOR THAT MUST BE considered in site selection is the topography of the site. Sites are classified as narrow-V, wide-V, narrow-U, or wide-U. Another factor to be considered in site selection is the quality of the rock foundation and the depth of excavation required to expose rock suitable for founding the structure. A third factor is the storage capacity of the reservoir provided by each different site investigated. All these factors must be considered in the economic comparison of each site to the others.

7.3 GEOLOGICAL INVESTIGATIONS OF SELECTED DAM SITE. Very detailed geological investigations must be performed at the selected dam site location to provide a thorough interpretation and analysis of foundation conditions. These investigations must completely define the rock mass characteristics in each abutment and the valley bottom to include accurate mapping of rock types, statistical analysis of rock mass discontinuities (joints, bedding planes, schistosity, etc.), location of faults and shear zones, and zonation of the subsurface according to rock quality as it is controlled by weathering. In addition to these geologic studies, the potential for earthquake effects must be assessed based on a seismological investigation.

7.3.1 SURFACE INVESTIGATIONS. This stage of investigation frequently entails additional topographic mapping to a more detailed scale than was needed for the site selection investigations. This is followed by detailed geologic mapping of all surface exposures of rock. Frequently it is necessary to increase these exposures by excavating trenches and pits to reveal the rock surface in areas covered by soil and vegetation. The fracture pattern existing in the rock mass is of particular significance and must be carefully mapped and analyzed. Any evidence of faulting and shearing should be investigated. Linear and abnormal configurations of surface drainage features revealed by remote sensing or on topographic maps may be surface reflections of faults or shear zones and should be investigated if they are located in or near the dam foundation. They may also be of seismological concern if there is evidence that they could be active faults. This concern may require fault trenching and age dating of gouge materials as well as establishing the relationship of the soil cover to last fault displacement to evaluate the potential for future activity on the fault. The surface geologic mapping should provide a sound basis for planning the subsurface investigations, which are the next step.

7.3.2 SUBSURFACE INVESTIGATIONS. The subsurface investigation program must be very thoroughly planned in advance to obtain all of the necessary information from each boring. This is a very expensive portion of the design effort and it can become much more expensive if the initial planning overlooks requirements which necessitate reboring or retesting of existing borings to obtain data which should have been obtained initially. The following paragraphs address procedures which must be considered in planning the subsurface investigation program for an arch dam.

7.3.2.1 CORE BORINGS MUST BE OBTAINED in order to provide hard data on foundation conditions. There are numerous decisions which must be made regarding the borings. The boring location plan is perhaps the first. This plan should contemplate a phased approach to the boring program so that future boring locations can be determined based upon data obtained from the earlier phase. The ultimate goal in locating borings is to provide sufficient coverage within the foundation to essentially preclude the possibility of adverse foundation features escaping detection. This can be accomplished by judicious spacing of borings along the dam axis utilizing both vertical and inclined orientations. Refer to Figures 10-1 and 10-2 for examples of an arch dam boring layout. Target depth for borings is another important consideration. Minimum depths should be established during planning with maximum depth left to the discretion of the geologist supervising the drilling as it is accomplished. Core diameter and type of core barrel are important considerations that affect both the cost of the investigation and the quality of the results. It may well be necessary to experiment with different combinations in order to determine the size and type of barrel that is most effective.

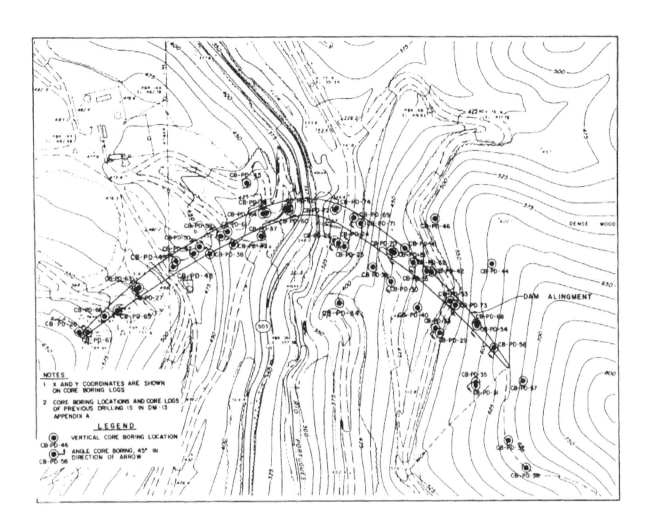

Figure 10-1

Arch dam boring layout plan from Portugues Dam Design

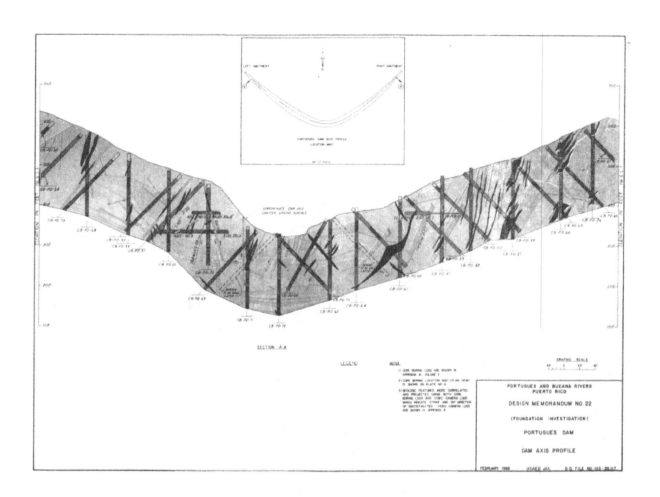

Figure 10-2

Arch dam boring profile from Portugues Dam Design

7.3.2.2 ROCK CORE LOGGING IS CRITICAL to the subsurface investigation. It is
essential that this be performed in considerable detail by a competent geologist, and it is
preferable that all rock core logging be done by the same individual, where feasible, for
the sake of consistency. The logging should include descriptions of rock type, rock quality
including degree of weathering, fractures, faults, shears, rock quality designation (RQD),
sufficient data to utilize the selected rock mass rating system, and should be supported
with photographs of all of the core taken while still fresh. It is important that the geologist
be present during drilling in order to log such occurrences as drill fluid losses, rod drops,
changes in drill fluid color, rod chatter, drilling rate, etc. These types of data used in
conjunction with the log of the rock core can greatly improve the interpretation of the
foundation encountered by a particular boring.

7.3.2.3 BORE HOLE LOGGING and testing should be utilized to enhance the amount of information obtained from each hole drilled. Certain techniques work better in some environments than in others; thus, the following techniques listed must be utilized discriminately according to their applicability to the site conditions. Bore hole logging systems include caliper logs, resistivity logs, SP logs, sonic logs, radioactive logs, etc. Bore hole TV cameras also provide important information on foundation conditions such as frequency and orientation of fractures and condition of rock in intervals of lost core. Bore hole pressure meters such as the Goodman Jack may provide valuable information on the rock mass deformation properties. Water pressure testing is important to develop data on the potential seepage characteristics of the dam foundation. All these techniques should be considered when planning the subsurface investigations. It is generally more efficient to perform these investigations at the time the hole is being drilled than to return to the hole at a later date.

7.3.2.4 LABORATORY TESTING OF CORE SAMPLES is necessary to provide design data on foundation conditions. Petrographic analysis is required to correctly identify the rock types involved. It is necessary to obtain shear strength parameters for each different rock type in order to analyze the stability of the foundation. The shear tests are normally run in a direct shear box and are performed on intact samples, sawed samples, and along preexisting fracture planes. This provides upper- and lower-bound parameters as well as parameters existing on natural fractures in the rock. The geologist and design engineer can then use these data to better evaluate and select appropriate shear and friction parameters for use in the foundation stability analysis. Another test performed on core samples is the unconfined compression test with modulus of elasticity determination. This provides an index of rock quality and gives upper-bound values of the deformation modulus of the rock for later comparison and correlation with in situ rock mass deformation tests.

7.3.2.5 GEOPHYSICAL SURVEYING TECHNIQUES can be utilized to improve the geological interpretation of the foundation conditions. These should be used in conjunction with the surface geological mapping and with the core borings to provide an integrated interpretation of subsurface conditions. Surface resistivity and refraction seismology are techniques which may provide usable data on depth of overburden and rock quality variations with depth as well as stratification. Cross-hole seismic surveys are sometimes successful in detecting large fault zones or shear zones trending between borings. Other geophysical techniques such as ground penetrating radar and electrical spontaneous potential are available, are being further refined and improved, and should be considered for environments where they have a likelihood of success. Seismic techniques are also appropriate for determining the dynamic modulus of deformation of the rock mass.

7.3.2.6 GROUND WATER INVESTIGATIONS and permeability testing are necessary for several reasons. These investigations provide the basis for design of any dewatering systems required during construction. They also provide the data to evaluate the reservoir's capability to impound water and to design seepage and uplift control required in the foundation beneath the dam and in the abutments. These data also provide the basis for making assumptions of uplift on rock wedges. Ground water levels, or their absence, should be carefully and accurately determined in all borings. Water pressure testing should be accomplished in most foundation and abutment borings to locate potential seepage zones and to provide data to help in designing the foundation grouting program. The literature is extensive concerning procedures for performing and evaluating bore hole pressure tests. Pumping tests are also very important in providing data for evaluation of the foundation seepage characteristics of the foundation. Reference is made to EM 1110-1-1804 for further guidance on both pressure testing and pump testing.

7.3.2.7 GROUT TESTING IS NECESSARY for multiple reasons. First, it is necessary to evaluate the groutability of the foundation. Water pressure testing alone can be very misleading in evaluating groutability because rock with very fine fractures may take significant quantities of water but be impervious to even a very thin cement grout. Grout testing is also required for determining the optimal size grout hole and the most effective means of drilling the hole. In some rock foundations, percussion drills with cuttings removed by air provide the best holes for grout injection, while in others, rotary drills utilizing water for cuttings removal are the most appropriate. A grout test provides the opportunity to experiment with multiple drilling techniques and various hole diameters to determine the most effective ones prior to entering into the main construction contract when changes are normally quite expensive. Another important reason for grout testing is to improve the estimates of quantity of grout take and length of holes likely to be required in the main contract. Perhaps the most important reason to perform a grout test is to provide an evaluation of the probable effectiveness of the grout curtain for consideration during design.

7.3.2.8 RIM TESTS AND EVALUATIONS are important in some reservoir areas where there may be concerns for excessive loss of water through rim leakage or where potentially large landslides may occur which could displace a significant volume of the lake causing over topping of the dam. These evaluations can be accomplished through topographic and geologic mapping of the reservoir rim followed by core boring, water table determination, and pressure testing in those areas of concern. Remedial measures may be required in areas which are susceptible either to excessive leakage or to significant landslides.

7.3.3 ABUTMENT ADITS. Adits provide excellent access for in situ observation and mapping of foundation conditions as well as large-scale rock mass testing. Information obtained from adit investigations provides a much higher level of confidence that all significant foundation defects have been detected than if borings and geophysical surveys alone are used. They also greatly improve the confidence level in the mapping and statistical evaluation of the rock mass fracture system. It is advisable to include adits in the subsurface investigation program for arch dam sites with difficult or unusual foundation conditions.

7.3.3.1 EXPLORATORY ADITS MAY BE of various sizes and shapes, however, 5 feet wide by 7 feet high is considered to be about the minimum size. A more practical size is 7 feet wide by 8 feet high in that it provides adequate space for the contractor's excavating equipment and for in situ rock mechanics testing. The horseshoe shape is a good configuration for an exploratory adit. It provides essentially vertical side surfaces and a horizontal floor surface which are more easily surveyed and mapped than curved configurations. The adit locations in the abutments should be selected with two factors in mind. First, it is desirable to locate them so that the in situ rock mechanics tests can be performed at or near the location of the maximum stress to be applied to the foundation by the dam. This is normally at about the one-third height of the dam. Another factor to be considered in the location is the geology of the abutments. If there are conditions of concern, such as faults, shear zones, etc., it may be necessary to locate the adits to provide access to these features for in-place inspection and testing. The adits should be oriented to provide maximum intersection of the fracture system and to provide access for in situ testing of the rock mass immediately below the founding level of the dam. It is prudent to construct an adit in each abutment if foundation conditions vary significantly from one abutment to the other. Geological conditions may require more than one adit in an abutment.

7.3.3.2 GEOLOGIC MAPPING IS REQUIRED FOR EACH ADIT. The preferred procedure to follow is the full periphery method. The results can be displayed in reports in both plan and isometric views. Refer to Figures 10-3 and 10-4 for examples.

7.3.2.3 SURVEYS OF JOINTS AND OTHER rock mass discontinuities should be accomplished while performing the adit mapping. The adits provide excellent exposures for obtaining data for a statistical analysis of the fracture system in the rock. It is important to perform bias checks to assure that the orientation of the adit is not resulting in over counting of some joint sets in relation to others. For instance, a joint set oriented perpendicular to the axis of the adit will be intersected much more frequently than one oriented parallel, thereby tending to bias the statistical analysis. It is also important to describe the spacing, frequency, extent or degree of separation, openness, roughness, joint filler, and wall rock condition of weathering of each fracture set across the width and height of the adit. This information is needed when determining shear strength parameters for individual fracture sets for use in stability analysis. Stereographic projection coupled with statistical analysis is an excellent method for determining the preferred orientation of the joint sets.

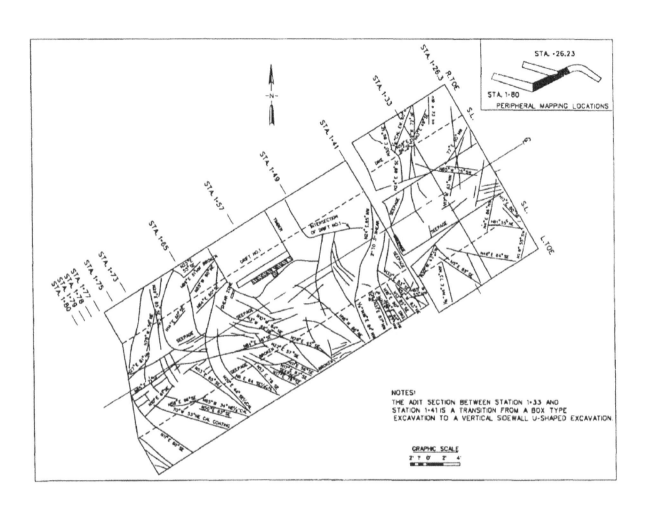

Figure 10-3

Adit periphery mapping plan view from Portugues Dam Design

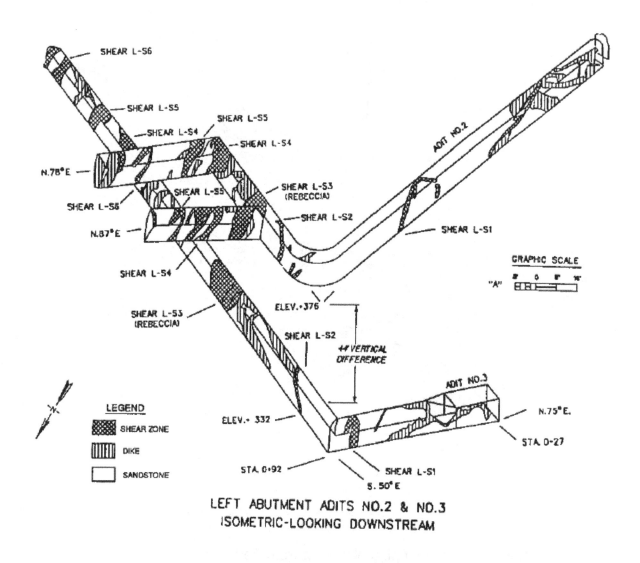

SHEAR L-S6

SHEAR L-S5

SHEAR L-S5

SHEAR L-S4

SHEAR L-S4

ADIT NO.2

SHEAR L-S4

N.78°E

SHEAR L-S3
(REBECCIA)

SHEAR L-S5

SHEAR L-S5

SHEAR L-S3
(REBECCIA)

SHEAR L-S2

N.87°E

SHEAR L-S1

GRAPHIC SCALE

"A"

SHEAR L-S4

ELEV.+376

SHEAR L-S3
(REBECCIA)

SHEAR L-S2

44 VERTICAL
DIFFERENCE

ADIT NO.3

ELEV.+ 332

N.75°E.

STA. 0+27

LEGEND

SHEAR ZONE

DIKE

SANDSTONE

STA. 0+92

SHEAR L-S1

S. 50°E

LEFT ABUTMENT ADITS NO.2 & NO.3
ISOMETRIC-LOOKING DOWNSTREAM

Figure 10-4

Adit periphery mapping isometric view from Portugues Dam

7.3.2.4 ADITS PROVIDE AN OPPORTUNITY to obtain samples for laboratory testing as well as providing access for in situ testing. Shear zones and faults with their gouge and brecciated intervals may be sampled for laboratory testing. The tests may include mineral identification, shear strength determination, Atterberg limits of soil-like materials, etc. In situ testing may include uniaxial jacking tests to determine the rock mass deformation properties, shear testing of discontinuities, and measurement of in situ rock stress.

7.3.3 TEST EXCAVATIONS. Test excavations are valuable aids in a dam site investigation.

7.3.3.1 TEST PITS AND TRENCHES can provide the field geologist with strategically located exposures of the bedrock for mapping purposes. A bulldozer is usually necessary to prepare access roads to core drill locations. Excavation for abutment access roads can provide very good exposures of the rock surface. These roads can be extended as necessary to provide the field geologist with exposures where they are needed. Normally, it is wise to expose the bedrock for mapping in trenches or road cuts that zig zag across the dam site from valley bottom to the top of dam on each abutment. This additional exposure of the rock will normally significantly improve the geologic interpretation of foundation conditions.

7.3.3.2 LARGE DIAMETER CALYX BORINGS may in some instances be required where individual foundation features are of such concern that it is necessary for the field geologist to examine them in place. They are very expensive but are less costly than excavating an adit.

7.3.3.3 THE DAM FOUNDATION IN THE VALLEY BOTTOM and on the abutments may be excavated by separate contract prior to the main dam contract as a means of fully exposing the foundation for examination by the geologists and the dam designers. At this point in time, changes in the dam's design can still be made without incurring excessive delay costs from the main dam contractor. The foundation should be carefully mapped and the geologic interpretation formalized as a part of the foundation design memorandum. It should also be incorporated into the final foundation report. Final foundation preparation and cleanup, including some additional excavation, should be left for the main dam contract because most rock surfaces will loosen and weather when left exposed to the elements for a significant period of time.

7.3.4 ROCK MASS RATING SYSTEM. An important consideration in the geological investigations of an arch dam foundation is to obtain sufficient data to allow the quality of the foundation to be quantitatively compared from one area to another. In order to do this, it is necessary to adopt a rock mass rating system for use throughout the geological investigations. Several such systems have been developed. Bieniawski (1990) provides a survey of the more widely accepted systems. These systems were for the most part developed to provide a means of evaluating rock mass quality for tunneling; however, they can be adapted to provide a meaningful comparison of rock quality in a foundation. The geomechanics classification proposed by Bieniawski (1973) is particularly useful. This system assigns numerical values to six different rock parameters which can be obtained in the field and from core borings. The rock mass rating (RMR) is calculated as follows:

$$R = A + B + C + D + E - F \qquad (10\text{-}1)$$

where

```
A = Compressive strength of intact rock
B = Deere's RQD
C = Spacing of joints
D = Condition of joints
E = Ground water conditions
F = Adjustment for adverse joint orientation
```

Factor F is very important in assessing rock quality in a tunnel but is not necessarily appropriate in a classification system for assessing the rock mass strength of an arch dam foundation, since it is taken into account in the foundation stability analysis. For this reason it may be advisable to consider altering the system for individual arch dam foundation evaluations. Table 10-1 from Bieniawski (1990) provides the geomechanics classification of jointed rock masses. It is important when logging rock core or when performing geologic mapping to assure that all data necessary for the rock mass rating system are collected.

7.3.5 STEREONET ANALYSIS OF ROCK FRACTURE SYSTEM. An analysis must be made of the fracture system in each abutment and the valley section for use in the rock mechanics analysis of foundation and abutment stability. The Schmidt equal area stereonet utilizing the lower hemisphere projection is the conventional system normally used. Individual fractures (joints) are located on the stereonet by plotting the point on the lower hemisphere where a pole constructed normal to the plane of the fracture would pierce the hemisphere. After all fractures being analyzed are plotted on the stereonet, an equal area counting procedure is used to determine the percentage of poles which fall in each area. These are then contoured similar to the contouring procedure for a topographic map. The contoured stereonet can then be readily evaluated to determine the orientation of the primary, secondary, tertiary, etc. joint sets. Refer to Figure 10-5 for an example of an equal area joint polar diagram.

7.4 ROCK MECHANICS INVESTIGATIONS. The foundation of an arch dam must function as an integral part of the structure. It is very important that the dam designer fully appreciate and understand the mechanical properties of the foundation. To fully describe the foundation conditions so that they may be quantified for incorporation in the dam design, it is first necessary to accurately and completely define the geologic conditions as described previously, and then define the rock mass mechanical properties. A thorough rock mechanics investigation of the geologic environment of the foundation is necessary to provide the quantification of foundation properties necessary for the dam foundation analysis. It is extremely important that the engineering geologist, geotechnical engineer, and the structural designer work closely and cooperatively during the rock mechanics investigation and foundation analysis to assure that the concerns of each are adequately addressed. Each has a unique perspective on the design which can contribute to the success of the rock mechanics investigation and foundation analysis.

A. CLASSIFICATION PARAMETERS AND THEIR RATINGS

	PARAMETER		RANGES OF VALUES						
1	Strength of intact rock material	Point-load strength index	> 10 MPa	4 - 10 MPa	2 - 4 MPa	1 - 2 MPa	For this low range – uniaxial compressive test is preferred		
		Uniaxial compressive strength	>250 MPa	100 - 250 MPa	50 - 100 MPa	25 - 50 MPa	5-25 MPa	1-5 MPa	<1 MPa
		Rating	15	12	7	4	2	1	0
2	Drill core quality RQD		90% - 100%	75% - 90%	50% - 75%	25% - 50%	< 25%		
	Rating		20	17	13	8	3		
3	Spacing of discontinuities		>2 m	0.6 - 2 m	200 - 600 mm	60 -200 mm	< 60 mm		
	Rating		20	15	10	8	5		
4	Condition of discontinuities		Very rough surfaces. Not continuous No separation Unweathered wall rock.	Slightly rough surfaces. Separation < 1 mm Slightly weathered walls	Slightly rough surfaces. Separation < 1 mm Highly weathered walls	Slickensided surfaces OR Gouge < 5 mm thick OR Separation 1-5 mm Continuous	Soft gouge > 5 mm thick OR Separation > 5 mm. Continuous		
	Rating		30	25	20	10	0		
5	Ground water	Inflow per 10 m tunnel length	None	<10 litres/min	10-25 litres/min	25 - 125 litres/min	> 125 litres/min		
			OR	OR	OR	OR	OR		
		Ratio $\frac{\text{joint water pressure}}{\text{major principal stress}}$	0	0.0-0.1	0.1-0.2	0.2-0.5	>0.5		
			OR	OR	OR	OR	OR		
		General conditions	Completely dry	Damp	Wet	Dripping	Flowing		
	Rating		15	10	7	4	0		

B. RATING ADJUSTMENT FOR JOINT ORIENTATIONS

Strike and dip orientations of joints		Very favourable	Favourable	Fair	Unfavourable	Very unfavourable
Ratings	Tunnels	0	-2	-5	-10	-12
	Foundations	0	-2	-7	-15	-25
	Slopes	0	-5	-25	-50	-60

C. ROCK MASS CLASSES DETERMINED FROM TOTAL RATINGS

Rating	100 ← 81	80 ← 61	60 ← 41	40 ← 31	< 20
Class No	I	II	III	IV	V
Description	Very good rock	Good rock	Fair rock	Poor rock	Very poor rock

D. MEANING OF ROCK MASS CLASSES

Class No	I	II	III	IV	V
Average stand-up time	10 years for 15 m span	6 months for 8 m span	1 week for 5 m span	10 hours for 2.5 m span	30 minutes for 1 m span
Cohesion of the rock mass	> 400 kPa	300 - 400 kPa	200 - 300 kPa	100 - 200 kPa	< 100 kPa
Friction angle of the rock mass	> 45°	35° - 45°	25° - 35°	15° - 25°	< 15°

TABLE 10-1

Geomechanics Classification of Jointed Rock Masses.

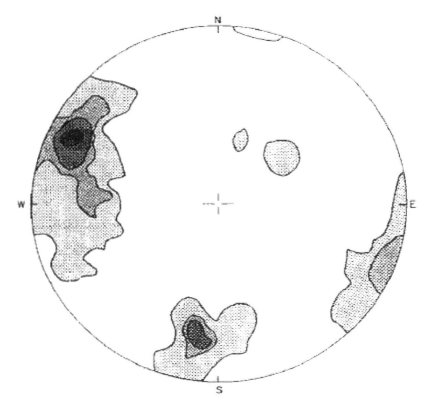

JOINT POLE CONTOUR DIAGRAM

172 MEASUREMENTS

LOWER HEMISPHERE
SCHMIDT NET
STEREOGRAPHIC PROJECTION

PERCENTAGE VALUES, PER 1% AREA

☐	≤ 2%	▓	> 6%, ≤ 8%
☐	> 2%, ≤ 4%	▓	> 8%, ≤ 10%
▓	> 4% ≤ 6%	■	> 10%

Figure 10-5

Equal area joint pole contour diagram from Portugues Dam Design

7.4.1 LABORATORY TESTING OF SAMPLES. Laboratory testing is much less

expensive than in situ testing and can provide very good data. The laboratory testing must

be evaluated and the design assumptions made by a competent and experienced

geotechnical engineer in full coordination with an engineering geologist who fully

appreciates the geologic conditions in the foundation which could adversely affect the

performance of the dam. The laboratory testing should incorporate unconfined

compression tests including determination of modulus of elasticity on a suite of samples from each rock type and each rock quality to be found below the founding level of the dam. These tests will provide an index of rock strength and will provide data on the upper bound deformation modulus of the rock. Direct shear testing should be performed on a suite of samples from each rock type and rock quality to provide data for use in determining shear strength values to be used in foundation analyses. This type of test should be performed under three separate conditions: on intact samples to provide an upper-bound set of data on shear strength parameters, on sawed surfaces to provide a lower-bound or residual strength set of data, and finally, on natural fractures to provide data on the shear friction parameters resisting sliding on these features. In those cases where clay or silt material exists as gouge in fault zones, in shear zones, or as in-filling in open joints, it is advisable to obtain samples of this material and perform shear tests on preferably undisturbed samples or on remolded samples where undisturbed samples are not feasible. Both triaxial and direct shear tests are appropriate for these samples. When interpreted conservatively and with a full appreciation of the geologic conditions, test data obtained as described will usually provide a sound basis for estimating the shear strength parameters for use in the foundation stability analyses.

7.4.2 ABUTMENT ADITS. Adits are excavated in the abutments for two primary purposes. First, they are constructed to provide access to map geologic conditions and to expose geologic features of structural concern. Of equal importance is the access provided to perform in situ rock mechanics testing of foundation conditions.

7.4.2.1 IN SITU DEFORMATION TESTING IS PERFORMED in the adits as a means of determining the stiffness of the foundation. From these tests the static modulus of deformation of the foundation can be calculated at that particular location. Several different techniques have been developed for performing deformation testing. These include the uniaxial jacking test, the radial jacking test, and the pressure chamber jacking test. The uniaxial jacking test is the one most commonly used primarily because it is less costly and easier to set up while still providing satisfactory results. Figure 10-6 is a diagram of a typical uniaxial jacking test setup. It consists of the following: a load frame which transfers load from one wall of the adit to the other, two flat jacks which apply the loads to the rock surfaces, two multiposition borehole extensometers which measure the deflection or deformation of

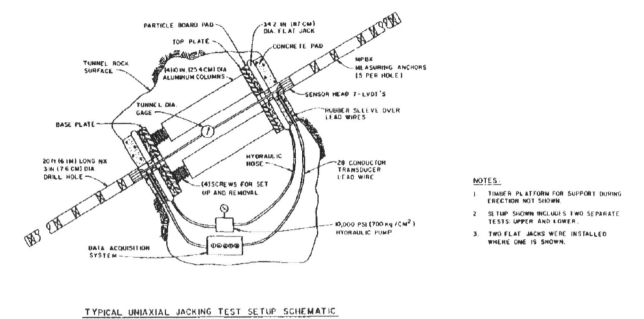

TYPICAL UNIAXIAL JACKING TEST SETUP SCHEMATIC

Figure 10-6

Uniaxial jacking test diagram from Portugues Dam

the rock mass as it is loaded or unloaded, a tunnel diameter gauge which measures changes in tunnel dimensions as load is applied or released, and a very high-pressure hydraulic jack. The axial orientation of the test is selected to coincide with the resultant of the forces applied to the foundation by the dam. The test as depicted in Figure 10-6 provides a measure of the rock mass deformation on two opposing surfaces of the adit. This in effect provides two rock mass modulus of deformation tests at this location. The test is performed by loading the rock in predetermined increments for a specific period of time. Incremental increases of 200 psi held for periods of 24 hours with complete unload between each pressure increase is commonly used. The unloaded increment of time is also commonly 24 hours. The maximum load applied should exceed the maximum pressure to be applied by the dam. A maximum test pressure of 1,000 psi is commonly used. Deformation measurements should be made at frequent intervals during both loading and unloading cycles. These measurements can provide data for evaluating the creep potential and initial set after loading in the rock mass in addition to the modulus of deformation. The following publication of the American Society for Testing and Materials

(ASTM) provide detailed test and analytical procedures for computing the rock mass modulus of deformation D4395-84 (ASTM 1984b). The following equations taken from these references may be used for computing the modulus of deformation:

<u>Flexible loading system</u>

(Surface deflection at center of circularly loaded area)

$$E = \frac{2(1-\mu^2)QR}{W_c} \qquad (10\text{-}2)$$

(Surface deflection at center of annularly loaded area)

$$E = \frac{2Q(1-\mu^2)(R_2-R_1)}{W_c} \qquad (10\text{-}3)$$

<u>Rigid loading system</u>

(surface deflection)

$$E = \frac{(1-\mu^2)(P_1)}{2W_a R} \qquad (10\text{-}4)$$

where

E = modulus of deformation
μ = Poisson's ratio
Q = pressure
R = radius of loaded area or radius of rigid plate
W_c = deflection at center of loaded area

R_2 = outside radius of annulus
R_1 = inside radius of annulus
W_a = average deflection of the rigid plate
P = total load on the rigid plate

The ASTM references contain other equations for calculating the modulus of deformation for deflections measured within the rock mass.

7.4.2.2 PROCEDURES ARE AVAILABLE FOR DETERMINING the elastic properties of rock utilizing seismic techniques. These are described in some detail in ASTM publication STP402 (ASTM 1965a). This technique provides a Poisson's ratio and the dynamic modulus of deformation of the rock mass. Use of the technique for arch dam foundation evaluation must be tempered with considerable practical knowledge and judgement, because the dynamic modulus of deformation so determined is usually significantly higher than the static modulus determined by the uniaxial or radial jacking tests. The open fractures in the rock mass are a major factor in this discrepancy. The rock mass behaves more in the static mode than the dynamic mode under loading from a dam, therefore, the static modulus is more appropriate for analyzing the foundation for an arch dam, except in evaluating dynamic earthquake response.

7.4.2.3 IN SITU SHEAR TESTING CAN BE PERFORMED in an adit to test the shearing resistance of an individual feature in the foundation if conditions demand this information. The test, however, is very expensive and provides information only on the feature tested. In many cases it is more practical to perform extensive laboratory tests of shear strength and to use this information along with engineering experience as the basis for arriving at the proper shear strength parameters for use in the foundation stability analyses.

7.4.2.4 IN THOSE CASES WHERE ABNORMALLY high in situ stress may exist in the foundation or abutment rock mass, it may be advisable to perform in situ testing to measure the in-place stress regime for consideration during the foundation analysis. Several instruments have been developed for measuring the strain release in an over cored bore hole. These instruments are suitable for determining the in situ stress existing within about 25 feet of a free face or surface from which a boring can be drilled. Flat jacks may be used for measuring stress existing immediately beneath a free face or surface. Bore hole hydraulic fracturing techniques are appropriate for measuring in situ stress at locations remote from a drilling surface. This procedure can measure in-place stress

© J. Paul Guyer   2017   239

hundreds of feet from the surface. One clue to the existence of a high in situ stress field is the appearance of disking in rock core samples. This disking is caused by stress release in the core occurring after it is freed from the restraint of the surrounding rock by the coring action. Numerous publications available in the literature describe the various in situ stress determination techniques. ASTM publication STP402 (ASTM 1965a) describes various instruments and techniques.

7.4.2.5 SAMPLING AND LABORATORY TESTING have been discussed previously. In addition to samples obtained from core borings, the adits are also good locations for obtaining samples for laboratory testing. This is particularly true of samples of hard-to-retrieve materials such as fault breccia and fault gouge. These materials must be sampled very carefully to minimize disturbance, and immediately after exposure to retain near natural moisture content. Where possible, planning and preparation for this sampling should be done prior to adit excavation to enhance the likelihood of obtaining usable samples.

7.5 ROCK MECHANICS ANALYSES. The dam foundation, and in particular its abutments, must be carefully analyzed to evaluate resistance to shear failure, deformation characteristics, and permeability. The foundation must respond as an integral part of the dam and must be fully considered in the design of the dam. Of particular importance is the analysis of the fracture system within the rock mass, including the joints, faults, shears, bedding, schistosity, and foliation. These features must be considered in relation to each other because intersecting fractures can sometimes form potential failure wedges which are more susceptible to sliding than either fracture alone. Since the elastic properties of the dam and its foundation are significant factors in the performance of the dam, it is necessary to estimate the deformation properties of the foundation and abutments within a reasonable degree of accuracy. The permeability of the foundation and uplift pressures on potential failure wedges must be evaluated and incorporated into the design. The engineering geologist, geotechnical engineer, and structural engineer responsible for the dam design must work in full coordination and cooperation in the performance of these analyses to assure that the concerns and objectives of each discipline are satisfied to the maximum extent possible.

7.5.1 ROCK MASS PROPERTY DETERMINATION. Methods of testing the rock mass to define its physical properties have previously been discussed. These discussions are continued here and include the processes necessary to arrive at values to be used in the foundation and abutment analyses.

7.5.1.1 IN ORDER TO PERFORM STABILITY analyses of the foundation and the abutments, it is necessary to select appropriate values of the shear strength of each fracture set, shear, fault, or other discontinuity which could form a side of a kinematically capable failure wedge. Laboratory shear tests normally will provide the basic data required for selecting these shear strength values. As stated previously, direct shear testing should be performed on a suite of samples from each rock type and rock quality. This test should be performed under three separate conditions to provide upper- and lower-bound rock strength and shear friction as described.

7.5.1.1.1 THE DATA PROVIDED BY THIS SERIES of tests should provide part of the basis for determining reliable shear friction values resisting movement on discontinuities. There are other factors which must also be considered in arriving at acceptable shear strength values on a discontinuity. Roughness measures such as the angle of the asperities have a significant effect upon the shear strength of a joint because, for movement to take place, the rock mass must either dilate by riding up and over the asperities or it must shear through them. Either mechanism takes considerable additional energy. The condition (degree of weathering) of the wall rock is another factor to be considered as is the continuity or extent of the open joints which can significantly affect shear resistance to movement. Fracture filling material, such as clay or silt, can dramatically reduce the shear friction strength of a fracture and must be considered.

7.5.1.1.2 A CONSERVATIVE EVALUATION OF SHEAR strength for rock-to-rock contact on a fracture under relatively low normal load is provided by the following equation:

$$S = N \tan(\phi_r + i) \qquad (10\text{-}5)$$

where

S = shear strength of fracture
N = normal stress on fracture
ϕ_r = residual friction angle
i = asperity angle

In this equation, the value of cohesion is omitted under the assumption that failure occurring under a low normal stress would likely result in the wedge overriding the asperities rather than shearing through them. The residual friction angle is derived from the direct shear test on sawed surfaces and should be taken at about the lower one-third point of the range of test values. The asperity angle is developed from actual measurements of the roughness on typical joints in the foundation. It is important to measure the asperity angle in the same direction that movement would likely occur since the degree of roughness varies considerably from one orientation to another. The angles are measured from a string line oriented in the likely direction of movement. (Refer to Figure 10-8 for an example of a field measurement setup.) Rock shear test results obtained from sawed surfaces are more consistent and amenable to interpretation than those obtained from shearing of intact rock or those obtained from shear testing along natural fractures. A more reliable shear strength value is thus developed by adding the angle of the asperities to the residual shear friction angle in this equation.

7.5.1.1.3 THE SELECTED SHEAR STRENGTH values for a particular joint may require modification depending upon the other factors noted previously which affect shear strength such as continuity or extent of the fracture, condition of its wall rock, and in-filling material, if any. If fracture continuity is less than 100 percent, then added strength can be allowed for shear through intact rock. The test data obtained from samples sheared through intact rock provide a basis for assigning shear strength values to the portion of the failure plane which is not part of the natural fracture. If the wall rock is weakened by weathering, the strength must be reduced. The suite of tests performed on weathered rock will provide data on which to base this reduction in shear strength. If soft in-filling material is present, rock-to-rock contact is diminished, and the strength must be reduced as a compensation.

7.5.1.1.4 WHERE GOUGE MATERIAL EXISTS along shears or faults or where joint filling is present, it is necessary to obtain samples of this material for testing. This can best be done from an adit because it is often impossible to obtain an adequate quantity for testing from a bore hole. Undisturbed samples are preferred, but often they are not feasible to obtain. Remolded samples tested at in situ density and moisture content will normally provide satisfactory results. The strength properties of the gouge or in-filling material can be tested utilizing conventional soil property tests.

Figure 10-8

Picture of asperity angle "i" measured in the field utilizing a string line for orientation from
Portugues Dam

The consolidated, undrained triaxial test is appropriate for testing this material. In the
stability analysis, an estimate must be made of the percentage of the potential failure
surface which could pass through these soil-like materials. The strength of that portion of
the potential failure surface must be assessed based on the tests of the gouge or in-filling
material.

7.5.1.2 THE DEFORMABILITY OR STIFFNESS OF an arch dam foundation must be estimated for incorporation in the stress analysis of the dam. The modulus of deformation of the rock mass is a measure of the foundation's deformation characteristics. The modulus can be expected to vary significantly from the valley bottom to the abutments and from one rock quality or rock type to another. The methods of testing and measuring the modulus of deformation have previously been described. It is necessary to translate the test results from a few specific locations to an interpretation of the deformation characteristics of the entire foundation. In order to do this on a quantitative basis, a rock mass rating system is required which permits quantitative evaluation of the rock mass quality over the entire foundation area. There are several rock mass rating systems currently in use worldwide. The geomechanics classification system developed by Bieniawski (1990) and described previously in paragraph 10-3e has proven very useful for foundation analysis. A relationship has been suggested between the geomechanics classification system RMR and the in situ modulus of deformation of the rock mass. This relationship is expressed in the following equation:

$$E = 10^{(\frac{RMR-10}{40})} \qquad (10\text{-}6)$$

where

E = modulus of deformation measured in gigapascals (GPa)

1 GPa = 145,037.7 psi

This equation was used in the rock mechanics analysis of the Portugues Dam Foundation and was a valid predictive model of the foundation deformation properties of the dam. Once this model is validated for a particular site, it is possible to compare the entire site conditions to the in situ tests of modulus of deformation. This is accomplished by determining the RMR for the segment of rock of concern in each core boring made in the foundation and then comparing these borings with the RMR of the core borings made for installation of the extensometers at the location of the in situ modulus of deformation tests. This comparison used in concert with the relationship noted then allows the assignment of modulus of deformation values to each major portion of the foundation.

These values can then be applied in the finite element analysis of the dam and its foundation.

7.5.1.3 THE PERMEABILITY OF THE FOUNDATION and abutments must be determined for several reasons. As stated earlier in paragraph 10-3b(6), these data are required to evaluate the reservoir's capability to impound water, to provide a basis for design of construction dewatering systems, to provide a basis for design of uplift relief systems, and to serve as the basis for estimating the amount of uplift that must be considered in the stability analysis. The permeability of the foundation can be estimated based primarily on bore hole pressure test data supported by a limited amount of pump test data. A pertinent equation is as follows:

$$K_e = \frac{Q \ln (R/r)}{2\pi LH} \tag{10-7}$$

where

K_e = equivalent coefficient of permeability
r = radius of bore hole in feet
Q = volume of flow rate in cfs
H = excess pressure head at center of test in feet
R = radius of influence in feet (0.5L to 1.0L)
L = length of test section of bore hole
π = pi

7. 5.1.4 IN CASES WHERE ABNORMAL in situ stress conditions are indicated, it may be necessary to perform tests to measure the in situ stress existing in the rock mass. These stresses may be significant in the foundation stability analysis. There are several techniques and variations of these techniques available for measuring in situ stress in rock masses. The overcoring procedure is commonly used to measure stresses within a relatively short distance (±25 feet) from an exposed surface or free face. The hydraulic fracturing procedure is used to measure stresses existing at locations remote from an exposed surface or free face.

7.5.1.5 THE POISSON'S RATIO OF THE ROCK MASS must be estimated for some foundation analyses. A satisfactory method of doing this is to obtain the Poisson's ratio at the same time that the modulus of elasticity is determined while measuring the unconfined compression strength of intact rock core samples from each different rock type and quality in the foundation. Mean values obtained from each rock type and quality will provide values that are satisfactory for this purpose.

7.5.2 ABUTMENT STABILITY ANALYSIS. Much of the previous narrative was intended to provide data and information necessary to perform the abutment stability analyses. Abutment stability is critical to the overall stability of an arch dam. The following subparagraph describes the analytical procedures, the first of which is the use of the stereonet for slope stability analysis.

7.5.2.1 THE PROCEDURE FOR PERFORMING A STATISTICAL representation of the rock mass fracture system utilizing the equal area stereonet has been described. This concept has also been adapted for use in the stability analysis of the foundation and abutment. The first step in analyzing the stability of the dam foundation is to locate any fracture, fracture set, or combination of fractures which could form a wedge kinematically capable of failure, either as a result of foundation excavation or under the forces applied to the foundation by the dam. For a wedge to be kinematically capable of failure, the dip of the potential failure plane or the plunge of the intersection of two fracture planes along which sliding could occur must intersect or "daylight" on the rock slope or free face. This must also occur in a location which would accommodate failure under the forces imposed by gravity and/or the dam. This step of the analysis is accomplished by plotting the great circle representation of each fracture or fracture set and the natural or cut slope (free face) on an equatorial equal angle stereonet, as demonstrated in Figure 10-9. If the great circle representation of the free face intersects the great circle of both fracture planes and the plunge of the wedge of rock is a flatter angle than the dip of the free face, then movement of the wedge is kinematically possible without the necessity for crossbed shear through intact rock. The same test can be applied to sliding on a single plane which strikes subparallel to the free face and dips at an angle flatter than the free face. All major fracture sets and unique fractures such as faults and shears must be analyzed for their kinematic capability of movement. Those with the potential for failure must be further analyzed taking into account shearing resistance on the failure planes and driving forces which contribute to the potential for sliding. Since the fractures within an identified joint set normally have a range of orientations, it is not adequate to consider only the average or median orientation. Orientations near the bounds of the range must also be evaluated since they do exist in the rock mass. This first step of determining those wedges which could kinematically fail will eliminate a great many wedges from the need for further analysis.

7.5.2.2 AFTER DETERMINING THOSE fractures and fracture combinations which are kinematically capable of allowing a wedge of rock to fail, it is then necessary to determine the geometry of a block which would be significant in the foundation of the dam. One conservative assumption that should be made in many cases is that the joint sets identified by the geological investigations are pervasive in the abutment on which they have been identified. By that it is meant that they can be expected to occur anywhere on the abutment. Next, a daylight point of the line of intersection of two fracture systems is chosen and the surface trace of the fractures is drawn. Once all the corners are located, the areas and volumes can be calculated. From this the volume of rock can be determined and converted into the weight of the wedge. The area of the fracture planes can be computed for use in determining the uplift forces acting on the wedge. The size of the wedge must be large enough to be significant to the stability of the dam; i.e., it should be large enough to cause catastrophic failure of the dam. For a wedge to be significant it must be possible for it to exist beneath the dam or immediately adjacent to the dam. Combinations of fractures which result in wedges that are above the top of the dam or outside the foundation of the dam need not be considered unless excavation will in some way make them a hazard to safety. A third fracture or

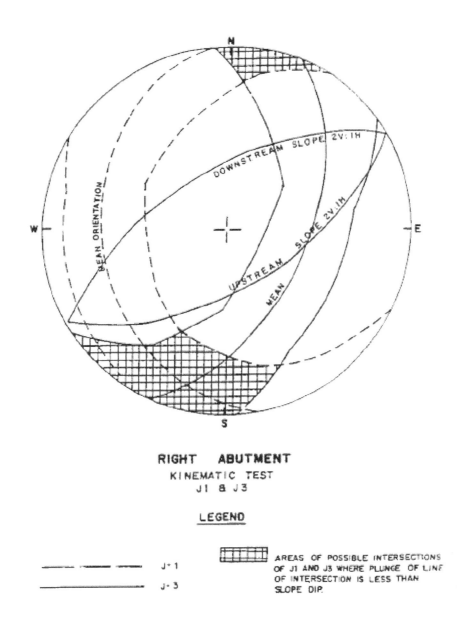

RIGHT ABUTMENT
KINEMATIC TEST
J1 & J3

LEGEND

J-1

J-3

AREAS OF POSSIBLE INTERSECTIONS
OF J1 AND J3 WHERE PLUNGE OF LINE
OF INTERSECTION IS LESS THAN
SLOPE DIP.

Figure 10-9

An equatorial angle stereonet plot showing two intersecting joint sets and two cut slopes

demonstrating kinematic capability for failure from Portugues Dam fracture set is required in most wedge geometries to cut the back of the wedge free. If no such fracture exists, it is conservative to assume that a tension fracture does exist in the position where such a feature is needed and no tensile strength exists across this feature. A wedge the size of the entire abutment is considered the most critical and is considered the definitive case for each set of intersecting fractures. Smaller wedges must also be considered but are less likely to result in catastrophic failure of the dam as they become smaller.

7.5.2.3 THREE DIFFERENT STATIC LOADING combinations must be analyzed. These cases have been described in Chapter 4 as static usual, the static unusual, and the static extreme. In addition, the dynamic loadings from earthquakes must be incorporated for dams in areas where there is earthquake potential. Refer to Chapter 4 for description of each different loading case. The loads which must be included in the analysis of potential rock failure wedges are: weight of the rock wedge; driving force applied by the dam; uplift applied by hydrostatic forces acting against the boundaries of the wedge; and dynamic forces generated by the design earthquake. The weight of the rock wedge is determined by calculating the volume of the wedge times the unit weight of the rock. The forces applied by the dam are obtained from the structural analysis of the dam. The computation of uplift forces is based upon the following assumptions:

7.5.2.3.1 FRACTURES ARE OPEN OVER 100 percent of the wedge area and are completely hydraulically connected to the surface.

7.5.2.3.2 HEAD VALUES VARY LINEARLY from maximum value at backplane to zero at daylight point.

7.5.2.3.3 BACK PLANES OR OTHER PLANES or segments of planes acted on directly by the reservoir receive full hydrostatic force. Dynamic loads attributed to the design earthquake are based upon the design earthquake studies which develop ground motions for both an OBE and the MCE. These studies provide values for the magnitude, distance, peak acceleration, peak velocity, peak displacement, and duration of the earthquakes.

7.5.2.3.4 THE FOLLOWING SLIDING FACTORS of safety (FS) should be used for the
different loading cases:

- (a) Static usual loads ------------- FS = 2.0
- (b) Static unusual loads ----------- FS = 1.3
- (c) Static extreme loads ----------- FS = 1.1
- (d) Dynamic unusual loads ---------- FS = 1.3
- (e) Dynamic extreme loads ---------- FS = 1.1

These FSs are based on a comprehensive field investigation and testing program as
described previously in this chapter.

7.5.2.3.5 THE FIRST STEP IN PERFORMING A STABILITY analysis of an abutment is
to determine those wedges of significant size which are kinematically capable of moving.
This step has already been described in paragraph 10-5b(1). The next step is to
determine the FS against sliding of the blocks where movement is kinematically possible.
There are three different methods that can be used for performing this analysis. The
conventional 2-D procedure can be used for conditions where sliding will occur on a
single fracture plane. This is the most simple of the three procedures, however, it is not
appropriate for the more complicated wedge type conditions where sliding can occur on
two or more planes, on the intersection of two planes, or by lifting off one plane and
sliding or rotating on another. For more complicated failure mechanisms one of the
following procedures should be employed: graphical slope stability analysis utilizing
stereonets or vector analysis, as described in the following paragraphs.

7.5.2.3.5.1 THE GRAPHICAL SLOPE stability analysis is a continuation of the procedure
already described for utilizing an equal angle stereonet to determine those wedges where
failure is kinematically possible. This step involves plotting on the stereonet those
parameters which are involved with the stability of the block. The first of these is the
reaction which resists failure. This consists of a plot of the friction cone which exists on
each plane involved in the wedge. This establishes the stable zone on the stereonet. The
friction cone will plot as a circle on the stereonet. There will be a separate friction cone
plotted for each fracture involved in the boundary of the wedge. The next step requires
the determination of the resultant of the forces that are driving the wedge. These forces

may include the weight of the rock wedge, uplift resulting from hydrostatic pressure acting normal to all the planes which define the wedge boundaries, thrust of the dam, and where earthquake loading is of concern those inertial forces which could be imposed by an earthquake. The resultant of these forces is obtained by the graphical summation of the vectors representing each force. If the resultant of these forces falls within the cone of friction, then the wedge is stable. In other words, if the resultant acts at an angle to the normal of the failure plane which is less than the angle of friction, then failure will not occur. The FS against sliding may be determined by dividing the tangent of the friction angle by the tangent of the angle made by the resultant of forces and the normal examples are illustrated in Figure 10-10.

7.5.2.3.5.2 THE VECTOR ANALYSIS PROCEDURE REQUIRES that all fractures forming the boundary of the wedge be described vectorially relative to the orientation of the abutment face. Vectors must be developed for the strike, dip, normal, and lines of intersection of the boundary fractures. Applied force vectors must be developed for weight of wedge, uplift on all boundary fractures, thrust of the dam, and inertial force resulting from the design earthquake. All these must be combined to form a resultant relative to the abutment face. From this the mode of failure is determined, i.e., sliding on a single plane, sliding on the intersection of two planes, or lifting all planes. By employing the vector analysis procedures, the stability of the wedge can be calculated and an FS can be determined. Figure 10-10 also illustrates this procedure.

7.5.2.3.5.3 THE ANALYSIS OF THE STABILITY OF the abutments of an arch dam requires very careful application of both engineering geology and rock mechanics investigative and analytical techniques. When these procedures are properly applied and their results accounted for in the design, a high degree of confidence in the stability of the dam foundation is justified.

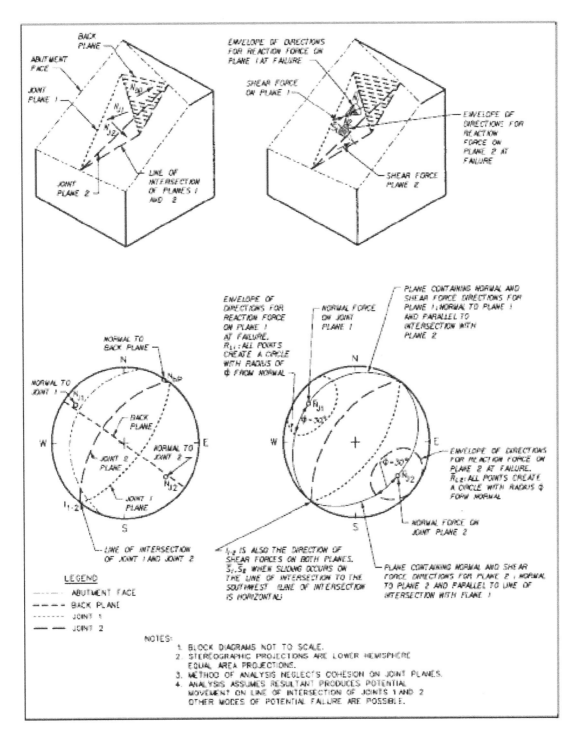

Figure 10-10

Illustration of both the vector and graphical stereonet technique of slope stability analysis

from Portugues Dam

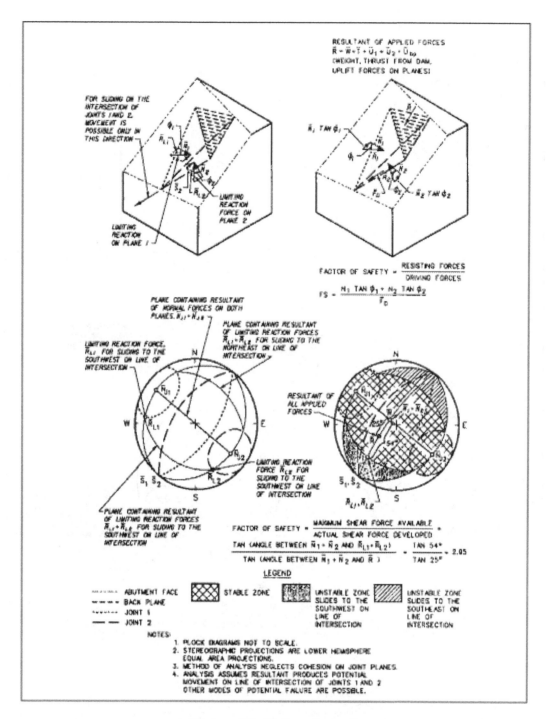

Figure 10-10 (continued)

Illustration of both the vector and graphical stereonet technique of slope stability analysis

from Portugues Dam

CHAPTER 8: ARCH DAM INSTRUMENTATION

8.1 INTRODUCTION. There are several reasons to incorporate instrumentation into any dam. One reason is so that the design and construction engineers can follow the behavior of the dam during its construction. Another reason is to compare the performance of the dam under operating conditions with that predicted by the designers. There is also the need to evaluate the overall condition and safety of the dam on a regular basis. A good instrumentation program can also increase the knowledge of the designer, and that knowledge can be applied to future projects. This chapter will discuss the instrumentation requirements that are needed to accomplish several of these points as they relate to arch dams.

8.1.1 GENERAL CONSIDERATIONS. In establishing any instrumentation program, it is important to understand the objectives of the program, the need for each type of instrument, the environment in which the instrument will be, the difficulty in gathering the data, and the time and effort in reducing and understanding the data generated. The wrong type of instrument may not measure the desired behavior. And while insufficient instrumentation may ignore important structural behavior, excessive instrumentation may bury the design engineer in a mound of paper that hides a serious condition. It is also important that some consideration be given to the initial and long-term cost of the program and the availability of trained personnel to collect the data. The cost of purchasing and installing the instruments will be up to 3 percent of the general construction cost (Fifteenth Congress on Large Dams 1985). A large instrumentation program will require a long-term commitment by management to maintain the program.

8.1.2 INSTRUMENT SELECTION. Instruments and associated equipment should be rugged and capable of long-term operation in an adverse environment. It is preferable to incorporate instruments of similar types (such as electric resistance versus vibrating wire types) in order to reduce the need for several types of readout equipment and training in numerous different types of equipment. The final decision of which type of instruments to use will be site specific. For example, most electric resistance instruments are restricted by the total length of the lead wires, while vibrating wire instruments are not. Therefore, if it is desirable to reduce the number of readout stations, then the length of the lead wires may dictate the use of vibrating wire instruments.

8.1.3 REDUNDANCY. Every instrumentation program should include some redundancy, especially with embedded instruments since it is usually not possible to repair damaged embedded instruments. The cost to retrofit replacement instruments will far exceed the cost of providing adequate redundancy. Redundancy, in this case, is more than only the furnishing of additional instruments to account for those that are defective or are damaged during installation. Redundancy includes providing different instruments which can measure similar behavior with different methods. For example, for measuring movement, a good instrumentation program should have both plumblines and a good trilateration system. Redundancy of instruments is especially critical in key areas and around special features.

8.2 AUTOMATIC DATA COLLECTION AND REMOTE MONITORING. The use of automatic data collection and remote monitoring is becoming more popular in recent years because electronic instruments lend themselves to this type of operation. Automatic data collection and remote monitoring will reduce the labor involved in gathering data and will reduce the time required to process and evaluate the data. However, automatic data collection and remote monitoring should not preclude the design engineer obtaining knowledge of onsite conditions existing when the data was obtained (Jansen 1988). It is important to remember that instruments do not ensure dam safety, they only document performance. Remote monitoring is not a substitute for a thorough inspection program. Regular visits to the dam by experienced design engineers are essential.

8.3 READINGS DURING CONSTRUCTION. Some of the most valuable information obtained in any instrumentation program is the information gathered prior to and during the construction of the dam and during initial reservoir filling. Because of the importance of gathering good data due to the effects of construction, efforts should be made to obtain readings as early and as frequently as possible during the construction. Special provisions should be made in the contract documents to allow access to the instrument readout points and/or stations at the regular intervals.

8.4 FABRICATION AND INSTALLATION. As noted, fabrication and installation of the instruments should be done by trained personnel, not by unskilled laborers. The initial readout schedule for the various types of instruments described.

8.5 INSTRUMENT TYPES. The basic types of instruments can be described by the items that they measure. The types discussed in this chapter include movement, stresses or strains, seepage, pressure, and temperature. Each of these types is discussed in the next few paragraphs, with recommendations for each. Other types such as seismic, water elevation gauges, etc. are not discussed herein.

8.6 MONITORING MOVEMENT. Because of the monolithic behavior of arch dams, displacement is probably the most meaningful parameter that can be readily monitored. Although displacements occur in all directions, the most significant displacements are usually the ones that take place in a horizontal plane. All concrete arch dams should have provisions for measuring these displacements, including relative movements between points within the dam and movement of the dam relative to a remote fixed point. In new construction, plumblines are still the preferred instrument to monitor the relative horizontal movements within an arch dam. In existing structures, it may be easier to install inclinometers or a series of tiltmeters or electrolevels. All arch dams (new or existing) should include trilateration or triangulation surveys.

8.6.1 PLUMBLINES. Plumblines and optical plummets measure bending, tilting, and deflections of concrete structures. Conventional plumblines are suspended from the top of the structure and extend down to the lowest readout gallery. If the curvature of the dam will not permit this type of installation, then a series of conventional plumblines can be installed as shown in Figure 12-1. Inverted plumblines are commonly used in conjunction with conventional plumblines to extend the total length of measurable deflections well into the foundation. The primary disadvantages to conventional plumblines are that they require trained personnel to obtain readings, the metal components are subject to corrosion, and no readings can be obtained until the structure is complete. An optical plummet can be used as a substitute for a conventional plumbline. The optical plummet is line-of-sight instrumentation that uses bubble levels or mercury reflectors to keep the reading line precisely vertical. Since it is an optical instrument, it is not susceptible to the corrosion problems of the conventional plumbline. However, as with the conventional plumbline, the optical plummet requires trained personnel to obtain readings, and no readings can be obtained until the structure is complete. Also, the optical plummet is susceptible to errors caused by refraction of light waves, as well as distortions due to atmospheric conditions.

8.6.2 INCLINOMETER. Inclinometers are used to measure angles from vertical. They can be used both in the concrete mass or extended into the foundation. Extending the inclinometer into the foundation can provide information on a potential sliding plane being investigated. Inclinometers consist of a metal or plastic casing embedded into the concrete, a probe, and a readout unit. The casing is grouted into a core hole, which makes it especially attractive for installation in existing structures. If an aluminum casing is used, the casing should be coated with epoxy to prevent corrosion. The casing has four grooves on the inside walls at 90 degrees from one another. The casing should be installed so that one set of groves are in the radial direction and one set in the tangential direction. As with the plumblines, inclinometers require trained personnel to obtain readings.

8.6.3 EXTENSOMETER. Extensometers are used to measure small (2- to 4-inch) deflections along the length of the instrument. Extensometers can be supplied with several (5 to 10) anchor heads. When inserted into a dam foundation, the heads can be positioned on either side of a zone of interest so that the deflection of that zone under load can be obtained. However, limitation on the amplitude of deflections that an extensometer can measure may require periodically resetting the reading heads. Extensometers can be installed at almost any angle. It is preferable to install some extensometers early in the concrete placement to determine deformations in the foundation during construction.

8.6.4 JOINT METER. Joint meters are used to measure the opening of monolith joints. Depending on the meter being used, the maximum opening that can be measured may range from 0.08 to 0.4 inches. Joint meters provide information about when the joints have begun to open and if there is adequate opening for grouting. They also give an indication of the effectiveness of the grouting and show whether any movement occurs in the joint during and after grouting.

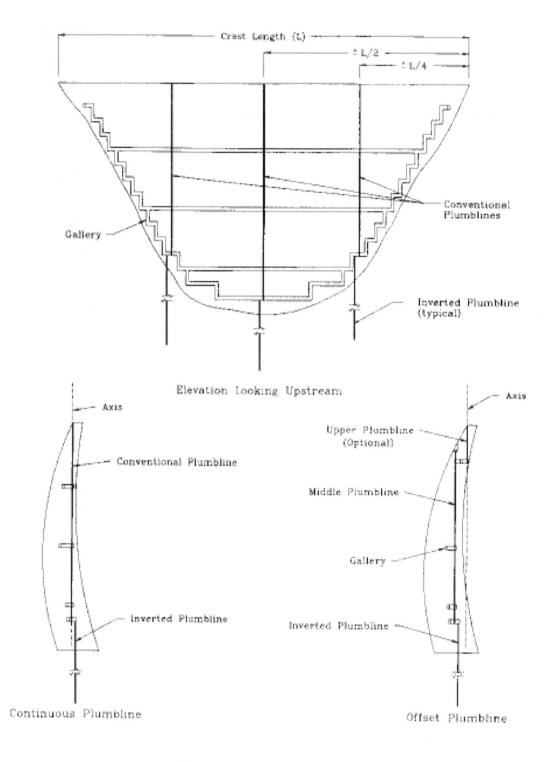

Figure 12-1

Typical layout for plumblines

8.7 TRILATERATION AND TRIANGULATION SURVEYS. These two methods of precise measurements utilize theories and equations of geometry to transform linear and angular measurements into deflections of alignment points. Triangulation is based on a single known side length (baseline) and three measured angles of a triangle to determine lengths of the remaining two sides. Trilateration uses electronic distance measurement (EDM) instruments to measure the lengths of the sides of the triangle and uses these lengths to calculate the angles. The accuracy of each of these methods is very dependent on the skill of the crew and equipment being used. It can also be costly and time consuming to gather and reduce the data.

8.8 MONITORING STRESSES AND STRAINS.

8.8.1 STRAIN METERS. Strain meters measure strain and temperature. Since they measure strain at one location and in one direction only, it is usually necessary to install strain meters in groups of up to 12 instruments with the help of a "spider" frame (Figure C-3 in EM 1110-2-4300). Since strain meters do not directly measure stress, it is usually necessary to convert strains to stresses. This will require a knowledge of the concrete materials which are changing with time, such as creep, shrinkage, and modulus of elasticity. These material properties, as well as the coefficient of thermal expansion and Poisson's ratio, are usually determined by laboratory testing prior to construction. However, if the mix design is revised due to field conditions, then the testing may need to be repeated to accurately determine stresses from the strain measurements.

8.8.2 "NO-STRESS" STRAIN METERS. These meters are identical to the strain meters discussed in the previous paragraph, except the method of installation isolates the meters within the mass concrete so that the volume changes in the concrete can be measured in the absence of external loading. This arrangement usually consists of one vertical and one horizontal meter.

8.8.3 STRESS METERS. Stress meters measure compressive stress independently of shrinkage, expansion, creep, or changes in modulus of elasticity. They are used for special applications such as determining vertical stress at the base of a section for comparison and checks of results from strain meters. They are also used in the archesfor determining horizontal stress normal to the direction of thrust in the thinner arch elements near the top of the dam. 12-5. Seepage Monitoring. Seepage through a dam or its foundation is visible evidence that the dam is not a perfect water barrier. Continued measurement of this seepage can provide an indication of progressive dissolution or erosion in a dam foundation or abutment. The types of instruments used to monitor seepage include weirs, flowmeters, and calibrated catch containers.

8.8.4 PRESSURE MONITORING. Although uplift pressure is not a critical issue in the stress analysis of arch dams, it is extremely important in the foundation stability analysis and should be included as part of the overall instrumentation program. However, measuring water pressures in a rock foundation is always a difficult problem since it can change over short distances because of jointing and fissuration.

8.8.5 OPEN PIEZOMETERS. Open piezometers are used to measure the average water level elevations in different zones of materials. Open piezometers include observation wells and slotted-pipe or porous-tube piezometers. Foundation drains can be considered a type of observation well. However, the use of drains or other open piezometers with long influence zones is not recommended for measuring uplift pressures in concrete dams. They are accessible to water over most, if not all, of their length and, as such, they tap water from different layers. The result is often a water level that is an average between the smallest and largest pressure heads crossed with no indications of what it truly represents.

8.8.6 CLOSED PIEZOMETERS. Closed piezometers measure pressure over a small influence zone (usually 3 feet). Typical closed piezometers are the electric resistance or vibrating wire piezometers. These piezometers are applicable for measuring pore pressures at the concrete/foundation contact as well as at several zones within the foundation rock.

8.9 STANDPIPE WITH BOURDON-TYPE GAUGE. A standpipe with a Bourdon-type gauge is the one of the simplest, and perhaps the cheapest, ways to measure uplift pressures at the dam/foundation contact. There are various ways of installing this uplift pressure system, but the two most common are as shown in Figure 12-2. This type of uplift pressure measuring system is an example of an open-system piezometer discussed in paragraph 12-6a, but with a small influence zone, usually only about 3 feet into the foundation. Closed piezometers can also be added at a few locations as verification of this uplift pressure measurement system. In dams without a foundation gallery, closed piezometers should be used to measure uplift pressures at the dam/foundation contact.

8.10 TEMPERATURE MONITORING. Temperature sensing devices are very important in arch dams since volume change caused by temperature fluctuation is a significant contributor to the loading on a arch dam. Thermometers are used to determine the temperature gradients and history of the concrete mass for use in evaluating thermal stresses which contribute to thermal cracking. They are also used to control the cooling process during the grouting operations and are used to determine the mean concrete temperatures due to reservoir and seasonal fluctuation. Thermometers are preferred over thermocouples because they have been more dependable, have greater precision, and are less complicated in their operation. Standpipes filled with water have also been used as a substitute for permanent thermometers. In using a standpipe, a thermometer is lowered into the standpipe to the desired elevation and held there until the reading stabilizes. Standpipes can be an effective way to obtain vertical temperature gradients but are not practical for measuring temperature variation between the upstream and downstream faces of the dam. The addition of several standpipes through the thickness of the dam will add extra complications to the overall construction process. Standpipes also tend to indicate higher temperatures than thermometers (Sixteenth Congress on Large Dams 1986).

8.11 GENERAL LAYOUT REQUIREMENTS. The general instrument layout recommended for arch dams is shown in Figures 12-3 and 12-4. Instruments, other than those needed specifically for construction, should be positioned at points that correspond to those used in the design and analysis. This is done for ease of comparison of the predicted and measured performance of the dam.

8.11.1 MOVEMENT INSTRUMENTS.

8.11.2 LOCATIONS THAT CORRESPOND TO THE MAXIMUM section of the dam (the crown cantilever) and to the midpoints to each abutment, as shown in Figure 12-1. If the curvature does not permit a continuous plumbline to be installed from the top of the dam, then a shortened plumbline can be installed. If an upper gallery is available, a staggered plumbline can be installed.

8.11.3 EXTENSOMETERS AND INCLINOMETERS. Extensometers and inclinometers should be installed into the foundation as early in the construction as practical, preferably before concrete placement, to determine deformations in the foundation due to construction activities. The location of the extensometers should correspond to the arch elevations for the other instrumentation groups. The total length of the extensometer should be between 25 and 50 percent of the height of the dam. Additional extensometers or inclinometers should be located wherever there is a change in foundation type or a potential slide plane. If inclinometers are to be used in lieu of plumblines, they should be installed at the general locations discussed.

8.11.4 JOINT METERS. One or two joint meters are required in every other monolith joint at the midheight elevation of alternate grout lifts (Figure 12-5).

8.11.5 TRIANGULATION AND TRILATERATION TARGETS. Targets should be placed at the crest and at one or more points on the downstream face, as shown in Figure 12-6. The targets should correspond to the location of the plumblines.

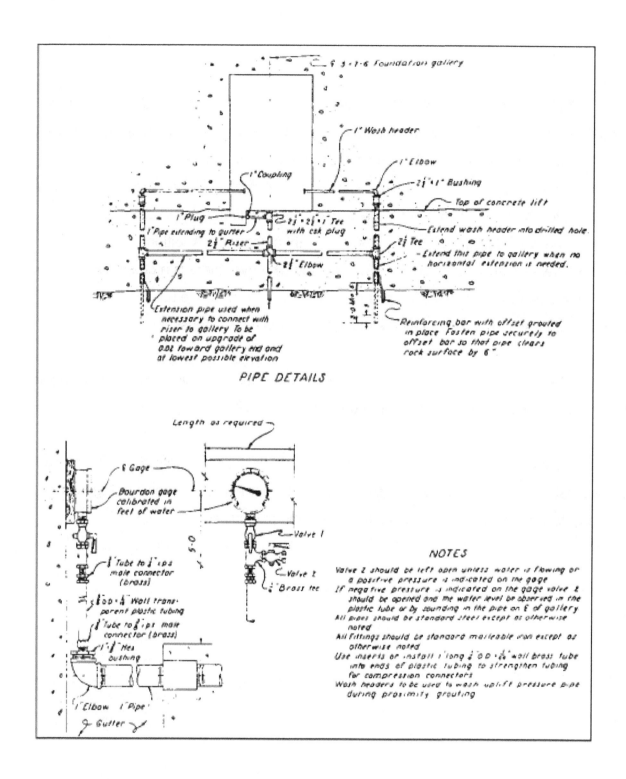

Figure 12-2

Typical Bourdon gauge installation

© J. Paul Guyer   2017

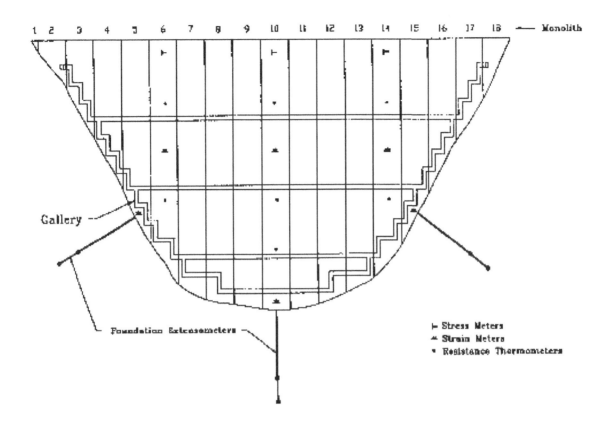

Figure 12-3

Recommended location of embedded instruments in arch dams (elevation looking upstream)

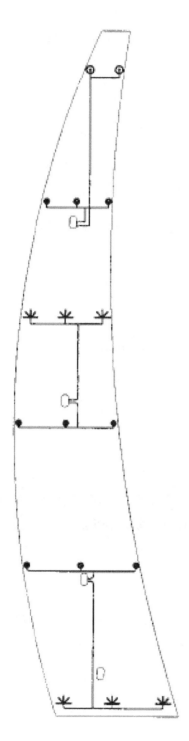

○ Stress Meters
⚹ Strain Meters
∘ Resistance Thermometers

Figure 12-4

Recommended location of embedded instrumentation in arch dam (section
at crown cantilever)

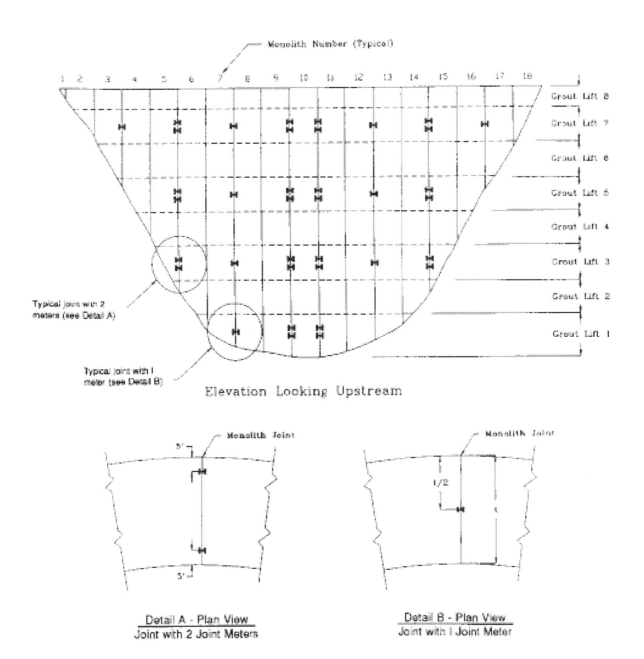

Figure 12-5

Layout for joint meters

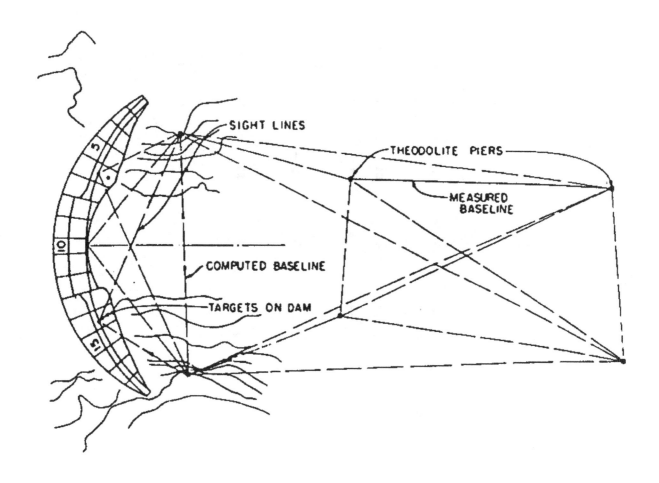

Figure 12-6

Layout for triangular or trilateration network

8.12 STRESS/STRAIN CLUSTERS. Clusters of strain meters, "no-stress" strain meters, and stress meters should be positioned along four or five arch elevations that correspond to arch elevations used in the design and analysis (see Figures 12-3 and 12-4). An additional set of instruments should be positioned at the base of the crown cantilever. At each of the instrumentation points, two clusters of instruments should be placed near the upstream and downstream faces or at three locations between the faces. Where the thickness of the dam permits and where more detailed information is desired about the stress distribution through the dam, five or more instrument groups can be placed between the upstream and downstream faces of the dam.

8.13 SEEPAGE. Initially, seepage through the joints and through the drains can be measured at two weirs, each located to collect seepage through the drains along each abutment. Measurements of individual drains should also be made on a regular basis. Additional seepage monitoring points can be added after the initial reservoir filling, if the need arises.

8.14 PRESSURE. Three uplift pressure groups (standpipe or closed piezometer groups) should be located in a similar manner as the plumblines. There should be at least four uplift pressure measuring points through the thickness of the dam in each group with a spacing between points of no more than 30 feet.

8.15 TEMPERATURE. Thermometers should be installed at locations shown in Figure 12-7 to verify the thermal gradient through the structure and to obtain the temperature history of the concrete. This information is used for comparison with the thermal studies discussed in Chapter 8. Thermometers are not typically located near the other instrument clusters when the instruments in these clusters also sense temperature. If the thermometer groups are not located at the center of the grout lifts, then additional thermometers may be required to monitor the temperature of the concrete from the time of placement through the time of joint grouting. These thermometers should be located at the center of each monolith at points that correspond to the midheight of each of the grout lifts and are used to assure that the concrete has reached the required grouting temperatures.

8.16 OTHER INSTRUMENT GROUPS. Instruments should also be positioned in various arrays near areas of special interest, such as around galleries or around any other openings through the dam. Conduits running through the dam should include strain meters on the metal conduits and on the reinforcing steel surrounding the conduits.

8.17 READOUT SCHEDULE. Each dam will have its own site-specific requirements for instrumentation and readout schedule. Table 12-1 shows some general guidelines that should be used in establishing the general readout schedule for each dam. Variations in the guidelines shown in Table 12-1 will occur when special conditions arise at the dam site, such as long periods of unusually high or low water levels, seismic activity, or unexplained behavior of the dam, such as cracking, increased seepage, etc.

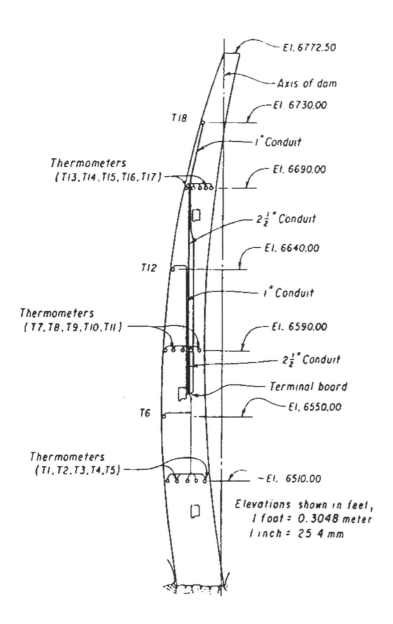

Figure 12-7

Layout for thermometer installation

Type of Instrument	During Construction[1]	During Initial Filling	First 2 years of Operation	Next 5 years of Operation	After 7 years of Operation
Plumblines and optical plummets	N/A	weekly	monthly	quarterly	semi-annually
Inverted Plumblines	weekly	weekly	monthly	quarterly	semi-annually
Inclinometers	weekly	weekly	monthly	quarterly	semi-annually
Extensometers	weekly	weekly	monthly	quarterly	semi-annually
Joint meters	weekly	weekly	biweekly	quarterly	semi-annually
Triangulation	N/A	weekly	monthly	quarterly	semi-annually
Trilateration	N/A	weekly	monthly	quarterly	semi-annually
Strain meters	weekly[2]	weekly	monthly	quarterly	semi-annually
"No-Stress" strain meters	weekly[2]	weekly	monthly	quarterly	semi-annually
Stress meters	weekly[2]	weekly	monthly	quarterly	semi-annually
Weirs, etc.	N/A	weekly	weekly	monthly	quarterly
Open piezometers	monthly	monthly	quarterly	quarterly	semi-annually
Closed piezometers	weekly[4]	weekly	monthly	quarterly	semi-annually
Uplift pressure gauges	weekly	weekly	biweekly	monthly	quarterly
Thermometers	weekly[4]	weekly	monthly	monthly	quarterly

[1] Initial readings for all embedded instruments should be made within 3 hours after embedment.

[2] After initial readings are obtained, readings are continued every 12 hours for 15 days, daily for the next 12 days, twice weekly for 4 weeks, and weekly thereafter.

[3] Daily during curtain grouting.

[4] After initial readings are obtained, readings are continued every 6 hours for 3 days, every 12 hours for 12 days, daily for the next 12 days, twice weekly for 4 weeks, and weekly thereafter.

12-14

TABLE 12-1

Recommended Readout Schedule

CHAPTER 9: MANUAL LAYOUT OF ARCH DAMS

9.1 GENERAL DESIGN PROCESS. Design of an arch dam involves the layout of a
tentative shape for the structure, preliminary static stress analysis of this layout,
evaluation of the stress results, and refinement of the arch dam shape. Several
iterations through the design process are necessary to produce a satisfactory design
which exhibits stress levels within an acceptable range. The final dam layout that
evolves from the iterative design process is then statically analyzed by the finite element
method. "Static analysis" refers to the analysis performed after a layout has been
achieved through the design process. "Preliminary stress analysis" refers to the method
of analysis performed during the iterative design process to investigate the state of
stress for tentative layouts. The computer program ADSAS (Arch Dam Stress Analysis
System) (USBR 1975) is used for the preliminary stress analysis and is discussed in
more detail in paragraph 5-5. GDAP (Graphics-based Dam Analysis Program) (Ghanaat
1993a) is a special purpose finite element program, specifically developed for the
analysis of arch dams. GDAP is discussed in more detail in Chapter 6. Preliminary
stress analyses are relatively quick and inexpensive to run compared to the static
analysis which is more detailed, both in its input as well as its output. Although past
history has shown that results from both procedures are comparable, an arch dam
layout that reaches the static analysis phase may still require refinement, pending
evaluation of static analysis results.

9.2 LEVELS OF DESIGN. There are three phases of the life cycle process of a project
for which layouts are developed; reconnaissance, feasibility, and preconstruction
engineering and design (PED). The degree of refinement for a layout is determined by
the phase for which the design is developed.

9.2.1 RECONNAISSANCE PHASE. Of the three phases mentioned, the least amount
of engineering design effort will be expended in the reconnaissance phase. Examination
of existing topographic maps in conjunction with site visits should result in the selection
of several potential sites. When selecting sites during the reconnaissance phase,

emphasis should be placed on site suitability, i.e., sites with adequate canyon profiles and foundation characteristics.

9.2.1.1 RECONNAISSANCE LEVEL LAYOUTS FOR DIFFERENT sites can be produced quickly from the empirical equations discussed. Empirical equations to determine concrete quantities from these layouts are presented. Alternatively, the structural engineer may elect to base reconnaissance level layouts on previous designs which are similar in height, shape of profile, loading configuration, and for which stresses are satisfactory. However, it should be pointed out that most arch dams which have been constructed to date are single-center. Because the technology exists today which simplifies the layout of more efficient, multicentered arch dams, most dams that will be designed in the future will be of the multicentered variety. Therefore, basing a tentative, reconnaissance level layout on an existing single-center arch dam may result in conservative estimates of concrete quantity.

9.2.1.2 TOPOGRAPHIC MAPS OR QUAD SHEETS that cover an adequate reach of river provide sufficient engineering data for this phase. From this region, one or more potential sites are selected. The areas around these sites are enlarged to 1:50 or 1:100 scale drawings. These enlarged topographic sheets and the empirical formulas will produce the geometry and concrete volume for a reconnaissance level layout.

9.2.2 FEASIBILITY PHASE. Designs during the feasibility phase are used in the selection of the final site location and as a basis for establishing the baseline cost estimate. Feasibility designs are made in greater detail than reconnaissance designs since a closer approximation to final design is required.

9.2.2.1 AS A RESULT OF THE WORK PERFORMED during the reconnaissance phase, the structural engineer should now have available one or more potential sites to evaluate during the feasibility phase. Using the iterative layout process discussed, tentative designs will be plotted, analyzed, evaluated, and refined for each potential site until a proposed layout evolves that provides the best balance between minimal concrete volume and minimum stress level. From the sites evaluated and their respective layouts, the structural engineer will select the most economical design, and this will be carried into the PED phase. A baseline cost estimate will be developed for the final layout.

9.2.2.2 THE ITERATIVE LAYOUT PROCESS requires a certain amount of topographic and subsurface information. However, these data should be obtained with the knowledge that funds for feasibility studies are limited. Aerial topographic surveys of potential sites are required as well as a few core borings to determine an approximation of the depth of overburden. Loading conditions, as discussed, should be defined.

9.2.3 PRECONSTRUCTION ENGINEERING AND DESIGN. Design work during this phase is presented in the FDM which is also used to develop contract plans and specifications. The final design layout produced during the feasibility stage is subjected to further static and dynamic analysis during the PED phase. Any remaining load cases that have not been analyzed during the feasibility phase should be analyzed and evaluated at this time. This may require that any missing data (operating conditions, thermal loads, etc.) be finalized prior to the analysis. If results from all of the preliminary stress analysis load cases indicate the final layout is acceptable, design may proceed to the static FEM analysis. Otherwise, the layout requires refinement.

9.3 PROCEDURE. A single-center, variable-thickness, arch dam is assumed for the purpose of discussion. The procedure for laying out other types of arch dams differs only in the way the arches are defined.

9.4 MANUAL LAYOUT. Although the term "Layout of an arch dam" implies a single procedure, layout actually consists of an iterative, refining process involving several layouts, each successive one improving on the previous. The first of these layouts require the structural designer to assume some initial parameters which will define the shape of the arch dam. As stated, a 1:50- or 1:100-scale topographic map of a dam site is required before layout begins. If possible, the contours should represent topography of foundation rock; however, in most instances, only surface topography is available at this stage of design. The structural designer must then assume a reasonable amount of overburden, based on core borings or sound judgement, to produce a topo sheet that reflects the excavated foundation.

9.4.1 AXIS. The crest elevation required for the dam should be known at this time from hydrologic data and this, in conjunction with the elevation of the streambed (or assumed foundation elevation) at the general location of the dam, determines the dam height, H (feet). The structural designer should select a value for the radius of the dam axis (RAXIS). For the initial layout where the engineer would have no reasonable estimate for the value of RAXIS from a previous layout, the following empirical relationship has been derived by the USBR based on historical data from existing dams:

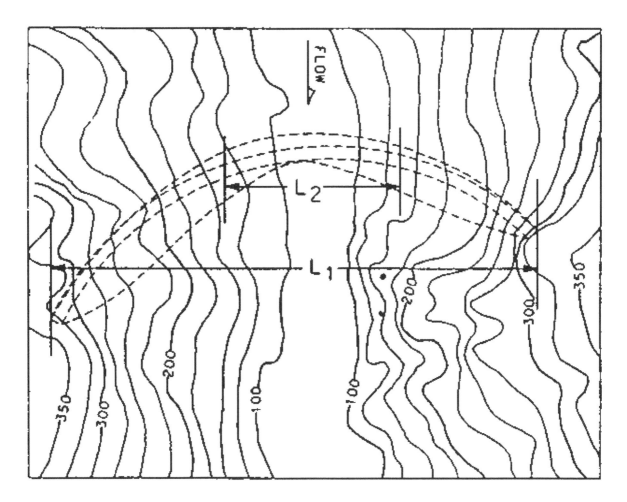

Figure 5-1

Determination of empirical values L1 and L2

9.4.1.1 ON A SHEET OF VELLUM or transparent paper, an arc is drawn with a radius equal to RAXIS at the same scale as the topo sheet. This arc represents the axis of the dam. The vellum is then overlaid and positioned on the topo sheet so as to produce an optimum position and location for the dam crest; for this position, the angle of incidence to the topo contour at the crest elevation (b in Figure 2-1) should be approximately equal on each side. As shown in Figure 5-2, RAXIS may require lengthening if the arc fails to make contact with the abutments or if the central angle exceeds 120 degrees.

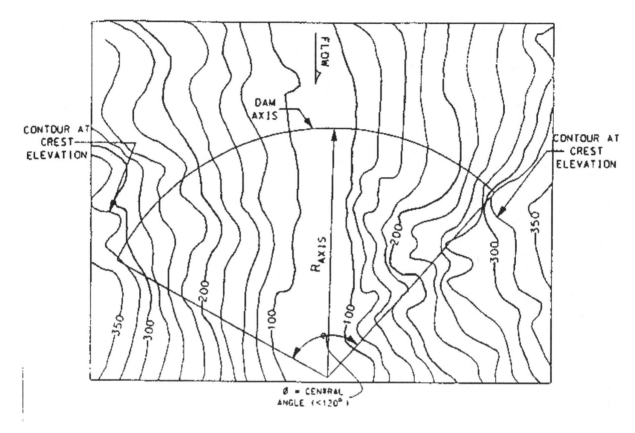

Figure 5-2

Layout of dam axis

9.4.1.2 THE MAGNITUDE OF THE CENTRAL ANGLE of the top arch is a controlling value which influences the curvature of the entire dam. Objectionable tensile stresses will develop in arches of insufficient curvature; such a condition often occurs in the lower elevations of a dam having a V-shape profile. The largest central angle practicable should be used considering that the foundation rock topography may be inaccurately mapped and that the arch abutments may need to be extended to somewhat deeper excavation than originally planned. Due to limitations imposed by topographic conditions and foundation requirements, for most layouts, the largest practicable central angle for the top arch varies between 100 and 120 degrees.

9.4.2 LOCATION OF CROWN CANTILEVER AND REFERENCE PLANE. On the overlay, locate the crown cantilever at the intersection of the dam axis and the lowest point of the site topography (i.e., the riverbed). This corresponds to the point of maximum depth of the dam. A vertical plane passing through this point and the axis center represents the reference plane (or plane of centers). On the overlay plan, this plane is shown as a line connecting the crown cantilever and the axis center. Later, when arcs representing arches at other elevations are drawn, they will be located so that the centers of the arcs will be located on the reference plane. Ideally, the reference plane should be at the midpoint of the axis. This seldom occurs, however, because most canyons are not symmetrical about their lowest point.

9.4.3 CROWN CANTILEVER GEOMETRY. The geometry of the crown cantilever controls the shape of the entire dam and, as a result, the distribution and magnitude of stresses within the body. The empirical equations which follow can be used to define thicknesses of the crown at three locations; the crest, the base, and at el 0.45H above the base:

$$T_c = 0.01(H + 1.2L_1) \qquad (5-2)$$

$$T_B = \sqrt[3]{0.0012HL_1L_2\left(\frac{H}{400}\right)^{\frac{H}{400}}} \qquad (5-3)$$

$$T_{0.45} = 0.95T_B \qquad (5-4)$$

9.4.3.1 IN ADDITION, UPSTREAM AND DOWNSTREAM PROJECTIONS of the extrados (upstream) and intrados (downstream) faces can also be arrived at empirically. Those relationships are:

$$USP_{CREST} = 0.0 \qquad\qquad (5-5)$$

$$USP_{BASE} = 0.67T_B \qquad\qquad (5-6)$$

$$USP_{0.45H} = 0.95T_B \qquad\qquad (5-7)$$

$$DSP_{CREST} = T_C \qquad\qquad (5-8)$$

$$DSP_{BASE} = 0.33T_B \qquad\qquad (5-9)$$

$$DSP_{0.45H} = 0.0 \qquad\qquad (5-10)$$

(Note: These empirical equations were developed by the USBR and are based on historical data compiled from existing dams. However, the engineer is not restricted to using the parameters derived from the empirical equations; they are presented as an aid for developing initial parameters and only for the first layout. Sound engineering judgement resulting from experience obtained in arch dam layout may also be utilized when defining an initial layout or refining a previous one. Values for subsequent layouts will consist of adjustments, usually by engineering evaluation of stress analysis, of the values used in the previous iterations.)

9.4.3.2 AS SHOWN, THE UPSTREAM AND DOWNSTREAM projections at the crest, base, and at el 0.45H above the base can now be plotted in elevation in reference to the dam axis. This plot is referred to as the "plane of centers" view. The next step is to define the upstream and downstream faces of the crown cantilever using a circular arc (or combinations of straight lines and circular arcs) which passes through the upstream and downstream projection points as shown in Figure 5-4. With the faces defined in this

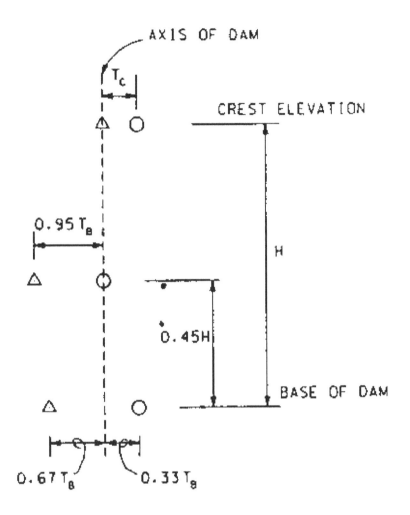

Figure 5-3

Empirically derived projections of the crown cantilever

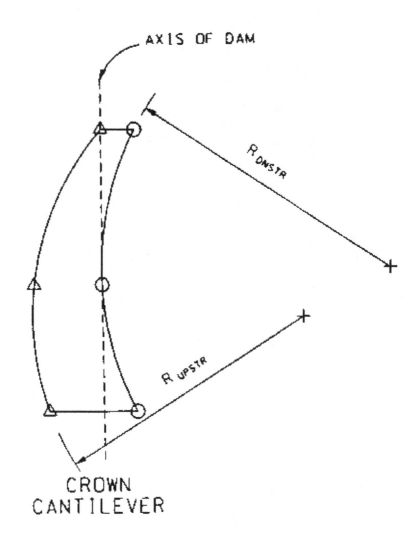

Figure 5-4

Definition of upstream and downstream faces

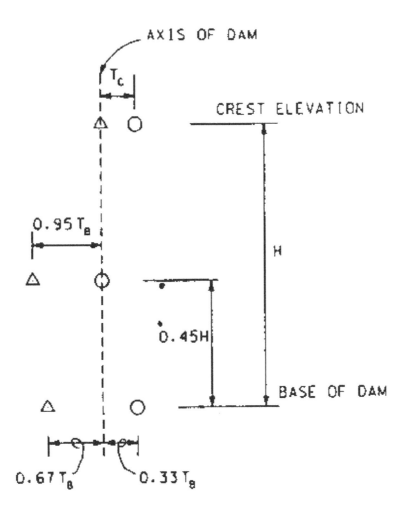

Figure 5-3

Empirically derived projections of the crown cantilever

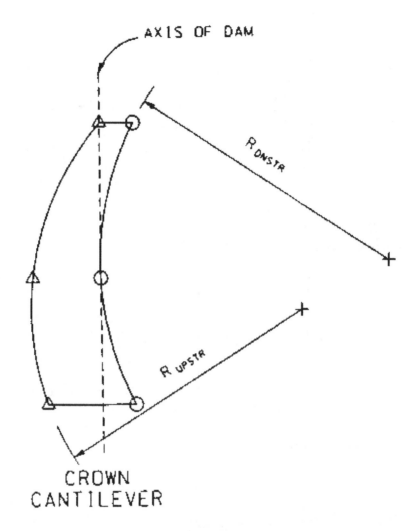

AXIS OF DAM

R DNSTR

R UPSTR

CROWN
CANTILEVER

Figure 5-4

Definition of upstream and downstream faces

manner, upstream and downstream projections at any elevation can be obtained. This information will be necessary when laying out the arches.

9.4.4 ESTIMATING THE DAM FOOTPRINT. The axis of the dam on the topographic overlay corresponds to the upstream face of the dam at the crest. An arc representing the downstream face of the crest can be drawn with the center of the arc at the axis center and a radius equal to RAXIS reduced by the thickness at the crest, TC. On the plan overlay, three points are identified to aid in laying out the contact line between the foundation and the upstream face of the dam. Two of the points are the intersection of the axis of the dam with the foundation contour at the crest elevation at each abutment (points A and B). The third point is the upstream projection of the crown cantilever at the base. This point can be plotted in reference to the axis of the dam based on information taken from the plane of centers view (Figure 5-5). Using a french curve, a smooth curve is placed beginning at the upstream face of the crest on one abutment, passing through the upstream projection of the crown cantilever at the base (point C), and terminating at the upstream face of the crest at the other abutment (points A and B in Figure 5-6).

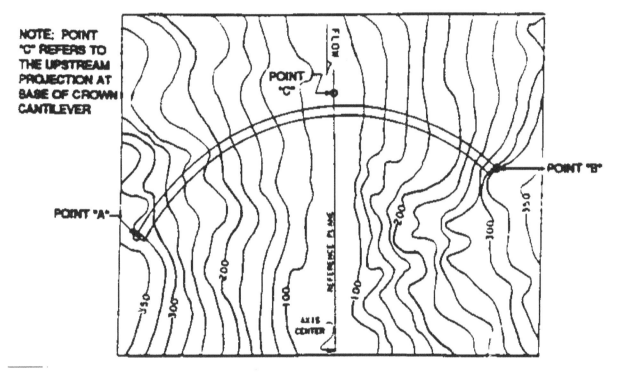

Figure 5-5

Contact points between dam and foundation at crest and crown cantilever

9.4.5 LAYOUT OF THE ARCHES. Of all that is involved in arch dam layout, this step is possibly the most difficult. For shaping and analysis purposes, between 5 and 10 evenly spaced horizontal arches are drawn. These arches should be spaced not less than 20 feet nor greater than 100 feet apart. The lowest arch should be 0.15H to 0.20H above the base of the crown cantilever.

9.4.5.1 BEGINNING AT THE ARCH IMMEDIATELY below the crest, determine, from the plane of the centers view, the upstream and downstream projections of the crown cantilever at that specific arch elevation. These projections are then

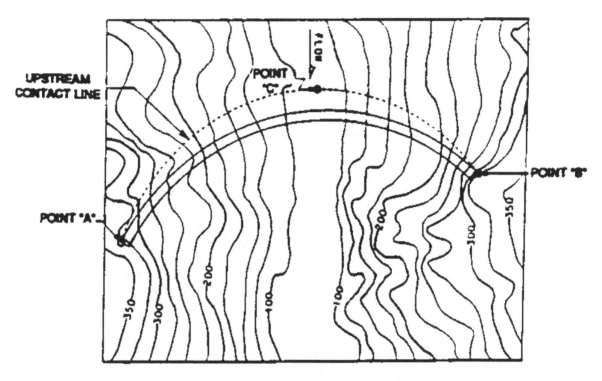

Figure 5-6

Upstream contact line at dam-foundation interface

plotted on the plan view along the reference plane. Using a beam compass, trial arcs representing the upstream face of the dam at that specific arch elevation are tested until one is found which meets the following criteria:

9.4.5.1.1 THE ARC CENTER must lie along the reference plane.

9.4.5.1.2 THE ARC MUST PASS through the upstream projection of the crown cantilever as plotted on the plan view.

9.4.5.1.3 BOTH ENDS OF THE ARC MUST TERMINATE on the upstream contact line at a foundation elevation equal to or slightly deeper than the arch elevation.

9.4.5.2 LOCATING AN ARCH WHICH SATISFIES all of these criteria is a trial and error process which may not be possible with a single-center layout. This is generally the case when dealing with unsymmetrical canyons where different lines of centers are required for each abutment. Figure 5-7 shows an example of an arch that meets the criteria. Of particular importance is that the ends of the arch must extend into the abutments and not fall short of them. This ensures that a "gap" does not exist between the dam and foundation.

9.4.5.3 THIS PROCEDURE IS REPEATED to produce the downstream face of the arch. Similar to what was performed for the upstream face, the downstream projection of the crown cantilever is determined from the plane of centers view and plotted on the plan view. The beam compass is then used to locate an arc that meets the three criteria with the exception that the arc must pass through the downstream projection of the crown cantilever with ends that terminate on the radial to the extrados at the abutment (Figure 5-7). If the same arch center is used for the upstream and downstream faces, a uniform thickness arch is produced. If the arch centers do not coincide, the arch produced will vary in thickness along its length (variable thickness arch).

9.4.5.4 ONCE A SATISFACTORY ARCH HAS been produced, the locations of the arch centers for the upstream and downstream face are marked along the reference plane on the plan view. Standard practice is to identify the extrados (upstream) face with a triangle and the intrados (downstream) face with a circle. The corresponding arch elevation should be identified with each. Although these arch centers appear to lie on a straight line in plan, they all are positioned at their respective arch elevations, and it is highly unlikely that they are on a straight line in three dimensional (3-D) space.

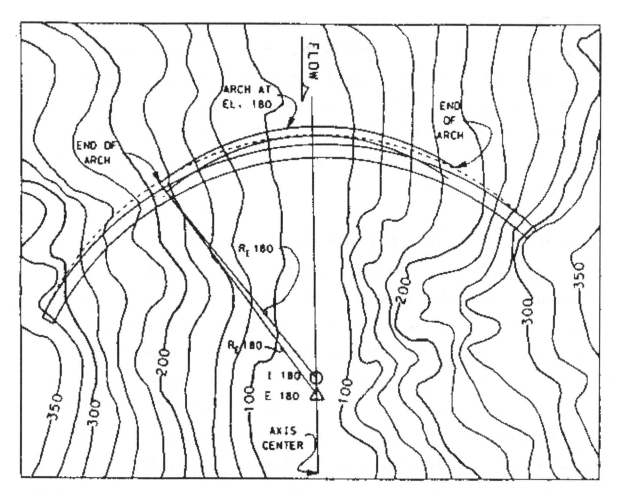

Figure 5-7

Typical layout of an arch section arch

9.4.5.5 ARCHES AT THE REMAINING elevations are plotted in similar fashion. Of particular importance to producing an acceptable plan view is to ensure that the footprint, when viewed in plan, is smooth and free flowing with no abrupt changes or reverse curvature. This requirement is usually met by the fact that a footprint is predetermined; however, endpoints of arches may not terminate exactly on the footprint. Revision of the footprint is necessary to ensure that it passes through all actual arch endpoints prior to checking it for smoothness.

9.4.6 REVIEWING THE LAYOUT. Layout of an arch dam includes the preparation of three different drawings. The first is the plan view, which begins with locating a crest and ends with plotting the arches. The second drawing is a section, in elevation, along the reference plane, called the plane-of-centers view. This view has been partially produced when the crown cantilever was created, but it requires expansion to include the lines of centers. The third drawing to be produced is a profile (looking downstream) of the axis of the dam and the foundation. Proper review includes examining all views for "smoothness," because abrupt changes in geometry will result in excessive stress concentrations. The term "smoothness" will be discussed in the following paragraph. Only when the plan view, plane of centers, and profile demonstrate "smoothness" and are in agreement is the layout ready for preliminary stress analysis. It should be pointed out that all three views are dependent on each other; when making adjustments to the geometry, it is impossible to change parameters in any view without impacting the others.

9.4.6.1 CREATING AND REVIEWING THE PLANE-OF-CENTERS VIEW. In addition to the crown cantilever, the plane of centers also includes the lines of centers for the upstream and downstream face. A section is passed along the reference plane in plan to produce the plane-of-centers view. Each arch center, upstream and downstream, is plotted in elevation in reference to the axis center, as shown in Figure 5-8. The lines of centers are produced by attempting to pass a smooth curve through each set of arch centers (Figure 5-9). These lines of center define the centers for all arches at any elevation. If the curve does not pass through the arch centers located during the arch layout procedure, those arch centers will be repositioned to fall on the appropriate line of centers. Those particular arches will require adjustment on the plan view to reflect the change in position of the arch center. The structural engineer should understand that this adjustment will involve either lengthening or shortening the radius for that particular arch which will impact where the ends of the arch terminate on the abutments. Lines of centers should be smooth flowing without abrupt changes and capable of being emulated using combinations of circular curve and straight line segments, as shown in the example on Figure 5-10. The circular arcs and straight line segments used to define the lines of centers will be input into ADSAS when performing the preliminary stress analysis.

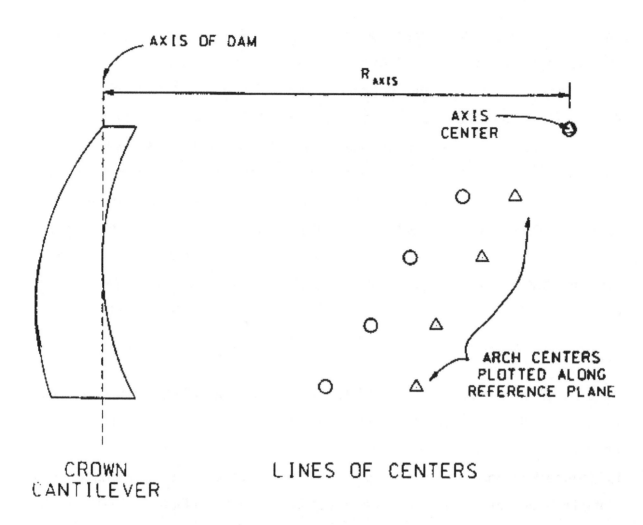

AXIS OF DAM

R_{AXIS}

AXIS CENTER

ARCH CENTERS PLOTTED ALONG REFERENCE PLANE

CROWN CANTILEVER

LINES OF CENTERS

Figure 5-8

Plotting of arch centers along reference

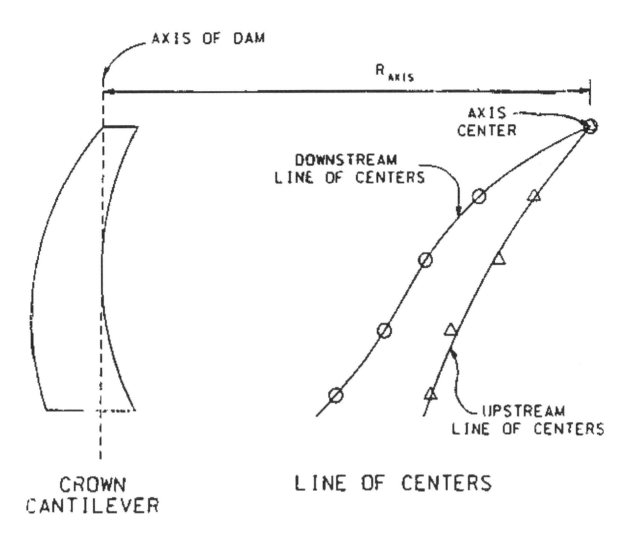

Figure 5-9

Development of the lines of centers

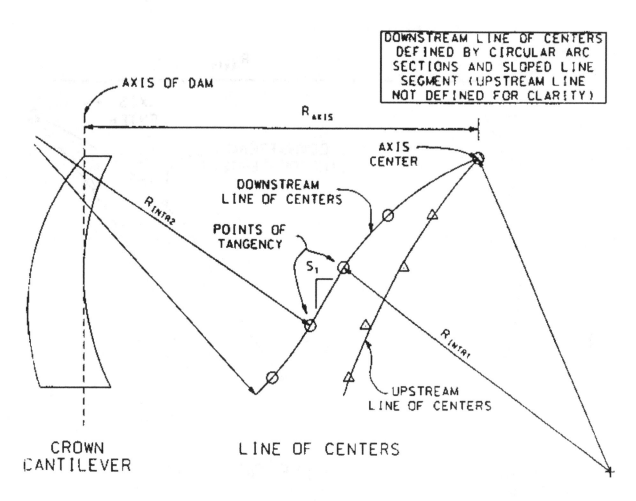

Figure 5-10

Defining the lines of centers

9.4.6.2 DEVELOPING THE PROFILE VIEW. Once a satisfactory plan view and plane-of-centers view has been obtained, the profile view is ready to be created. The profile view is used to examine the amount of excavation that a particular layout has induced. The profile view consists of a developed elevation of the upstream face of the dam (looking downstream) with the foundation topography shown. It should be noted that this is a developed view rather than projection of the upstream face onto a flat plane. This "unwrapping" results in a view in which no distortion of the abutments exists. Figures 5-11 and 5-12 show, respectively, examples of acceptable and unacceptable profiles.

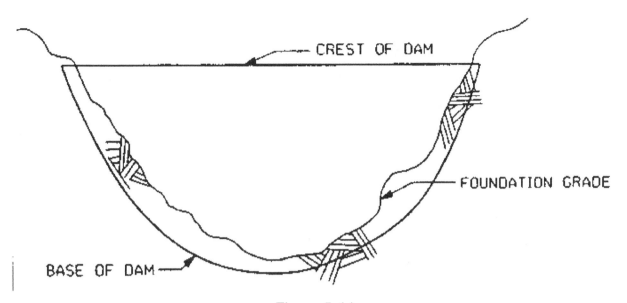

Figure 5-11

Example of an acceptable developed profile view

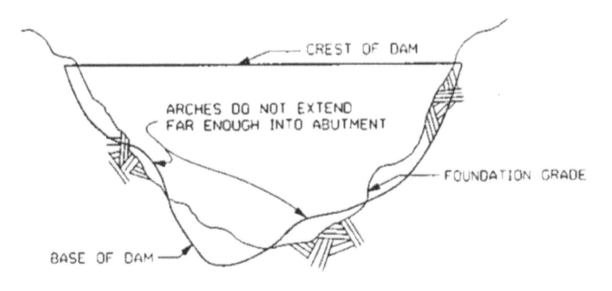

Figure 5-12

Example of an unacceptable developed profile view

9.4.6.3 FOUNDATION OF DAM. In general, the foundation of the dam should be as the lines of centers: smooth and free flowing with no abrupt changes in geometry. The base of the dam must also extend into the foundation; otherwise, an undesirable condition develops in that "gaps" will occur between the base of the dam and the foundation, requiring foundation treatment. Pronounced anomalies should be removed by reshaping the affected arches until a smooth profile is obtained.

9.5 PRELIMINARY STRESS ANALYSES. After a satisfactory layout has been obtained, a preliminary stress analysis is performed to determine the state of stress of the dam under various loading conditions. The computer program ADSAS was developed by the USBR (1975) for this purpose. The Corps of Engineers adapted ADSAS for use on a microcomputer. ADSAS is based on the trial load method of analysis. A discussion on the theory of the trial load method is beyond the scope of this document; however, the USBR (1977) addresses this topic in detail.

9.5.1 ADSAS INPUT. While an exact description of the steps necessary to prepare an input data file for ADSAS is documented in the operations manual (USBR 1975) and a Corps of Engineers' manual. A brief description is included here. A typical ADSAS input data file contains four groups of information: a geometry definition section, material properties, loading conditions, and output control cards.

9.5.1.1 GEOMETRY DEFINITION. Critical to the success of obtaining an accurate analysis is the ability to convey to the program the geometry which defines the shape of the arch dam. This geometry consists of:

9.5.1.1.1 CROWN CANTILEVER GEOMETRY. Base elevation, crest elevation, projections of the upstream and downstream faces at the crest and the base, X and Y coordinates, and radii of all circular arcs used in defining the upstream and downstream faces, and slopes of any straight line segments used in defining the upstream and downstream faces.

9.5.1.1.2 LINES-OF-CENTERS GEOMETRY. Axis radius, X and Y coordinates and radii of any circular arcs used in defining all lines of centers, slopes of any straight line segments used in defining all lines of centers, elevations at intersections between segments defining lines of centers, and horizontal distances from axis center to intrados and extrados lines of centers.

9.5.1.1.3 ARCH GEOMETRY. Elevations of all arches, angles to abutments for all arches, and angles of compound curvature. All data required for the crown cantilever and lines of centers geometry are taken from the planes-of-centers view while data required for the arch geometry should be available from the plan view.

9.5.1.2 MATERIALS PROPERTIES. ADSAS analysis also requires material properties of both the concrete and the foundation rock. These data include modulus of elasticity of the concrete and foundation rock, Poisson's ratio for the concrete and foundation rock, coefficient of thermal expansion of the concrete, and unit weight of concrete.

9.5.1.3 LOADING CONDITIONS. During layout, only static loading conditions are analyzed. ADSAS is capable of analyzing hydrostatic, thermal, silt, ice, tailwater loads, and dead weight.

9.5.1.4 OUTPUT CONTROL. ADSAS provides the user with the ability to toggle on or off different portions of the output to control the length of the report while capturing pertinent information.

9.6 EVALUATION OF RESULTS. Evaluation requires a thorough examination of all the analytical output. Types of information to be reviewed are the crown cantilever description, intrados and extrados lines of centers, geometrical statistics, dead load stresses and stability of blocks during construction, radial and tangential deflections and angular deformations, loading distributions, arch and cantilever stresses, and principal stresses. If any aspect of the design is either incorrect or does not comply with established criteria, modifications must be made to improve the design.

9.6.1 RESULTANT COMPONENTS. Evaluation of the arch dam may also include examination of the resultants along the abutments. These resultants are separated into three components; radial, tangential, and vertical. The combined radial and tangential resultant should be directed into the abutment rock. In the lower arches, that abutment may tend to parallel the surface contours or daylight into the canyon. Prudent engineering suggests that the resultant be turned into the abutment. The solution may be a combination of increasing stiffness in the upper arches or flexibility in the lower arches. The effect is mitigated by including the vertical component which then directs the total resultant downward into the foundation.

9.6.2 VOLUME OF CONCRETE. One major factor of a layout that requires evaluation is the volume of concrete that is generated. ADSAS computes this volume as part of its output. If a quantity is desired without proceeding through a preliminary stress analysis (as for a reconnaissance layout), that value can be arrived at empirically by the following equation.

$$v = 0.000002H^2L_2 \frac{(H + 0.8L_2)}{L_1 - L_2} + 0.0004HL_1[H + L_1] \qquad (5-11)$$

The volume of concrete calculated in ADSAS or derived from Equation 5-11 does not reflect mass concrete in thrust blocks, flip buckets, spillways, or other appurtenances.

9.7 IMPROVEMENT OF DESIGN. The best of alternative designs will have stresses distributed as uniformly as possible within allowable limits combined with a minimum of concrete. Where to terminate a design and accept a final layout based on these criteria are difficult in some dams with widely varying loading conditions, such as with a flood control dam which has periods of low and high reservoir elevations. The primary means of effecting changes in the behavior of the dam is by adjusting the shape of the structure. Whenever the overall stress level in the structure is far below the allowable limits, concrete volume can be reduced, thereby utilizing the remaining concrete more efficiently and improving the economy. Following are some examples of how a design can be improved by shaping.

9.7.1 LOADS AND DEFLECTIONS. Load distribution and deflection patterns should vary smoothly from point to point. Often when an irregular pattern occurs, it is necessary to cause load to be shifted from the vertical cantilever units to the horizontal arches. Such a transfer can be produced by changing the stiffness of the cantilever relative to the arch.

9.7.2 RESHAPING ARCHES. If an arch exhibits tensile stress on the downstream face at the crown, one alternative would be to reduce the arch thickness by cutting concrete from the downstream face at the crown while maintaining the same intrados contact at the abutment. Another possibility would be to stiffen the crown area of the arch by increasing the horizontal curvature which increases the rise of the arch.

9.7.3 RESHAPING CANTILEVERS. When cantilevers are too severely undercut, they are unstable and tend to overturn upstream during construction. The cantilevers must then be shaped to redistribute the dead weight such that the sections are stable. Severe overhang will cause tension to develop on the upper upstream face, contraction joints in the affected area to close, and prevent satisfactory grouting.

9.7.4 FORCE-STRESS RELATIONSHIP. Shaping is the key to producing a complete and balanced arch dam design. The task of the designer is to determine where and to what degree the shape should be adjusted. Figure 5-13 should be used to determine the appropriate changes to be made to the structural shape. If an unsatisfactory stress condition is noticed, the forces causing those stresses and the direction in which they act can be determined by Figure 5-13. For example, the equations of stress indicate which forces combine to produce a particular stress. Knowing the force involved and its algebraic sign, it is possible to determine its direction from the sign convention shown on the figure. With that information the proper adjustment in the shape can be made so that the forces act to produce the desired stresses.

9.8 PRESENTATION OF DESIGN LAYOUT. Figures and plates that clearly show the results of the design layout and preliminary stress analyses should be included in the FDM. Plates that illustrate and describe the detailed geometry of the arch dam include:

9.8.1 PLAN VIEW. Arches, arch centers, angles to the abutments, axis center, dam-foundation contact line, and dam orientation angle are some items that are included in this view of the dam overlaid on the site topography.

9.8.2 SECTION ALONG REFERENCE PLANE. This plate includes all the information that defines the vertical curvature of the crown cantilever and the line(s) of centers (Figure 1-6).

9.8.3 CANTILEVER SECTIONS. All cantilevers generated during the preliminary stress analysis should be shown. Showing the thickness at the base and at the crest of each cantilever is also recommended as shown in Figure 5-14.

9.8.4 ARCH SECTIONS. Arch sections generated as a result of the preliminary stress analysis should be plotted. Appropriate thicknesses at the reference plane and at each abutment should be shown for each arch as shown in Figure 5-15.

9.8.5 PROFILE. A profile, developed along the axis of the dam, should be presented, showing locations of cantilevers and the existing foundation grade as shown in Figure 5-16.

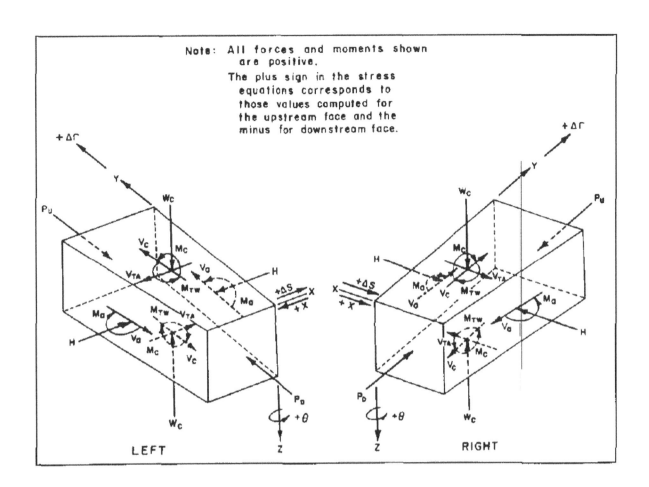

Figure 5-13

Sign convention for arch computations in ADSAS

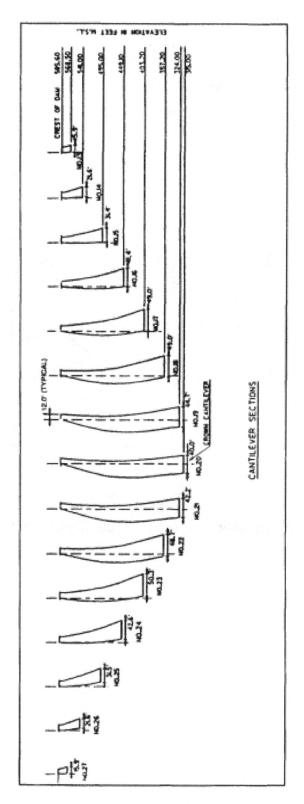

Figure 5-14

Cantilever sections

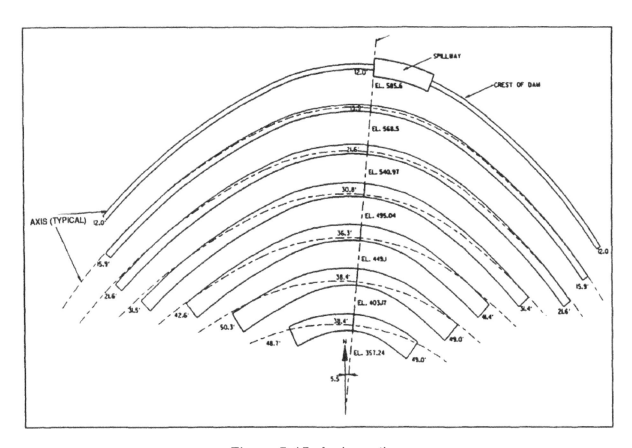

Figure 5-15. Arch sections

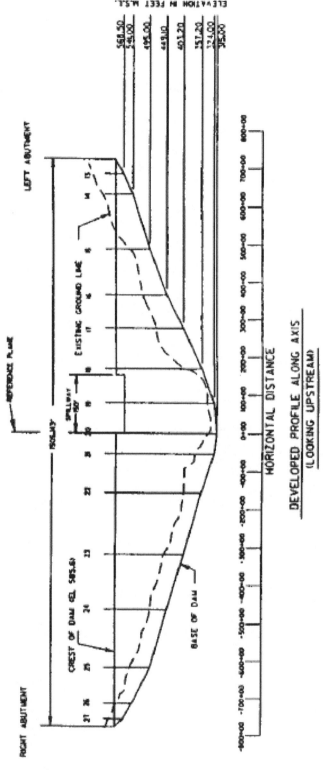

Figure 5-16

Developed profile

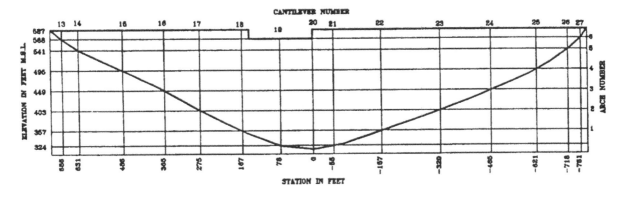

CANTILEVER NUMBER

STATION IN FEET

(PROJECTED LOOKING DOWNSTREAM)

Figure 5-17
ADSAS model

9.8.6 ADSAS MODEL. A plot showing the model of cantilever and arch units generated by ADSAS and the associated cantilever and arch numbering scheme is plotted and shown in the FDM as shown in Figure 5-17.

9.8.7 STRESS CONTOURS. Contour plots of arch and cantilever stresses on the upstream and downstream faces for all load cases are presented in the FDM as shown in Figure 5-18.

9.8.8 DEAD LOAD STRESSES. Stresses produced in the ungrouted cantilevers as a result of the construction sequence should be tabulated and presented in the FDM as shown in Figure 5-19.

9.9 COMPUTER-ASSISTED LAYOUTS. The procedures mentioned in this chapter involve manual layout routines using normal drafting equipment. As mentioned, the iterative layout procedure can be quite time consuming. Automated capabilities using desktop computers have been developed which enable the structural designer to interactively edit trial layouts while continuously updating plan, plane of centers, and profile views. When an acceptable layout is achieved, the program generates an ADSAS data file which is input into a PC version of ADSAS. In all, developing these tools on a desktop computer allows the structural designer to proceed through the layout process at a faster rate than could be achieved manually, thereby, reducing design time and cost.

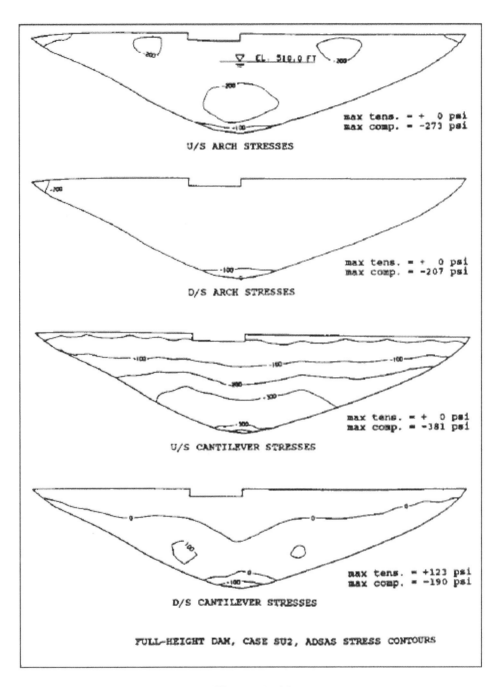

Figure 5-18

ADSAS stress contours

MINIMUM DEAD LOAD STRESSES (IN PSI) BY CANTILEVER				
CANTILEVER NUMBER	STRESS BASED ON CONCRETE PLACED ON	ELEVATION OF MINIMUM STRESS	MINIMUM DOWNSTREAM STRESS	UPSTREAM FACE STRESS
20	495.04	314.96	48	-406.
21	540.97	324.00	66.	-459.
22	587.20	357.24	142.	-483.
23	587.20	403.27	111.	-357.
24	587.20	449.10	80.	-265.
25	587.20	495.04	61.	-197.
26	587.20	540.97	18.	-95.
27	587.20	567.20	-2.	-35.
19	540.97	324.00	103.	-482.
18	587.20	357.24	156.	-496.
17	587.20	403.17	89.	-340.
16	587.20	449.10	37.	-227.
15	587.20	495.04	15.	-152.
14	587.20	540.97	5.	-81.
13	587.20	567.20	-6.	-32.
NOTE: MINUS SIGN INDICATES COMPRESSION				

Figure 5-19

Dead load stresses

CHAPTER 10: ARCH DAM OUTLETS

10.1 INTRODUCTION. This discussion describes the influence of voids through the arch dam and structural additions outside the theoretical limits of the arch dam. Voids through the dam in the radial direction are spillways, access adits, and outlet works, in the tangential direction are adits, galleries, and tunnels, and in the vertical direction are stairway wells and elevator shafts. Often associated with spillways and outlet works are blockouts or chambers for gate structures. External structures are restitution concrete which includes thrust blocks, pads, pulvino, socle, or other dental type concrete, spillway flip buckets on the downstream face, and corbels on the upstream face.

10.2 SPILLWAYS. Numerous types of spillways are associated with arch dams. Each is a function of the project purpose, i.e., storage or detention, or to bypass flood flows or flows that exceed diversion needs. Spillways for concrete dams may be considered attached or detached.

10.2.1 ATTACHED SPILLWAYS. Attached spillways are through the crest or through the dam. Through-the-crest spillways have a free fall which is controlled or uncontrolled; OG (ogee) types are shaped to optimize the nappe. In general, the usual crest spillway will be constructed as a notch at the crest. The spillway notch can be located either over the streambed or along one or both abutments as shown in Figure 3-1. A spillway opening can also be placed below the crest through the dam. Similar to the notch spillway, this opening can be located either over the streambed, as shown in Figure 3-2, or near one or both abutments. The location through the dam, whether at the crest or below the crest, is always a compromise between hydraulic, geotechnical, and structural considerations. Impact of the jets on the foundation rock may require treatment to avoid eroding the foundation. Spillways through the dam are located sufficiently below the crest so that effective arch action exists above and below the spillway openings.

10.2.1.1 SPILLWAY AT CREST. With this alignment, the spillway crest, piers, and flipbucket are designed to align the flow with the streambed to cause minimal possible bank erosion and/or to require minimal subsequent beneficiation. However, the notch reduces the arch action by the depth of the notch, i.e., the vertical distance between the dam crest and spillway crest which is normally pure arch restraint is nullified and replaced with cantilever action. To accommodate this reduced stiffness, additional concrete must be added below the spillway crest or the entire arch dam must be reshaped, thus complicating the geometry. Not that this is detrimental, but arch dam shapes work more efficiently when kept simple and smooth in both plan and elevation. Moving the spillway notch to either abutment as shown in Figure 3-1 or splitting the spillway crest length and locating half along each abutment will restore most of the arch action to the dam crest. Spillway notches through the crest near abutments interrupt arch action locally, but not significantly, as can be shown in numerous numerical analysis and scale model studies. The effect of abutment spillways is structurally less distressful on arch dams in wide

Figure 3-1

East Canyon Dam with spillway notch near left abutment

valleys where the top arch is long compared to the structural height, such as a 5:1 crest

length-to-height ratio, or in canyons where the climate fluctuates excessively (±50 oF)

between summer and winter. In this latter case, winter temperature loads generally

cause tensile stresses on both faces near the crest abutment, where the dam is thinnest

and responds more quickly and dramatically than thicker sections. Thus, locating the

spillway notches along the abutments is a natural structural location. The effects of a

center notch, in addition to reducing arch action, are to require that concrete above the

spillway crest support the reservoir load by cantilever action. Consequently, design of

the vertical section must not only account for stresses from dead load and reservoir but

meet stability requirements for shear. Temperature load in this portion of the dam is

usually omitted from structural analyses. Earthquake loads also can become a problem

and must be considered. On certain arch dams where the spillway width is a small proportion of the total crest length, some arch action will occur in the adjacent curved sections that will improve resistance to flood loads. Usually the beginning of this arch action is about the notch depth away from the pier. The more recent arch dams are thin efficient structures that make flood loads above the spillway crest of greater concern.

Figure 3-2

Through spillway below crest on Morrow Point Dam

10.2.1.2 SPILLWAYS BELOW CREST (THRU SPILLWAYS). Spillways are constructed through the arch dam at some optimal distance below the dam crest to reduce the plunge and provide for additional discharge. The spillway may be visualized as multiple orifices, either round or rectangular, and controlled with some type of gates.

The set of openings either may be centered over the streambed as shown in Figure 3-2 or split toward either or both abutments. The set is surrounded by mass concrete and locally reinforced to preserve, for the analyses, the assumption of a homogenous and monolithic structure. With this in mind, local reinforcement and/or added mass concrete must be designed so that the dam as a whole is not affected by the existence of the spillway. To minimize disruption of the flow of forces within the dam, the several openings should be aligned with the major principal compressive stresses resulting from the most frequent loading combination. Around the abutment, the major principal stresses on the downstream face are generally normal to the abutment. This alignment would tend to stagger the openings, thus creating design difficulties. In practice, however, all openings are aligned at the same elevation and oriented radially through the dam. If necessary, each orifice may be directed at a predetermined nonradial angle to converge the flows for energy dissipation or to direct the flow to a smaller impact area such as a stilling basin or a reinforced concrete impact pad. Between each orifice, within a set, are normal reinforced concrete piers designed to support the gravity load above the spillway and the water force on the gates.

10.2.1.3 FLIP BUCKET. The massive flip bucket, depending on site conditions, may be located near the crest to direct the jet impact near the dam toe or farther down the face to flip the jet away from the toe. In either case, the supporting structure is constructed of solid mass concrete generally with vertical sides. By judiciously limiting its width and height, the supporting structure may be designed not to add stiffness to any of the arches or cantilevers. To assure this result, mastic is inserted in the contraction joints to the theoretical limits of the downstream face defined before the flip bucket was added as shown in Figure 3-3. The mastic disrupts any arch action that might develop. For the same reason, mastic is inserted during construction in the OG corbel overhang on the upstream face. These features protect the smooth flow of stresses and avoid reentrant corners which may precipitate cracking or spalling. Cantilever stiffness is enhanced locally but not enough to cause redistribution of the applied loads. Reinforcement in the supporting structure and accompanying training walls will not add stiffness to the arch dam.

10.2.2 DETACHED SPILLWAYS. Detached spillways consist of side channel, chute, tunnel, and morning glory spillways. The selection is dependent upon site conditions.

10.2.2.1 SIDE CHANNEL. The side channel spillway is one in which the control weir is placed along the side of and parallel to the upper portion of the discharge channel as shown in Figure 3-4. While this type is not hydraulically efficient nor inexpensive, it is used where a long overflow crest is desired to limit the surcharge head, where the abutments are steep, or where the control must be connected to a narrow channel or tunnel. Consequently, by being entirely upstream from the arch dam, the spillway causes no interference with the dam and only has a limited effect on the foundation. Similarly, along the upstream abutment, any spillway interference is mitigated by the usual low stresses in the foundation caused by the dam loads. Usually, stresses near the crest abutment are much less than the maximum allowable or, quite possibly, are tensile stresses.

10.2.2.2 CHUTE SPILLWAY. Chute spillways shown in Figure 3-5 convey discharge from the reservoir to the downstream river level through an open channel placed either along a dam abutment or through a saddle. In either case, the chute is not only removed from the main dam, but the initial slope by being flat isolates the remaining chute from the eventual stressed foundation rock.

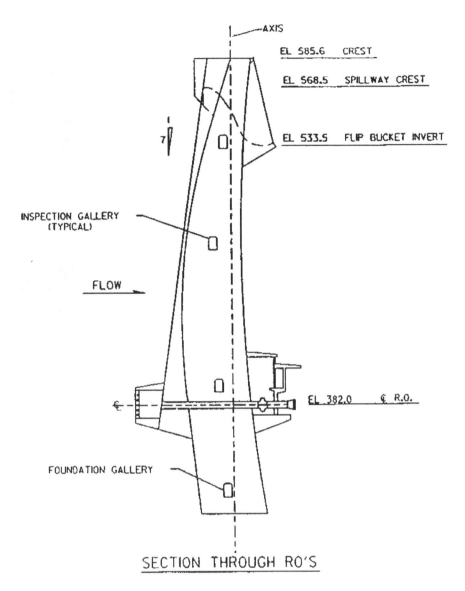

Figure 3-3

Typical section through spillway of a dam

Figure 3-4

Side channel spillway at Hoover Dam

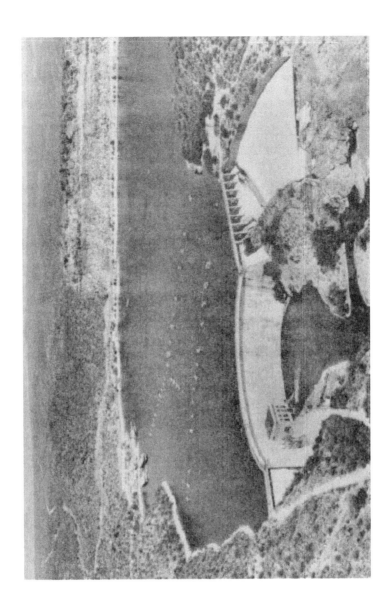

Figure 3-5

Chute Spillway at Stewart Mountain Dam

10.2.2.3 TUNNEL SPILLWAY. Tunnel spillways convey the discharge around the dam and consist of a vertical or inclined shaft, a large radius elbow, and a horizontal tunnel at the downstream end. Tunnel spillways may present advantages for damsites in narrow canyons with steep abutments or at sites where there is danger to open channels from snow or rock slides. The tunnel alignment usually traverses the stressed foundation and consequently should be at least one abutment thickness from the concrete to rock contact. Any need for increased spacing would be based on structural height and applied load combinations. Note that these tunnels are designed to flow up to 75 percent full, in which case in situ and superimposed stresses from the dam may govern the design. The corollary is that the tunnel walls must be strong enough to avoid creating a weakness in the foundation and subsequent stability problems with the dam.

10.2.2.4 MORNING GLORY SPILLWAY. A morning glory spillway such as that shown in Figure 3-6 is one in which the water enters over a horizontally positioned lip, which is circular in plan, and then flows to the downstream river channel through a horizontal or near horizontal tunnel. A morning glory spillway usually can be used advantageously at damsites in narrow canyons where the abutments rise steeply or where a diversion tunnel is available for use as the downstream leg. If a vertical drop structure is to be located upstream, it should not interfere structurally with the dam. A sloping tunnel offers the same concerns as previously discussed.

10.3 OUTLET WORKS. Outlet works are a combination of structures and equipment required for the safe operation and control of water released from a reservoir to serve downstream needs. Outlet works are usually classified according to their purpose such as river outlets, which serve to regulate flows to the river and control the reservoir elevation, irrigation or municipal water supply outlets, which control the flow of water into a canal, pipeline, or river to satisfy specified needs, or, power outlets, which provide passage of water to turbines for power generation. In general, outlet works do not structurally impact the design of an arch dam as shown in Figure 3-7. The major difficulty may lie in adapting the outlet works to the arch dam, especially a small double-curvature thin arch dam where the midheight thickness may be 25 feet or less. Taller and/or heavier dams will have a significant differential head between the intake and the valve or gate house, and the conduit may have several bends. All of the features can be designed and constructed but not without some compromise. Outlet works should be located away from the abutments to avoid interference with the smooth flow of stresses into the rock and smooth flow of water into the conduit. A nominal distance of 10 diameters will provide sufficient space for convergence of stresses past the conduit.

10.3.1 INTAKE STRUCTURES. Intake structures, in addition to forming the entrance into the outlet works, may accommodate control devices and the necessary auxiliary appurtenances such as trashracks, fish screens, and bypass devices (Figure 3-8). An intake structure usually consists of a submerged structure on the upstream face or an intake tower in the reservoir. Vertical curvature on the upstream face may result in more than usual massive concrete components as shown in Figure 3-9 to provide a straight track for the stop logs or bulkhead gate. A compromise to relieve some of the massiveness is to recess portions of the track into the dam face. This recess which resembles a rectangular notch in plan reduces the stiffness of those arches involved not only at the notch but for some lateral distance, depending on the notch depth.

Figure 3-6

Morning glory spillway at Hungry Horse Dam. Note upstream face of dam in upper right.

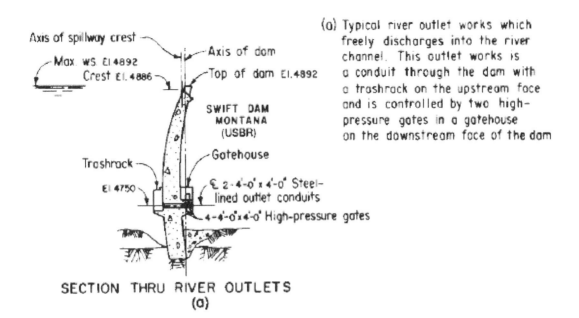

(a) Typical river outlet works which freely discharges into the river channel. This outlet works is a conduit through the dam with a trashrack on the upstream face and is controlled by two high-pressure gates in a gatehouse on the downstream face of the dam

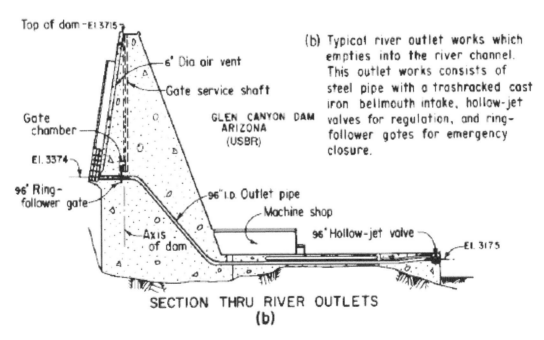

(b) Typical river outlet works which empties into the river channel. This outlet works consists of steel pipe with a trashracked cast iron bellmouth intake, hollow-jet valves for regulation, and ring-follower gates for emergency closure.

Figure 3-7

Typical river outlet works without a stilling basin

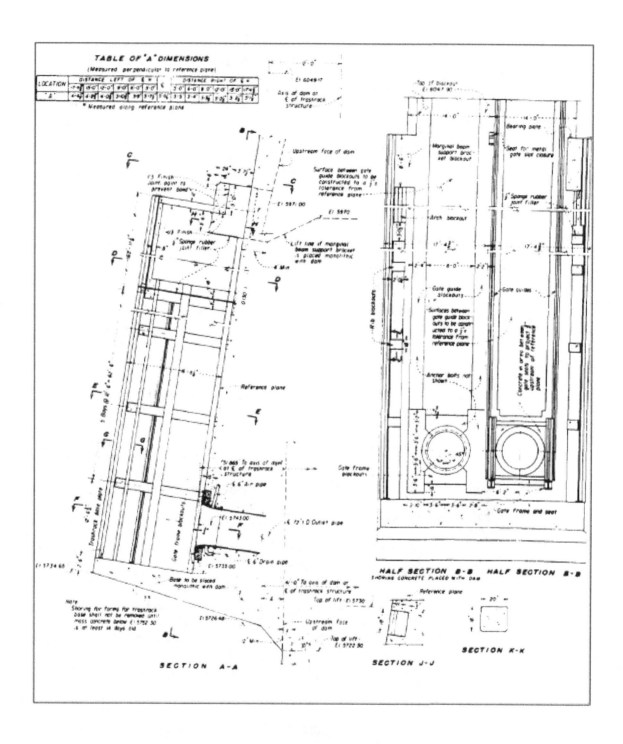

Figure 3-8

Typical river outlets trashrack structure

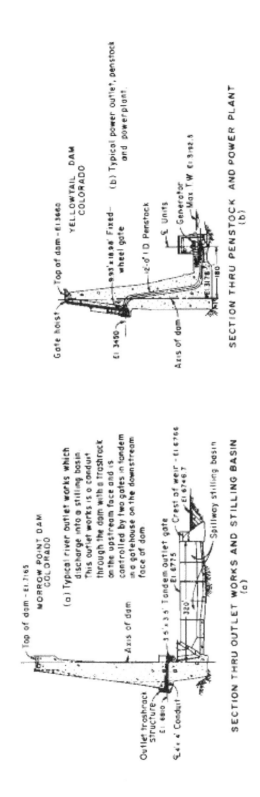

Figure 3-9

Typical river outlet works and power outlet

10.3.2 CONDUIT. The outlet works conduit through a concrete dam may be lined or unlined, as in a power outlet, (Figure 3-10) but when the conduit is lined it may be assumed that a portion of the stress is being taken by the liner and not all is being transferred to the surrounding concrete. The temperature differential between the cool water passing through the conduit and the warmer concrete mass will produce tensile stresses in the concrete immediately adjacent to the conduit. In addition, the bursting effect from hydrostatic pressures will cause tensile stresses at the periphery of the conduit. Such tensile stresses and possible propagation of concrete cracking usually extend only a short distance in from the opening of the conduit. It is common practice to reinforce only the concrete adjacent to the opening. The most useful method for determining the stresses in the concrete surrounding the outlet conduit is the finite element method (FEM) of analysis.

10.3.3 CONTROL HOUSE. The design of a control house depends upon the location and size of the structure, the operating and control equipment required, and the conditions of operation. The loadings and temperature conditions used in the design should be established to meet any situation which may be expected to occur during construction or during operation. In thin dams, the floor for the house is normally a reinforced concrete haunched slab cantilevered from the downstream face and designed to support the valves and houses. Neither the reinforcement nor the concrete should interfere with structural action of the arches and cantilevers. Emergency gates or valves are used only to completely shut off the flow in the outlet for repair, inspection maintenance or emergency closure. Common fixed wheel gates, such as shown in Figure 3-11, are either at the face or in a slot in the dam.

10.4 APPURTENANCES.

10.4.1 ELEVATOR TOWER AND SHAFT. Elevators are placed in concrete dams to provide access between the top of the dam and the gallery system, equipment and control chambers, and the power plant as shown in Figure 3-12. The elevator structure consists of an elevator shaft that is formed within the mass concrete and a tower at the crest of the dam. The shaft should have connecting adits which provide access into the gallery system and into operation and maintenance chambers. These adits should be located to provide access to the various galleries and to all locations at which monitoring and inspection of the dam or maintenance and control of equipment may be required. Stairways and/or emergency adits to the gallery system should be incorporated between elevator stops to provide an emergency exit such as shown in Figure 3-13. The design of reinforcement around a shaft can be accomplished by the use of finite element studies using the appropriate forces or stresses computed when analyzing the arch dam. In addition, stresses within the dam near the shaft due to temperature and other appropriate loads should be analyzed to determine if tension can develop at the shaft and be of such magnitude that reinforcement would be required. To minimize structural damage to the arches, the shaft should be aligned totally within the mass concrete and centered between the faces. However, the necessarily vertical shaft may not fit inside a thin arch dam. In such a case, the shaft could be moved to the abutment or designed to be entirely outside the dam but attached for vertical stability to the downstream face. This solution may not be esthetically pleasing, but it is functional, because if the shaft emerges through the downstream face it forms a rectangular notch that diverts the smooth flow of arch stresses and reduces arch stiffness.

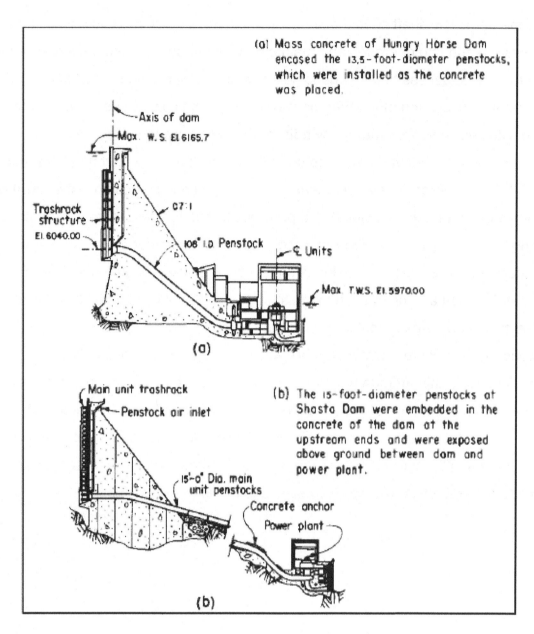

(a) Mass concrete of Hungry Horse Dam encased the 13.5-foot-diameter penstocks, which were installed as the concrete was placed.

(b) The 15-foot-diameter penstocks at Shasta Dam were embedded in the concrete of the dam at the upstream ends and were exposed above ground between dam and power plant.

Figure 3-10

Typical penstock installations

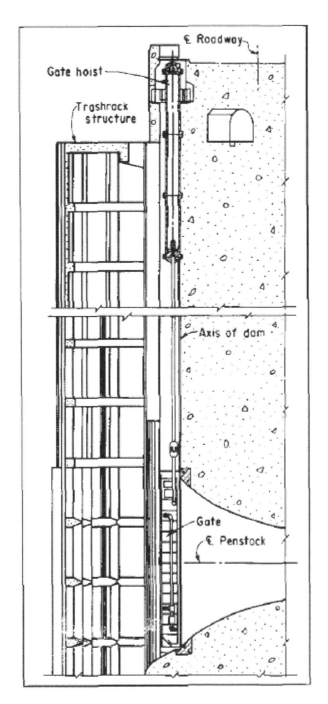

Figure 3-11

Typical fixed wheel gate installation at upstream face of dam

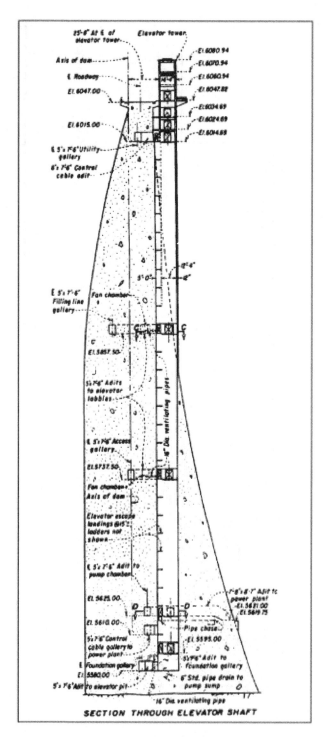

Figure 3-12

Structural and architectural layout of elevator shaft and tower in Flaming Gorge Dam

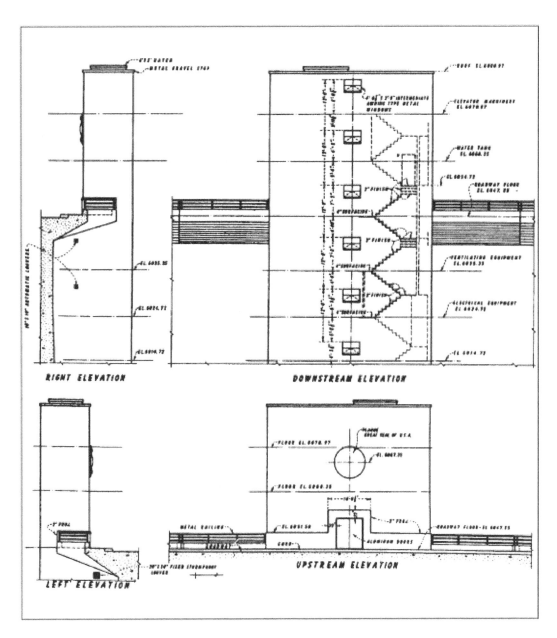

Figure 3-12 (continued)

Structural and architectural layout of elevator shaft and tower in Flaming Gorge Dam

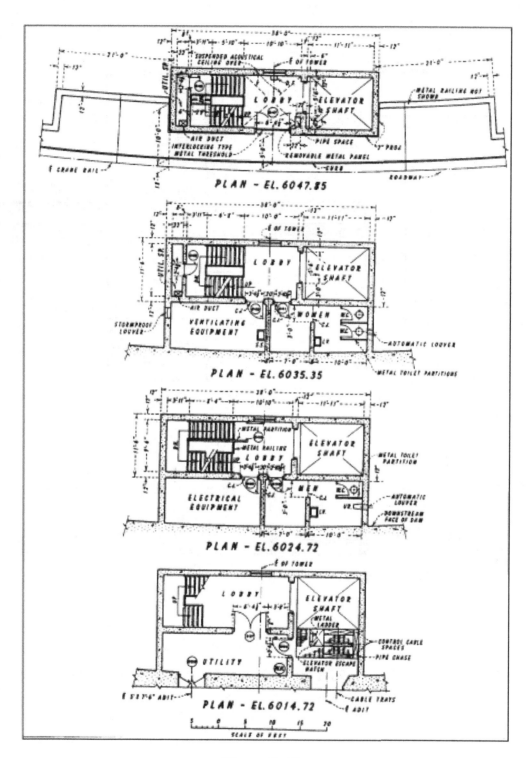

Figure 3-13

Details of layout of elevator shaft and tower in Flaming Gorge Dam

10.4.2 BRIDGES. Bridges may be required on the top of the dam to carry a highway over the spillway or to provide roadway access to the top of the dam at some point other than its end. Design criteria for highway bridges usually conform to the standard specifications adopted by the American Association of State Highway Officials and are modified to satisfy local conditions and any particular requirement of the project. Bridges, regardless of how heavy, whether of steel or concrete, or with fixed or pinned connections, are not considered sufficiently strong to transfer arch loads from one mass concrete pier to the other.

10.4.3 TOP OF THE DAM. The top of the dam may contain a highway, maintenance road, or walkway such as shown in Figure 3-14. If a roadway is to be built across the dam, the normal top of the dam can be widened with corbels which cantilever the road or walkway out from the upstream or downstream faces of the dam. The width of the roadway on the top of the dam is dependent upon the type and size of roadway, sidewalks, and maintenance and operation space needed to accomplish the tasks required. Parapets or handrails are required both upstream and downstream on the top of the dam and should be designed to meet safety requirements. The minimum height of parapet above the sidewalk should be 3 feet 6 inches. A solid upstream parapet may be used to increase the freeboard if additional height is needed. The design of the reinforcement for the top of the dam involves determining the amount of reinforcement required for the live and dead loads on the roadway cantilevers and any temperature stresses which may develop. Temperature reinforcement required at the top of the dam is dependent upon the configuration and size of the area and the temperature condition which may occur at the site. After the temperature distributions are determined by studies, the temperature stresses that occur can be analyzed by use of the FEM. If a roadway width greater than the theoretical crest thickness of the dam is required, the additional horizontal stiffness of the roadway section may interfere with the arch actions at the top if the dam. To prevent such interference, horizontal contraction joints should be provided in the roadway section with appropriate joint material. The contraction joints should begin at the edge of the roadway and extend to the theoretical limits of the dam face. Live loads such as hoists, cranes, stoplogs, and trucks are not added to the vertical loads when analyzing the arch dam; these loads may weigh less than 10 cubic yards (cu yd) of concrete, an insignificant amount in a concrete dam.

10.4.4 GALLERIES AND ADITS. A gallery is a formed opening within the dam to provide access into or through the dam. Galleries are either transverse or longitudinal and may be horizontal or on a slope as shown in Figure 3-15. Galleries connecting other galleries or connecting with other features such as power plants, elevators, and pump chambers are called adits. Some of the more common uses or purposes of galleries are to provide a drainageway for water percolating through the upstream face or seeping through the foundation, space for drilling and grouting the foundation, space for headers and equipment used in artificially cooling the concrete blocks and for grouting contraction joints, access to the interior of the structure for observing its behavior, access to and room for mechanical and electrical equipment, access through the dam for control cables and/or power cables, and access routes for visitors, as shown in Figure 3-16. The location and size of a gallery will depend on its intended use or purpose. The size is normally 5 feet wide by 7.5 feet high. In small, thin arch dams, galleries are not used where the radial thickness is less than five times the width. This gives the cantilevers sufficient section modulus to perform as intended. A distance of two diameters on either side of the gallery provides sufficient thickness to mitigate the high vertical stresses that concentrate along the gallery walls and to develop the necessary section modulus. In thin arch dams, access galleries can be designed smaller than 5 feet wide such as an elliptically shaped gallery 2 feet wide by 8 feet high that contains lighting across the top, a noncorrosive grating for pedestrian traffic, and provides space for drainage.

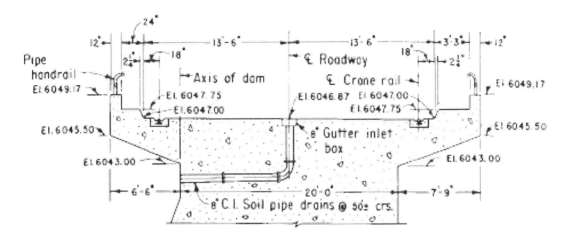

SECTION THRU TOP OF DAM
FLAMING GORGE DAM
(SHOWING ROADWAY AND CRANE RAILS)

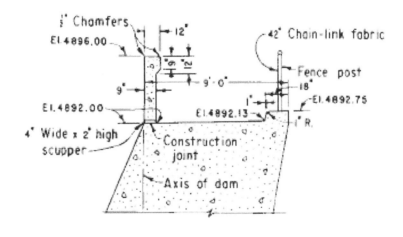

SECTION THRU TOP OF DAM
EAST CANYON DAM
(SHOWING WALKWAY)

Figure 3-14

Typical sections at top of an arch dam

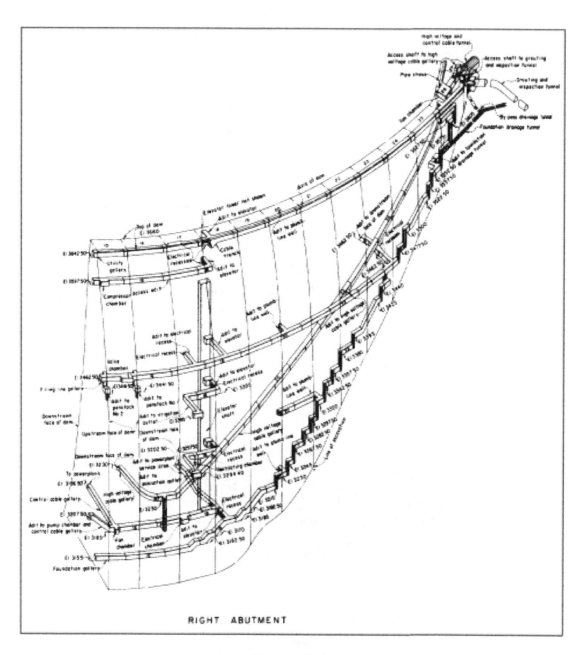

RIGHT ABUTMENT

Figure 3-15

Gallery system in right side of Yellowtail Dam

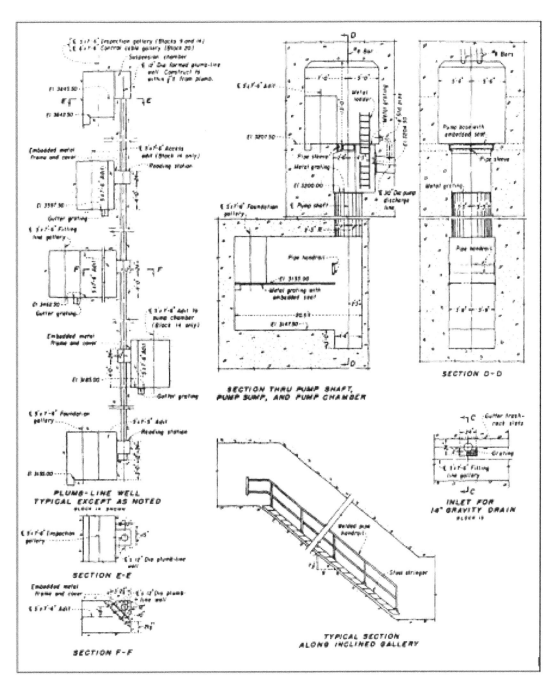

Figure 3-16

Details of galleries and shafts in Yellowtail Dam

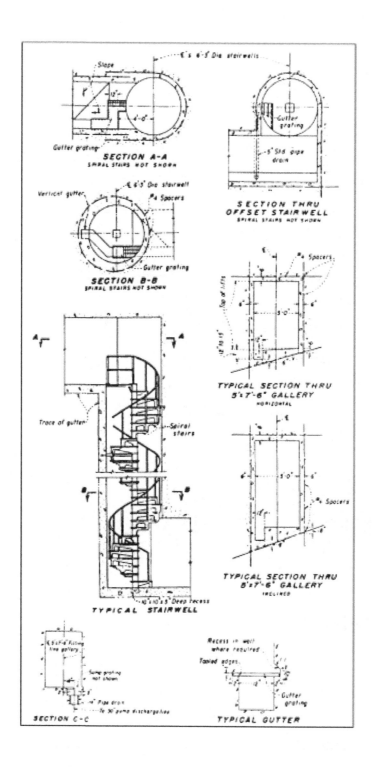

Figure 3-16 (continued)

Details of galleries and shafts in Yellowtail Dam

10.5 RESTITUTION CONCRETE. Because of topographical and geological features, all damsites are nonsymmetrical and have an irregular profile; however, during the design of an arch dam, a significant saving in keyway excavation may be achieved by building up certain regions of the footprint with mass concrete to form an artificial foundation and provide a smooth perimeter for the dam. At a particular site, restitution concrete may be local "dental" concrete, the more extensive "pad," or "thrust blocks" along the crest. In each case, the longitudinal and transverse shape is different for different design purposes, and, accordingly, restitution concrete may be extensive upstream, downstream, or around the perimeter of the dam. In keeping with the concept of efficiency and economy, each arch dam design should be made as geometrically simple as possible; the optimum is a symmetrical design. With this in mind, restitution concrete can be added to the foundation to smooth the profile and make the site more symmetrical, and/or provide a better distribution of stresses to the foundation. Restitution concrete is added to the rock contact either before or during construction; the concrete mix is the same mass concrete used to construct the dam.

10.5.1 DENTAL CONCRETE. Dental concrete is used to improve local geological or topographical discontinuities that might adversely affect stability or deformation as shown in Figure 3-17. Discontinuities include joints, seams, faults, and shattered or inferior rock uncovered during exploratory drilling or final excavation that make complete removal impractical. The necessary amount of concrete replacement in these weak geological zones is usually determined from finite element analyses in which geologic properties, geometric limits, and internal and/or external loads are defined. For relatively homogenous rock foundations with only nominal faulting or shearing, the following approximate formulas can be used for determining the depth of dental treatment:

```
d = 0.002bH + 5 for H greater than or equal to 150 feet
d = 0.3b + 5 for H less than 150 feet
```

where

```
H = height of dam above general foundation level, in feet
b = width of weak zone, in feet, and
d = depth of excavation of weak zone below surface of adjoining
    sound rock, in feet
```

10.5.2 PAD. A concrete pad is added to the foundation to smooth the arch dam profile, to make the site more symmetrical, to reduce excavation, and/or to provide a better distribution of stresses on the foundation. To smooth the profile, pad concrete is placed around the arch dam perimeter, made irregular due to topography or geology, such as in box canyons or bridging low-strength rock types. Such treatment is not unique to the United States. The

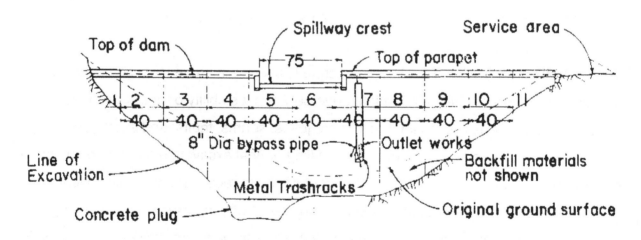

Figure 3-17

Upstream elevation of Wildhorse Dam. Note dental concrete (concrete plug)

additional thickness along the abutment is called a socle in Portugal. Arch
dams in nonsymmetrical sites can be designed more efficiently by constructing a pad
along the abutment of the long side. Reduced excavation is accomplished by filling in a
single but prominent depression with a pad rather than excavating the entire abutment

to smooth the profile. The pad, because it is analogous to a spread footing, will reduce pressure and deformation, especially on weak rock.

10.5.2.1 THE PAD IN CROSS SECTION has the geometric shape of a trapezoid. The top surface of the trapezoid then becomes the profile for the arch dam. To develop a smooth profile for the arch dam, the trapezoidal height will vary along the contact. The size of the footing is a function of the arch abutment thickness. At the contact of the arch and the pad, the pad thickness is greater than the arch thickness by a nominal amount of 5 feet, or greater, as determined from two-dimensional (2-D) finite element studies. The extra concrete thickness, analogous to a berm, is constructed on both faces. This berm provides geometric and structural delineation between the dam and the pad. Below the berm, the pad slopes according to the canyon profile. The slope is described in a vertical radial plane. In a wide valley where gravity action is predominant, a nominal slope of 1:1 is suggested on the downstream side. In narrow canyons, where arch action is the major structural resistance, the slope may be steeper. The upstream face may be sloped or vertical depending on the loading combinations. For example, reservoir drawdown during the summer coupled with high temperatures will cause upstream deflection and corresponding larger compressive stresses along the heel. Thus, to simulate the abutment the upstream face should be sloped upstream. Otherwise, a vertical or near vertical face will suffice. Normally, a constant slope will be sufficient, both upstream and downstream. A pad should be used to fill depressions in the profile that would otherwise cause overexcavation and to smooth the profile or to improve the site symmetry.

10.5.2.2 THE APPROPRIATE SHAPE SHOULD be evaluated with horizontal and vertical sections at points on the berm. Classical shallow beam structural analyses are not applicable because the pad is not a shallow beam. For completeness, check the shape for stress and stability at several representative locations around the perimeter. Loads to be considered in structural design of the pad include moment, thrust, and shear from the arch dam, as well as the reservoir pressure. The load on the massive footing for a cantilever should also include its own weight. Two-dimensional FEMs are ideally suited to shape and analyze the various loads and load combinations. Reshaping and reanalysis then can also be easily accomplished.

10.5.2.3 CONSTRUCTION OF THE PAD may occur before or during construction of the arch dam. For example, if the foundation rock is very hard and/or massive, higher-strength concrete can be placed months before the arch is constructed. In this way, the pad concrete has time to cure and perhaps more closely approximate the actual foundation conditions. Or, as with the socle where a slightly different geometric shape is required at the arch abutment than at the top of the footing, the more efficient method is to construct the footing monolithic with the arch dam, in blocks, and lifts. Special forming details are necessary at the berm; above and below the berm, normal slip forming is sufficient. Artificial cooling and contraction joint grouting is recommended to avoid radial crack propagation into the dam from future shrinkage and settlement. As with the arch dam, no reinforcing steel is necessary in the foundation shaping concrete. Longitudinal contraction joints should be avoided to prevent possible tangential crack propagation into the dam. If, during construction, a significant crack should appear in the foundation or dam concrete and continue to run through successive lifts, a proven remedy is to provide a mat of reinforcing steel on the next lift of the block with the crack but not necessarily across contraction joints.

10.5.3 THRUST BLOCKS. Thrust blocks are another type of restitution concrete. These components are constructed of mass concrete on foundation rock and form an extension of the arch dam crest. They are particularly useful in sites with steep side slopes extending about three-fourths the distance to the top and then rapidly flattening. In such a site, significantly additional water storage can be achieved by thrust blocks without a proportional increase in costs. For small and short extensions beyond the neat line, the cross-sectional shape can simply be a continuation of arch dam geometry as shown on Figure 3-18; thus, for some distance past the neat line, arch action will resist some of the applied load. Beyond that distance, cantilever action resists the water load and must be stable as with a gravity dam. The extension may be a straight tangent or curved as dictated by the topography, as shown in Figure 3-19. If curved, the applied load is distributed horizontally and vertically, in which case the section can be thinner than a straight gravity section. If straight, the tangent section will exhibit some horizontal beam action, but conservatively, none should be assumed unless artificial cooling and contraction joint grouting are utilized. For cases where the thrust block sections are shaped as gravity dams, the analysis approximates the thrust block stiffness by reducing the foundation modulus on those arches connected to the thrust blocks. A reasonable value for the crest abutments is 100 kips per square inch (ksi). The abutment/foundation modulus should be linearly interpolated at elevations between the crest and the first lower arch abutting on rock. The reliability of this assumption should be tested by performing parametric studies with several different assumed rock moduli, comparing arch and cantilever stresses on each face, and noting the stress differences in the lower half of the dam. In general, stress differences should be localized around the thrust blocks.

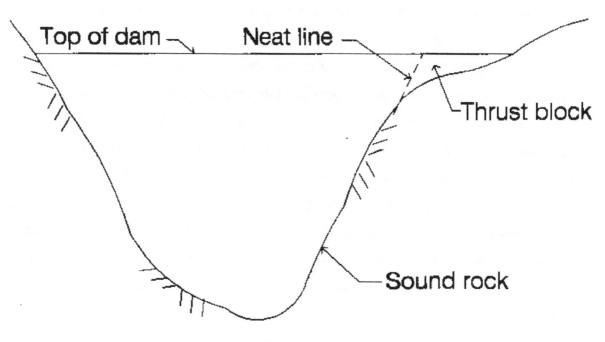

Figure 3-18

Schematic elevation of simple thrust block as a right abutment extension of an arch dam

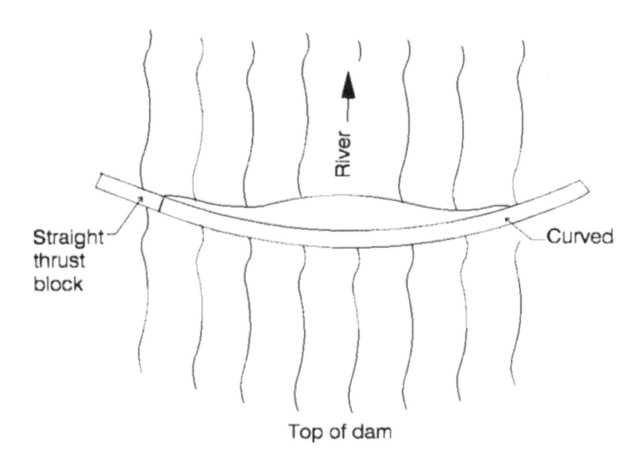

Figure 3-19

Schematic plan of straight and curved thrust blocks and water barrier

CHAPTER 11: STATIC ANALYSIS OF ARCH DAMS

11.1 INTRODUCTION. This discussion describes static analysis of concrete arch dams using the Finite Element Mesh (FEM) approach. The purpose of FEM analysis is to perform more accurate and realistic analysis by eliminating many assumptions made in the traditional methods. The main advantages of FEM are its versatility and its ability for exploring foundation conditions and representing the more realistic interaction of dam and foundation rock. In particular, nonhomogeneous rock properties, weak zones, clay or gouge seams, and discontinuities in the foundation may be considered in the analysis to evaluate their effects on the stress distribution. The cracked sections or open joints in the structure can be modeled; thrust blocks and the spillway openings in the crest are appropriately included in the mathematical models; and the stresses around the galleries and other openings can be investigated.

11.2 DESIGN DATA REQUIRED. Design data needed for structural analysis of a concrete arch dam are: Poisson's ratio, strength and elastic properties of the concrete, Poisson's ratio and deformation modulus of the foundation rock, unit weight and coefficient of thermal expansion of the concrete, geometric data of the dam layout, geometric data of spillway openings and thrust blocks, operating reservoir and tailwater surfaces, temperature changes within the dam, probable sediment depth in the reservoir, probable ice load, and the uplift pressure. A description of each data type is as follows:

11.2.1 CONCRETE PROPERTIES. The material properties of the concrete for use in static analysis are influenced by mix proportions, cement, aggregate, admixtures, and age. These data are not available beforehand and should be estimated based on published data and according to experience in similar design and personal judgment; however, actual measured data should be used in the final analysis as they become available. The concrete data needed for the analysis are:

- Sustained modulus of elasticity
- Poisson's ratio

- Unit weight

- Compressive strength

- Tensile strength

- Coefficient of thermal expansion

The sustained modulus of elasticity is used in the analyses of the static loads to account for the creep effects. In the absence of long-term test data, a sustained modulus of elasticity equal to 60 to 70 percent of the instantaneous modulus may be used.

11.2.2 FOUNDATION PROPERTIES. The foundation data required for structural analysis are Poisson's ratio and the deformation modulus of the rock supporting the arch dam. Deformation modulus is defined as the ratio of applied stress to resulting elastic plus inelastic strains and thus includes the effects of joints, shears, and faults. Deformation modulus is obtained by in situ tests (Structural Properties, Chapter 9) or is estimated from elastic modulus of the rock using a reduction factor (Von Thun and Tarbox 1971 (Oct)). If more than one material type is present in the foundation, an effective deformation modulus should be used instead. For nonhomogeneous foundations, several effective deformation modulus values may be needed to adequately define the foundation characteristics.

11.2.3 GEOMETRIC DATA. The necessary data for constructing a finite element mesh of an arch dam is obtained from drawings containing information defining the geometry of the dam shape. These include the plan view and section along the reference plane, as shown in Figures 1-5 and 1-6. In practice, arch dams are geometrically described as multicentered arches with their centers varied by elevation in addition to the arch opening angles and radii varying for each side with elevation. Elliptical arch shapes may be approximated for the various elevations by three-centered arches including central segments with shorter radii and two outer segments with equal but longer radii. The basic geometric data of a multicentered dam at each elevation for the upstream and downstream faces are as follows:

- Radius of central arcs

- Radius of outer arcs
- Angles to point of compounding curvatures
- Angles to abutments
- Location of centers of central arcs

Preparation of finite element mesh data from these geometric data is very time consuming because most general-purpose finite element programs cannot directly handle these data or the similar ADSAS input data; however, GDAP (Ghanaat 1993a), a specialized arch dam analysis program, can automatically generate coordinates of all nodal points, element data, element distributed loads, and the nodal boundary conditions from such limited geometric data or even directly from ADSAS input data for any arch dam-foundation system. Geometric data for modeling thrust blocks, spillway openings, and other structural features are obtained directly from the associated design drawings.

11.2.4 STATIC LOADS. The basic loads contributing to the design or safety analysis of arch dams are gravity, reservoir water, temperature changes, silt, ice, uplift, and earthquake loads. The data needed to specify each individual static load are described in this section.

11.2.4.1 GRAVITY LOADS. Gravity loads due to weight of the material are computed from the unit weight and geometry of the finite elements. The dead weight may be applied either to free-standing cantilevers without arch action to simulate the construction process or to the monolithic arch structure with all the contraction joints grouted. Although the first assumption usually is more appropriate, a combination of the two is more realistic in situations where the vertical curvature of the cantilevers is so pronounced that it is necessary to limit the height of free-standing cantilevers by grouting the lower part of the dam. In those cases, a gravity load analysis which closely follows the construction sequence is more representative. The weight of the appurtenant structures that are not modeled as part of the finite element model but are supported by the dam, if significant, are input as external concentrated loads and are applied to the supporting nodal points.

11.2.4.2 RESERVOIR WATER. Most finite element programs such as the GDAP and SAP-IV handle hydrostatic loads as distributed surface loads. The surface loads are then applied to the structure as concentrated nodal loads. Therefore, hydrostatically varying surface pressure can be specified by using a reference fluid surface and a fluid weight density as input.

11.2.4.3 TEMPERATURE. Temperature data needed in structural analysis result from the differences between the closure temperature and concrete temperature expected in the dam during its operation. Temperature changes include high and low temperature conditions and usually vary: by elevation, across the arch, and in the upstream-downstream direction. Temperature distribution in the concrete is determined by temperature studies (Chapter 8) considering the effects of transient air and water temperatures, fluctuation of reservoir level, and the solar radiation. The nonlinear temperature distribution calculated in these studies is approximated by straight line distribution through the dam thickness for the use in structural analysis performed in using shell elements. However, if several solid elements are used through the thickness, a nonlinear temperature distribution can be approximated.

11.2.4.4 SILT. Arch dams often are subjected to silt pressures due to sedimentary materials deposited in the reservoir. The saturated silt loads are treated as hydrostatically varying pressures acting on the upstream face of the dam and on the valley floor. A silt reference level and the weight density of the equivalent fluid are needed to specify the silt pressures.

11.2.4.5 ICE. Ice pressure can exert a significant load on dams located at high altitudes and should be considered as a design load when the ice cover is relatively thick. The actual ice pressure is very difficult to estimate because it depends on a number of parameters that are not easily available. In that case, an estimate of ice pressure as given in Chapter 4 may be used. (6) Uplift. The effects of uplift pressures on stress distribution in thin arch dams are negligible and, thus, may be ignored; however, uplift can have a significant influence on the stability of a thick gravity-arch dam and should be considered in the analysis. For more discussion on the subject, refer to "Theoretical Manual for Analysis of Arch Dams" (Ghanaat 1993b).

11.3 METHOD OF ANALYSIS. The static analysis of an arch dam should be based on the 3-D FEM. The FEM is capable of representing the actual 3-D behavior of an arch dam-foundation system and can handle any arbitrary geometry of the dam and valley shape. Furthermore, the method can account for a variety of loads and is equally applicable to gravity arch sections as well as to slender and doubly curved arch dam structures.

11.3.1 THE FEM IS ESSENTIALLY a procedure by which a continuum such as an arch dam structure is approximated by an assemblage of discrete elements interconnected only at a finite number of nodal points having a finite number of unknowns. Although various formulations of the FEM exist today, only the displacement-based formulation which is the basis for almost all major practical structural analysis programs is described briefly here. The displacement based FEM is an extension of the displacement method that was used extensively for the analysis of the framed and truss type structures before the FEM was developed. Following is an outline of the finite element computer analysis for static loads, as a sequence of analytical steps:

11.3.1.1 DIVIDE THE DAM STRUCTURE and the foundation rock into an appropriate number of discrete subregions (finite elements) connected at joints called nodal points.

11.3.1.2 COMPUTE THE STIFFNESS MATRIX of each individual element according to the nodal degrees of freedom and the force-displacement relationships defining the element.

11.3.1.3 ADD THE STIFFNESS MATRICES of the individual elements to form the stiffness matrix of the complete structure (direct stiffness method).

11.3.1.4 DEFINE APPROPRIATE BOUNDARY conditions and establish equilibrium conditions at the nodal points. The resulting system of equations for the assembled structure may be expressed as:

$$ku = p \qquad\qquad (6\text{-}1)$$

where

```
k = stiffness matrix
u = displacement vector
p = load vector
```

11.3.1.5 SOLVE THE SYSTEM OF EQUATIONS for the unknown nodal displacements u.

11.3.1.6 CALCULATE ELEMENT STRESSES from the relationship between the element strains and the nodal displacements assuming an elastic strain-stress relationship.

11.3.2 MOST GENERAL-PURPOSE FINITE ELEMENT COMPUTER programs follow these above analytical steps for static structural analysis, but their applicability to arch dams may be judged by whether they have the following characteristics:

11.3.2.1 AN EFFICIENT GRAPHICS-BASED preprocessor with automatic mesh generation capabilities to facilitate development of mathematical models and to check the accuracy of input data.

11.3.2.2 EFFICIENT AND APPROPRIATE finite element types for realistic representation of the various components of the dam structure.

11.3.2.3 EFFICIENT PROGRAMMING methods and numerical techniques appropriate for the solution of large systems with many degrees of freedom.

11.3.2.4 POSTPROCESSING CAPABILITIES providing graphics for evaluation and presentation of the results.

11.3.3 SAP-IV (BATHE, WILSON, AND PETERSON 1974) AND GDAP (GHANAAT 1993A) ARE TWO WIDELY used programs for the analysis of arch dams. These programs are briefly described here. SAP-IV is a general-purpose finite element computer program for the static and dynamic analysis of linearly elastic structures and continua. This program has been designed for the analysis of large structural systems. Its element library for dam analysis includes eight-node and variable-number-node, 3-D solid elements. The program can handle various static loads including hydrostatic pressures, temperature, gravity due to weight of the material, and concentrated loads applied at the nodal points. However, the program lacks pre- and postprocessing capabilities. Thus, finite element meshes of the dam and foundation must be constructed manually from the input nodal coordinates and element connectivities. Also, the computed stress results are given in the direction of local or global axes and cannot be interpreted reliably unless they are transformed into dam surface arch and cantilever stresses by the user.

11.3.4 GDAP HAS BEEN SPECIFICALLY designed for the analysis of arch dams. It uses the basic program organization and numerical techniques of SAP-IV but has pre- and postprocessing capabilities in addition to the special shell elements. The thick-shell element of GDAP, which is represented by its midsurface nodes, uses a special integration scheme that improves bending behavior of the element by reducing erroneous shear energy. The 16-node shell is the other GDAP special dam analysis element; this retains all 16-surface nodes and uses incompatible modes to improve the bending behavior of the element. In addition to the shell elements, eight-node solid elements are also provided for modeling the foundation rock. The preprocessor of GDAP automatically generates finite element meshes for any arbitrary geometry of the dam and the valley shape, and it produces various 3-D and 2-D graphics for examining the accuracy of mathematical models. The postprocessor of GDAP displays nodal displacements and provides contours of the dam face arch and cantilever stresses as well as vector plots of the principal stresses acting in the faces.

11.3.5 OTHER GENERAL-PURPOSE FEM PROGRAMS, such as ABAQUS and GTSTRUDL can also be used in the analysis of arch dams. Special care should be used to assure that they have the characteristics identified. Also, the stress results from general-purpose FEM programs may be computed in local or global coordinates and, therefore, may need to be translated into surface arch and cantilever stresses by the user prior to postprocessing.

11.4 STRUCTURAL MODELING. Arch dams are 3-D systems consisting of a concrete arch supported by flexible foundation rock and impounding a reservoir of water. One of the most important requirements in arch dam analysis is to develop accurate models representative of the actual 3-D behavior of the system. A typical finite element idealization of a concrete arch dam and its foundation rock is shown in Figure 6-1. This section presents general guidelines on structural modeling for linear-elastic static analysis of single arch dams. The guidelines aim to provide a reasonable compromise between the accuracy of the analysis and the computational costs. They are primarily based on the results of numerous case studies and not on any rigorous mathematical derivation. The procedures and guidelines for developing mathematical models of various components of an arch dam are as follows:

11.4.1 DAM MODEL. An appropriate finite element mesh for an arch dam can only be achieved by careful consideration of the dam geometry and the type of analysis for which the dam is modeled. For example, the finite element model of a double-curvature thin-shell structure differs from the model of a thick gravity-arch section. Furthermore, a structural model developed solely for a linear-elastic analysis generally is not appropriate for a nonlinear analysis.

11.4.1.1 NUMBER OF ELEMENT LAYERS. Arch dam types may be divided, according to the geometry of their cross sections, into thin, moderately thin, and thick gravity-arch sections. Table 6-1 identifies each of these types with regard to crest thickness (tc) and base thickness (tb), each expressed as a ratio to the height (H). Also shown is the ratio of base-to-crest thickness. Each of these dam types may be subject to further classification based on the geometry of arch sections. The GDAP element library contains several elements for modeling the dam and foundation, as described previously and shown in Figures 6-2, 6-3, and 6-4. The body of a thin arch dam, usually curved both in plan and elevation, is best represented by a combination of special-purpose shell elements available in the GDAP program (Figures 6-2b, 6-4c and d). The general 3-D solid element shown in Figure 6-4b, which may have from 8 to 21 nodes, can also be used, but these are not as accurate as the GDAP shell elements in representing bending moments and shear deformations of thin shell structures. In either case, a single layer of solid elements which use quadratic displacement and geometry interpolation functions in the dam face directions and linear interpolation in the dam thickness direction is sufficient to accurately represent the body of the dam (Figure 6-1).

11.4.1.1.1 MODERATELY THIN ARCH DAMS are modeled essentially similar to the thin arch dams, except that 3-D solid elements should be used near the base and the abutment regions where the shell behavior assumption becomes invalid due to excessive thickness of the arch.

11.4.1.1.2 GRAVITY-ARCH DAMS SHOULD be modeled by two or more layers of solid elements in the thickness direction depending on their section thickness. Any of the solid elements shown in Figures 6-4a, b, or d may be used to model the dam. It is important to note that multilayer element meshes are essential to determine a detailed stress distribution across the thickness and to provide additional element nodes for specifying nonlinear temperature distributions.

11.4.1.2 SIZE OF THE DAM MESH. There are no established rules for selecting an optimum mesh size for subdividing an arch dam in the surface directions. The best approach, however, is to define and analyze several meshes of different element types and sizes and then select the one that is computationally efficient and provides reasonably accurate results. The main factors to consider in choosing the mesh include the size and geometry of the dam, type of elements to be used, type and location of spillway, foundation profile, as well as dynamic characteristics of the dam, and the number of vibration modes required in the subsequent earthquake analysis.

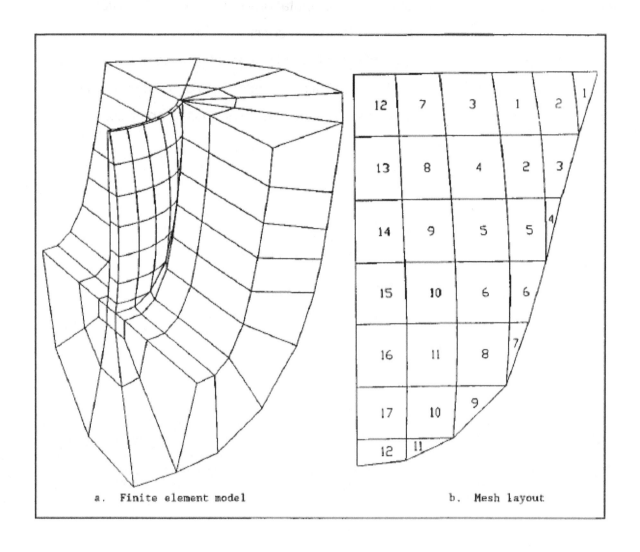

Figure 6-1

Finite element model of Morrow Point Dam and foundation

	t_c/H	t_b/H	t_b/t_c
Thin arch	0.025-0.05	0.09-0.25	2.9-5
Moderately thin	0.025-0.05	0.25-0.4	5-10
Thick gravity-arch	0.05 -0.10	0.5 -1.0	8-15

TABLE 6-1

Arch Dam Types

The size of the finite elements should be selected so that the mesh accurately matches the overall geometry, the thickness, and the curvature of the dam structures. As the dam curvature increases, smaller elements are needed to represent the geometry. The element types used to model a dam affect not only the required mesh size but greatly influence the results. For example, idealization of arch dams with flat face elements requires the use of smaller elements and, thus, a larger number of them, and yet such elements cannot reproduce the transverse shear deformations through the dam which may not be negligible. On the other hand, the same dam can be modeled with fewer curved thick-shell elements such as those available in GDAP and thus obtain superior bending behavior and also include the transverse shear deformations. Figure 6-2 shows an example of three finite element meshes of Morrow Point Arch Dam with rigid foundation rock. Downstream deflections of the crown cantilever due to hydrostatic loads (Figure 6-5) indicate that normal and fine meshes of shell elements provide essentially identical results, and the coarse mesh of shell elements underestimates the deflections by less than 1 percent at the crest and by less than 10 percent at lower elevations. Similar results were obtained for the stresses but are not shown here. This example indicates that the normal mesh size provides accurate results and can be used in most typical analysis. If desired, however, the coarse mesh may be used in preliminary analyses for reasons of economy. For the thick-shell elements used in this example, various parameters such as the element length along the surface (a), the ratio of the thickness to the length (t/a), and the ratio of the length to the radius of curvature (f = a/R) for the coarse and normal meshes are given in Table 6-2 as a reference. These data indicate that the GDAP shell elements with an angle of curvature less than 20 degrees and a length equal or less than 150 feet, provide sufficient accuracy for practical analysis of arch dams with simple geometry and size comparable to that of Morrow Point Dam. For other arch dams with irregular foundation profile, or with attached spillway, or when the lower-order solid elements are used, a finer mesh than that shown in Table 6-2 may be required.

11.4.2 FOUNDATION MODEL. An ideal foundation model is one which extends to infinity or includes all actual geological features of the rock and extends to a very large distance where boundary effects on the stresses in the dam become negligible. In practice, however, these idealized models are not possible because analytical techniques to deal with infinite foundation models are not yet sufficiently developed, and very extensive models are computationally prohibitive, even if the necessary geological data were available. Instead, a simplified foundation model is used which extends a sufficiently large distance that boundary effects are insignificant; the effects of the geological formation are partly accounted for by using modulus of deformation rather than the modulus of elasticity of the rock. In general, the geometry of the rock supporting an arch dam is completely different for different dams and cannot be represented by a single rule; however, simplified prismatic foundation models available in the GDAP program (Figure 6-1) provide adequate models that can conveniently be adapted to different conditions. The foundation mesh types available in the GDAP are shown in Figure 6-3. All three meshes are

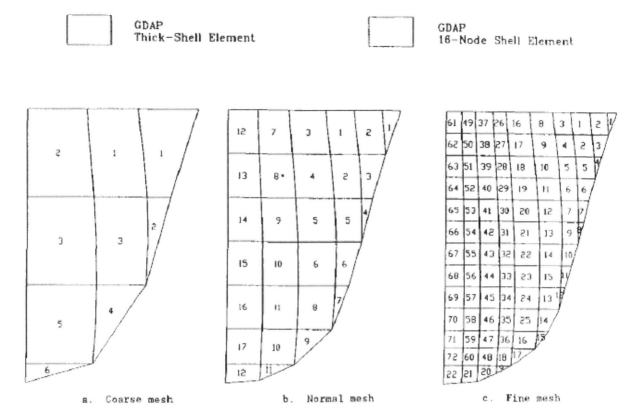

GDAP
Thick-Shell Element

GDAP
16-Node Shell Element

a. Coarse mesh   b. Normal mesh   c. Fine mesh

Figure 6~2

Alternative meshes for Morrow Point Dam

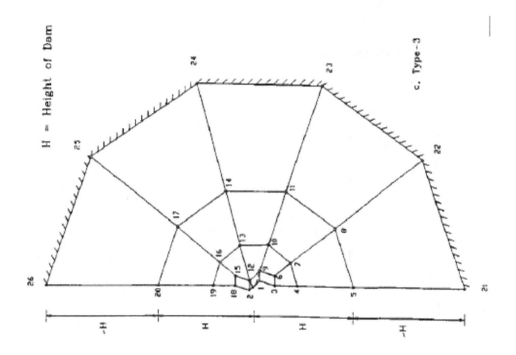

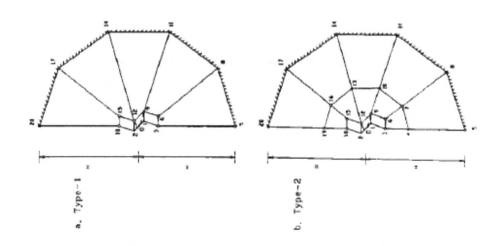

Figure 6-3

GDAP Foundation mesh types

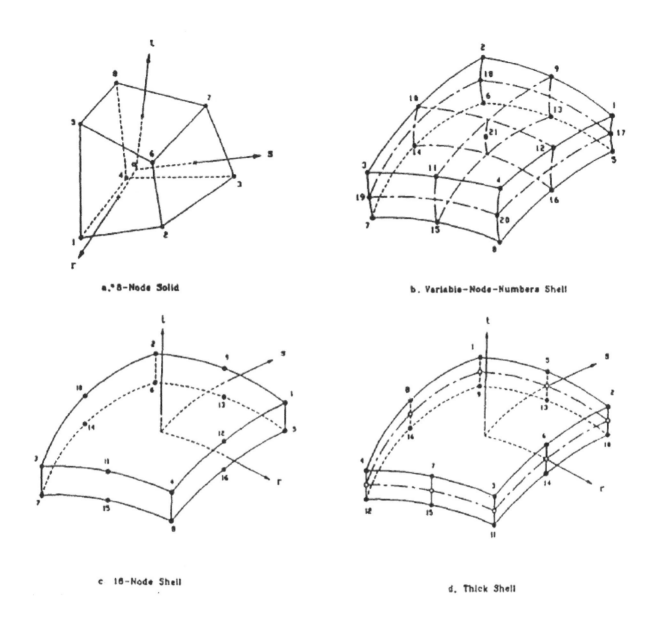

Figure 6-4

Shell and 3-D elements for arch dams

a very large distance where boundary effects on the stresses in the dam become negligible. In practice, however, these idealized models are not possible because analytical techniques to deal with infinite foundation models are not yet sufficiently developed, and very extensive models are computationally prohibitive, even if the necessary geological data were available. Instead, a simplified foundation model is used which extends a sufficiently large distance that boundary effects are insignificant; the effects of the geological formation are partly accounted for by using modulus of deformation rather than the modulus of elasticity of the rock. In general, the geometry of the rock supporting an arch dam is completely different for different dams and cannot be represented by a single rule; however, simplified prismatic foundation models available in the GDAP program (Figure 6-1) provide adequate models that can conveniently be adapted to different conditions. The foundation mesh types available in the GDAP are shown in Figure 6-3. All three meshes are constructed on semicircular planes cut into the canyon walls and oriented normal to the rock-concrete interface as indicated in Figure 6-1a; they differ only with respect to the extent of the rock and the number of elements in each semicircular plane. Eight-node solid elements with anisotropic material properties (Figure 6-4a) are most commonly used for modeling the foundation rock. The foundation mesh is arranged so that smaller elements are located adjacent to the dam-foundation contact surface and the elements become larger toward the boundaries of the model. The size of elements used near the interface is controlled by the dam thickness, and the size of the larger elements depends on the extent of foundation mesh and the number of elements to be used in each section.

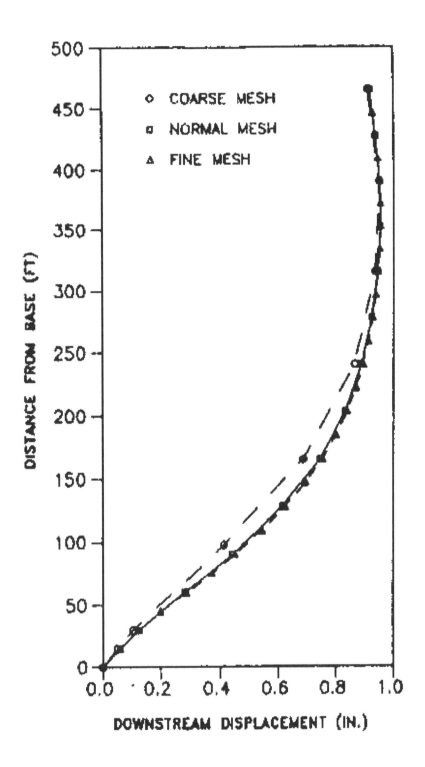

Figure 6-5

Crown section displacements of Morrow Point Dam for alternative meshes

Mesh	a ft	t/a crest	t/a base	φ
Coarse	150	0.08	0.34	20
Normal	75	0.15	0.70	10

TABLE 6-2

Element Mesh Parameters

11.4.2.1 EFFECTS OF FOUNDATION DEFORMABILITY. The importance of foundation interaction on the displacements and stresses resulting from loading an arch dam has long been recognized. The results of a parametric study of Morrow Point Dam, presented in Figures 6-6 through 6-8, demonstrate qualitatively the relative importance of the foundation modulus on the dam response. Three values of the rock modulus were considered: (a) rigid, (b) the same modulus as the concrete, and (c) one-fifth (1/5) the modulus of concrete. The analyses were made only for hydrostatic loads, and the effect of water load acting on the flexible foundation at the valley floor and on the flanks was also investigated. Figure 6-6 shows the deformation patterns while Figure 6-8 compares the arch and cantilever stresses along the crown cantilever section. Deformations clearly are strongly affected by the rock modulus. The rotation of foundation rock, caused by the reservoir water, results in a slight rotation of the dam section in the upstream direction which is more pronounced for weaker rocks. Stresses also are considerably affected by foundation flexibility as compared with the rigid foundation assumption and are further increased by the weight of the impounded water which causes deformations of the foundation rock at the valley floor and flanks. It is important to realize that actual foundations are seldom uniform and may have extensive weak zones. In such cases different values of rock modulus should be assigned to different zones so that the variability effects may be assessed.

11.4.2.2 SIZE OF THE FOUNDATION-ROCK REGION. To account for the flexibility effects of the foundation rock, an appropriate volume of the foundation should be included in the dam-foundation model to be analyzed; however, the amount of flexibility that is contributed by the foundation rock in actual field conditions has not been established. Larger foundation meshes can provide greater flexibility; however, if more finite elements are used to subdivide the foundation rock, greater data preparation and computational efforts are required. Moreover, the increased flexibility also can be obtained by using a reduced foundation modulus. Therefore, the foundation idealization models presented in Figure 6-3 may be sufficient to select the minimum mesh extent (i.e., radius of semicircle R_f) which adequately represents the foundation flexibility effects. In static analysis, flexibility of foundation affects displacements and stresses induced in the dam. For practical analysis, the minimum R_f is selected as a distance beyond which increasing R_f has negligible effects on the displacements and stresses in the dam. The static displacements along the crown cantilever of Morrow Point Dam for two concrete to rock modulus ratios and for three foundation mesh types are shown in Figure 6-7. These results and the stress results (not shown here) suggest that foundation mesh type-1 is adequate for most practical analyses and especially for foundations in which the rock modulus is equal to or greater than the concrete modulus. For very flexible foundation rocks, however, mesh type-3 with an R_f equal to two dam heights should be used.

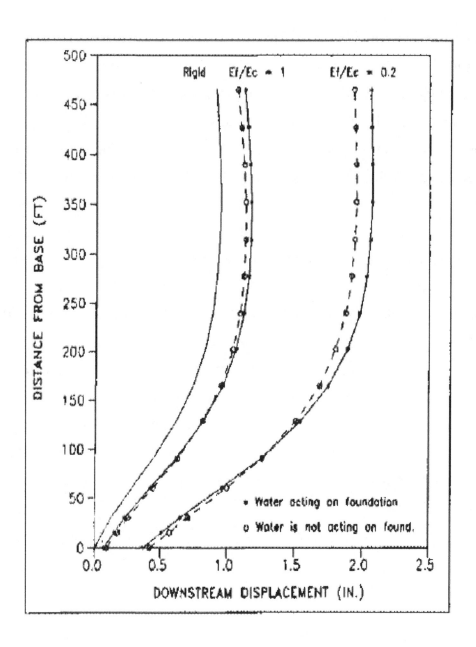

Figure 6~6

Crown section displacements of Morrow Point Dam for different foundation~to-concrete

modulus ratios with and without water acting on valley floor

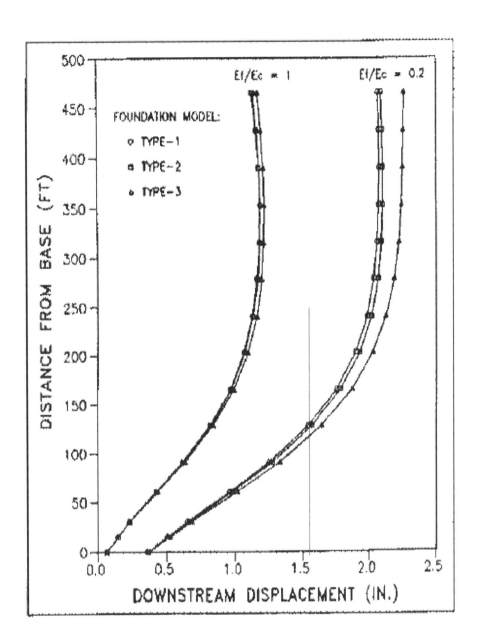

Figure 6-7

Crown section displacements of Morrow Point Dam for two foundation-to-concrete

modulus ratios and for three foundation mesh types

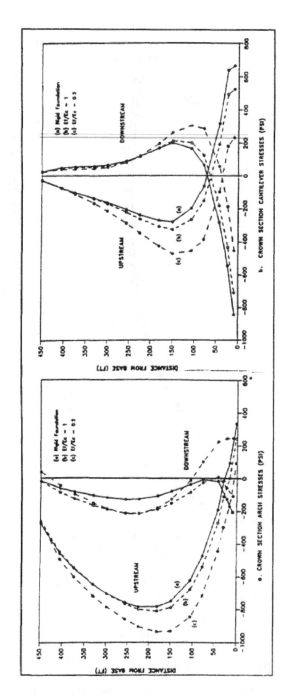

Figure 6-8

Crown section arch and cantilever stresses of Morrow Point Dam for different

foundation-to-concrete modulus ratios with (c & d) and without (a & b) water loads

acting on valley floor (Continued)

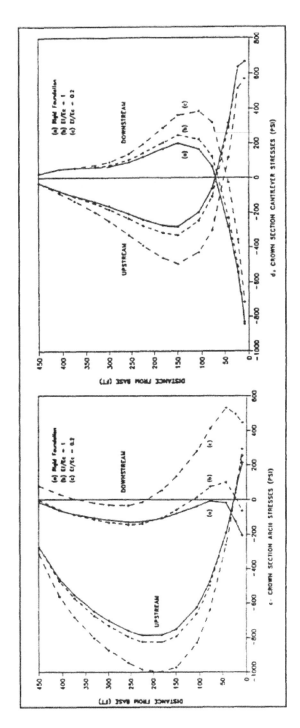

Figure 6-8 concluded)

Crown section arch and cantilever stresses of Morrow Point Dam for different foundation-to-concrete modulus ratios with (c & d) and without (a & b) water loads acting on valley floor

11.4.3 APPURTENANT STRUCTURES. All modern dams include a number of appurtenant structures and devices such as thrust blocks, spillways, galleries, and other openings. The effects of such appurtenances, if significant, should be considered in the analysis by including them in the finite element model of the dam structure.

11.4.3.1 THRUST BLOCKS. Thrust blocks are often used as an artificial abutment where the foundation rock does not extend high enough to support the arches. Their main function is to resist the forces transmitted by the upper arches and transfer them to sound foundation rock at their base. They are a critical component of an arch dam design and should be included appropriately as part of the finite element model of the dam-foundation system. Thrust blocks may be adequately modeled using 8-node elements or any variation of 8-to-20 node solid elements shown in Figures 6-4a and 6-4b. Several element layers are usually required to match the arch mesh at the junction and to account for excessive thickness of the thrust block.

11.4.3.2 SPILLWAYS, GALLERIES, AND OTHER OPENINGS. Arch dams may be designed to accommodate spillways and various other openings such as galleries, sluiceways, and river outlets. Stresses usually tend to concentrate excessively in the area of such openings, and care should be taken to reduce their effects. The large cuts made at the crests of arch dams to provide openings for overflow spillways should be included in the finite element model of the dam structure in order to assess their effects on the stress distribution. If necessary, the design of the dam should be modified to transfer load around the opening in the crest or to proportion the dam thickness to reduce the resulting stress concentrations. Spillways provided by tunnels or side channels that are independent of the arch dam are analyzed and designed separately; thus, they are not included in the finite element model of the dam.

11.4.3.3 OTHER OPENINGS SUCH AS THE GALLERIES, Sluiceways, and River Outlets. These openings introduce a local disturbance in the prevalent stress field and, in general, weaken the structure locally; however, the size of these openings usually does not have a significant influence on the overall stiffness of the dam structure, and their effect on the stress distribution may be ignored if adequate reinforcing is provided to carry the forces around the openings. Therefore, such openings need not be considered in the finite element mesh provided that the openings are small and adequately reinforced.

11.5 PRESENTATION OF RESULTS. An important aspect of any finite element analysis is that of selecting and presenting essential information from the extensive results produced. It is extremely helpful to have the results presented in graphical form, both for checking and evaluation purposes. The results should contain information for the complete structure to make a judgement regarding the dam safety, as well as to determine whether the boundary locations are suitable or whether there are inconsistencies in the stress distribution.

11.5.1 THE BASIC RESULTS OF A TYPICAL static analysis of an arch dam consist of nodal displacements and element stresses computed at various element locations. As a minimum, nodal displacements and surface stresses for the design load combinations specified in Chapter 4 should be presented. Additional displacement and stress results due to the individual load pattern are also desirable because they provide basic information for interpretation of the indicated dam behavior.

11.5.2 NODAL DISPLACEMENTS ARE computed in most computer analyses and are directly available. They are simply presented as deflected shapes across selected arches and cantilevers or for the entire dam structure in the form of 3-D plots. However, consideration should be given to whether the displacements should be indicated in global (x,y) coordinates, or in terms of radial and tangential components for each surface node. The stresses usually are computed with respect to a global coordinate system but they should be transformed to surface arch, cantilever, and principal stress directions to simplify their interpretation. The arch and cantilever stress quantities usually are plotted as stress contours on each dam face, while the principal stresses on each face are presented in the form of vector plots as shown in Figure 6-9. In addition, plots of the arch and cantilever stresses determined across the upper arch section and along the cantilever sections are desirable for further detailed study of the stresses.

11.6 EVALUATION OF STRESS RESULTS.

11.6.1 EVALUATION OF THE STRESS results should start with careful examination of the dam response to assure the validity of the computed results. Nodal displacements and stresses due to the individual loads are the most appropriate data for this purpose. In particular, displacements and stresses across the upper arch and the crown cantilever sections are extremely helpful. Such data are inspected for any unusual distributions and magnitudes that cannot be explained by intuition and which differ significantly from the results for similar arch dams. Once the accuracy of the analytical results has been accepted, the performance of the dam for the postulated loading combinations must be evaluated.

11.6.2 THIS SECOND STAGE OF EVALUATION involves comparing the maximum calculated stresses with the specified strength of the concrete according to the criteria established in Chapter 11. The analysis should include the effects of all actual static loads that will act on the structure during the operations, in accordance with the "Load Combinations" criteria presented. The largest compressive and tensile stress for each load combination case should be less than the compressive and tensile strength of the concrete by the factors of safety specified for each design load combination. When design criteria for all postulated loads are met and the factors of safety are in the acceptable range, the design is considered satisfactory, or, in the case of an existing dam, it is considered safe under the static loads. However, if calculated tensile stresses exceed the cracking strength of the concrete or the lift joints or if tensile stresses are indicated across the vertical monolith joints, the possibility of tension cracking and joint opening must be considered and judgement is required to interpret the results.

11.6.2.1 UNDER THE STATIC LOADS, a well-designed arch dam should develop essentially compressive stresses that are significantly less than the compressive strength of the concrete; however, tensile stresses may be developed under multiple loading combinations, particularly when the temperature drop is large and other conditions are unfavorable. Although unreinforced concrete can tolerate a limited amount of tensile stress, it is important to keep the tension to a minimum so that the arch has sufficient reserve strength if subjected to additional seismic loads. Vertical (cantilever) tensile stresses can be minimized by vertical arching and overhanging of the crest, but the amount by which this can be done is limited by the stress and stability of individual cantilever blocks during the construction process. When the design limits are reached or, as in the case of many existing dams, when the dam is not designed for severe loading conditions, some cracking could occur at the base and near the abutments. Linear-elastic analyses often indicate large stresses near the geometric discontinuity at the foundation contact. However, it is important to note that the tensile stresses indicated at the base of the arch dams by linear-elastic analyses are partly fictitious because these analyses do not take into account the limited bond between the concrete and foundation rock as well as the joints in the rock that could open when subjected to tensile forces. In this situation, a more realistic estimate of static stresses at the base of the dam may be obtained by a linear-elastic analysis that uses a reduced foundation deformation modulus to decrease the tension in the fractured rock.

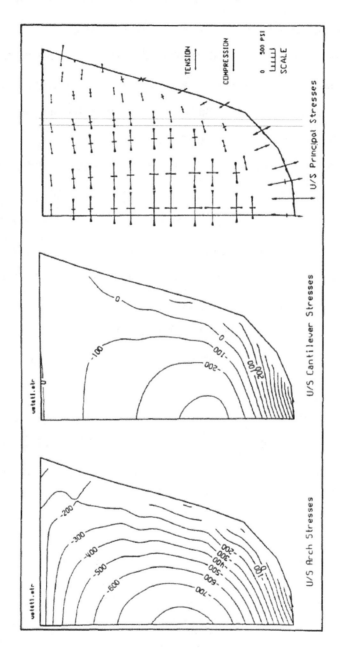

Figure 6-9

Arch and cantilever stress contours and vector plot of principal stresses for upstream face of Morrow Point Dam

11.6.2.2 ARCH DAMS RELY SIGNIFICANTLY on arch action to transfer horizontal loads to the foundation. Therefore, in general, compressive arch stresses are expected throughout the dam; however, the analyses of monolithic arch dams with empty reservoirs, with low water levels, or with severe low temperatures have indicated that zones of horizontal tensile stresses can develop on the upstream and downstream dam faces. These tensile stresses combined with additional tensile stresses due to temperature drop tend to open the vertical contraction joints which are expected to have little or no tensile strength. It is apparent that joint opening will relieve any indicated arch tensile stresses, and the corresponding loads can be redistributed to cantilever action provided that tensile arch stresses are limited to only a small portion of the dam.

11.6.2.3 SHEAR STRESSES ARE RARELY A PROBLEM in an arch dam; nevertheless, they should be checked to make sure that they remain within the allowable limits.

11.6.3 IN CONCLUSION, THE RESULTS OF a linear elastic analysis are valid only if the cracking or joint openings that occur in the dam are minor and the total stiffness of the structure is not affected significantly. Therefore, it is necessary to evaluate the extent of cracking and to judge whether or not a state of no tension can safely be achieved in the dam and its foundation. If appreciable cracking is indicated, it is desirable to investigate its extent and its effects on actual stresses and deflections by analytical procedures. An approximate investigation based on a simplified nonlinear analysis may be made by eliminating the tension areas by iteration and reanalyzing the arch.

CHAPTER 12: TEMPERATURE STUDIES FOR ARCH DAMS

12.1. INTRODUCTION. Temperature studies for arch dams fall into two distinct categories. The first category is the operational temperature study which is used to determine the temperature loading in the dam. This study is performed early in the design process. The second category includes the construction temperature studies which are usually performed after an acceptable layout has been obtained. The construction temperature studies are needed to assure that the design closure temperature can be obtained while minimizing the possibility of thermally induced cracking. The details of each of these studies are discussed in this chapter. Guidance is given on when the studies should be started, values that can be assumed prior to completion of the studies, how to perform the studies, and what information is required to do the studies.

12.2 OPERATIONAL TEMPERATURE STUDIES.

12.2.1 GENERAL. The operational temperature studies are studies that are performed to determine the temperature distributions that the dam will experience during its expected life time. The shape of the temperature distribution through the thickness of the dam is, for the most part, controlled by the thickness of the structure. Dams with relatively thin sections will tend to experience temperature distributions that approach a straight line from the reservoir temperature on the upstream face to the air temperature on the downstream face as shown in Figure 8-1. Dams with a relatively thick section will experience a somewhat different temperature distribution. The temperatures in the center of a thick section will not respond as quickly to changes as temperatures at the faces. The temperatures in the center of the section will remain at or about the closure temperature,1 with fluctuations of small amplitude caused by varying environmental conditions. The concrete in close proximity to the faces will respond quickly to the air and water temperature changes. Therefore, temperature distributions will result that are similar to those shown in Figure 8-2.

12.2.1.1 BEFORE DESCRIBING HOW THESE DISTRIBUTIONS can be obtained for analysis, a description of how the temperatures are applied in the various analysis tools is appropriate. During the early design stages, when a dam layout is being determined, the trial load method is used. The computerized version of the trial load method which is widely used for the layout of the dam is the program ADSAS. ADSAS allows for temperatures to be applied in two ways. The first represents a uniform change in temperature from the grout temperature. The second is a linear temperature load. This linear load can be used to describe a straight line change in temperature from the upstream to downstream faces. These two methods can be used in combination to apply changes in temperature from the grout temperature as well as temperature 1 The terms grout temperature and closure temperature are often used interchangeably. They represent the concrete temperature condition at which no temperature stress exists. This is also referred to as the stress-free temperature condition differences through the section. The resulting distribution will be a straight line distribution.

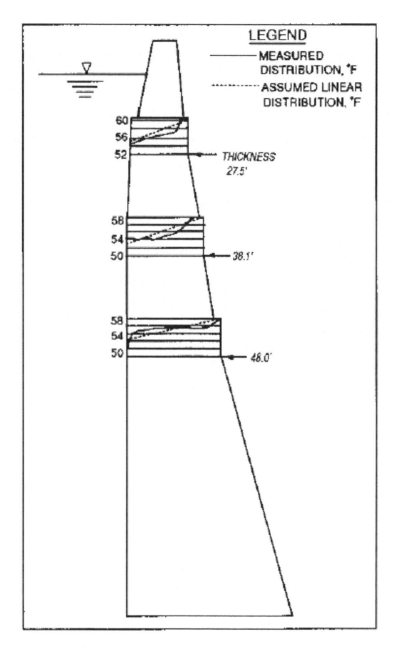

Figure 8-1

Measured temperature distributions in March for a relatively thin dam

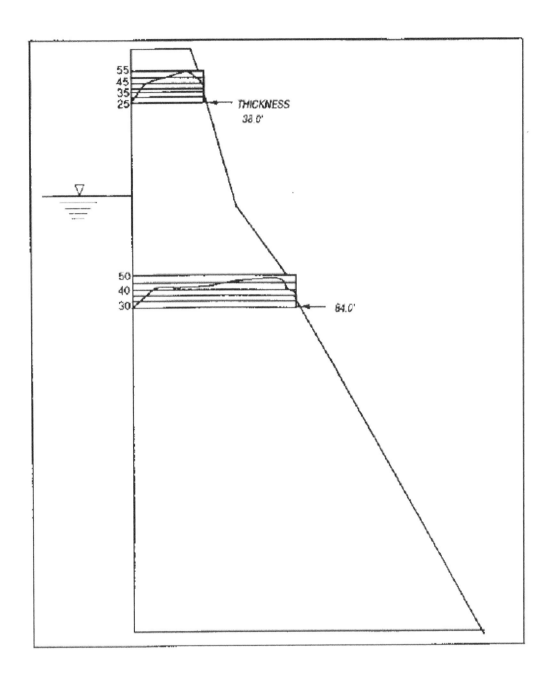

Figure 8-2

Measured temperature distributions in December for a relatively thick dam

12.2.1.2 DURING LATER STAGES of analysis, usually after the final shape of the dam has been determined, the FEM is used to analyze both the static and dynamic conditions. In most general-purpose finite element programs, temperatures are applied at nodal points. This allows for the application of temperature distributions other than linear if nodes are provided through the thickness of the dam as well as at the faces.

12.2.1.3 KEEPING IN MIND the method of stress analysis to be used, one can now choose the method of determining the temperature distributions. There are two methods available for determining the distributions. The first method involves determining the range of mean concrete temperatures that a slab of concrete will experience if it is exposed to varying temperatures on its two faces. This method can be performed in a relatively short time frame and is especially applicable when the trial load method is being used and when the dam being analyzed is relatively thin. When the dam being analyzed is a thick structure, the FEM can be used to determine the temperature distributions.

12.2.1.4 THE TEMPERATURE DISTRIBUTIONS are controlled by material properties and various site specific conditions, including air temperatures, reservoir water temperatures, solar radiation, and in some instances, foundation temperatures. The remainder of this section will discuss how the site conditions can be estimated for a new site and how these conditions are applied to the various computational techniques to determine temperature distributions to be used in stress analysis of the dam.

12.2.2 RESERVOIR TEMPERATURE. The temperature of a dam will be greatly influenced by the temperature of the impounded water. In all reservoirs the temperature of the water varies with depth and with the seasons of the year. It is reasonable to assume that the temperature of the water will have only an annual variation, i.e., to neglect daily variations. The amount of this variation is dependent on the depth of reservoir and on the reservoir operation. The key characteristics of the reservoir operation are inflow-outflow rates and the storage capacity of the reservoir.

12.2.2.1 WHEN A STRUCTURE IS BEING designed there is obviously no data available on the resulting reservoir. The best source of this data would be nearby reservoirs. Criteria for judging applicability of these reservoirs to the site in question should include elevation, latitude, air temperatures, river temperatures and reservoir exchange rate.1 The USBR has compiled this type of information as well as reservoir temperature distributions for various reservoirs and has reported the data in its Engineering Monograph No. 34 (Townsend 1965). Figure 8-3 has been reproduced from that publication.

12.2.2.2 IF DATA ARE AVAILABLE ON RIVER FLOWS and the temperature of the river water, the principle of heat continuity can be used to obtain estimates heat transfer across the reservoir surface. Determination of this heat transfer requires estimates of evaporation, conduction, and absorption. The reservoir exchange rate is measured as the ratio of the mean annual river discharge to the reservoir capacity.reflection of solar radiation and reradiation, which are based on estimates of cloud cover, air temperatures, wind, and relative humidity. Since so many parameters need to be assumed, this method may be no better than using available reservoir data and adapting it to the new site.

12.2.2.3 THE DESIGNER SHOULD RECOGNIZE that the dam's temperatures will be influenced significantly by reservoir temperatures. Therefore, as additional data become available, the assumptions made during design should be reevaluated. Also, it is good practice to provide instrumentation in the completed structure to verify all design assumptions.

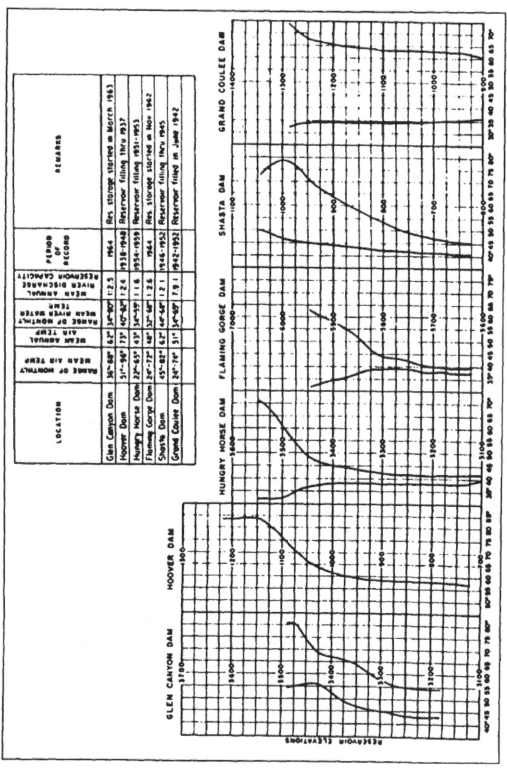

Figure 8-3

Typical reservoir temperature distributions

12.2.3 AIR TEMPERATURES. Estimates of the air temperatures at a dam sitewill usually be made based on the data at nearby weather stations. The U.S. Weather Bureau has published data for many locations in the United States, compiled by state. Adjustments of the data from the nearest recording stations to the dam site can be used to estimate the temperatures at the site. For every 250 feet of elevation increase there is about a 1^0 decrease in temperature. To account for a positive 1.4-degree latitude change, the temperatures can be reduced by 1° F. As with the reservoir temperatures, it is prudent to begin compiling air temperature data as early in the design process as possible to verify the assumed temperatures.

12.2.3.1 DURING THE DISCUSSION of reservoir temperatures, it was pointed out that daily water temperature fluctuations were not of significant concern; however, daily air temperature fluctuations will have a significant effect on the concrete temperatures. Therefore, estimates of the mean daily and mean annual air cycles are needed. A third temperature cycle is also used to account for the maximum and minimum air temperatures at the site. This cycle has a period of 15 days. During the computation of the concrete temperatures, these cycles are applied as sinusoidal variations. The air cycles are not truly sinusoidal, however, this assumption is an acceptable approximation. The pertinent data from the weather station required for the analyses are:

- Mean monthly temperatures (maximum, minimum, and average temperatures)
- Mean annual temperature
- Highest recorded temperature
- Lowest recorded temperature

12.2.4 SOLAR RADIATION. The effect of solar radiation on the exposed surfaces of a dam is to raise the temperature of the structure. Most concrete arch dams are subjected to their most severe loading in the winter. Therefore, the effect of solar radiation generally is to reduce the design loads. However, for cases where the high or summer temperature condition governs the design, the effect of solar radiation worsens the design loads. Also, in harsh climates where the dam is oriented in an advantageous direction, the effect of solar radiation on the low temperature conditions may be significant enough to reduce the temperature loads to an acceptable level.

12.2.4.1 THE MEAN CONCRETE temperature requires adjustments due to the effect of solar radiation on the surface of the dam. The downstream face, and the upstream face when not covered by reservoir water, receive an appreciable amount of radiant heat from the sun, and this has the effect of warming the concrete surface above the surrounding air temperature. The amount of this temperature rise has been recorded at the faces of several dams in the western portion of the United States. These data were then correlated with theoretical studies which take into consideration varying slopes, orientation of the exposed faces, and latitudes. Figures 8-4 to 8-7 summarize the results and give values of the temperature increase for various latitudes, slopes, and orientations. It should be noted that the curves give a value for the mean annual increase in temperature and not for any particular hour, day, or month. Examples of how this solar radiation varies throughout the year are given in Figure 8-8.

12.2.4.2 IF A STRAIGHT GRAVITY DAM is being considered, the orientation will be the same for all points on the dam, and only one value for each of the upstream and downstream faces will be required. For an arch dam, values at the quarter points should be obtained as the sun's rays will strike different parts of the dam at varying angles. The temperature rises shown on the graph should be corrected by a terrain factor which is expressed as the ratio of actual exposure to the sun's rays to the theoretical exposure. This is required because the theoretical computations assumed a horizontal plane at the base of the structure, and the effect of the surrounding terrain is to block out certain hours of sunshine. Although this terrain factor will actually vary for different points on the dam, an east-west profile of the area terrain, which passes through the crown cantilever of the dam, will give a single factor which can be used for all points and remain within the limits of accuracy of the method itself.

12.2.4.3 THE CURVES SHOWN in Figures 8-4 to 8-7 are based on data obtained by the USBR. The data are based on the weather patterns and the latitudes of the western portion of the United States. A USBR memorandum entitled "The Average Temperature Rise of the Surface of a Concrete Dam Due to Solar Radiations," by W. A. Trimble (1954), describes the mathematics and the measured data which were used to determine the curves. Unfortunately, the amount of time required to gather data for such studies is significant. Therefore, if an arch dam is to be built in an area where the available data is not applicable and solar radiation is expected to be important, it is necessary to recognize this early in the design process and begin gathering the necessary data as soon as possible.

12.2.5 PROCEDURE. This section will provide a description of the procedures used to determine the concrete temperature loads.

12.2.5.1 THE FIRST METHOD involves the calculation of the range of mean concrete temperatures. This method will result in the mean concrete temperatures that a flat slab will experience if exposed to: a) air on both faces or

12.2.5.2 WATER ON BOTH FACES. These two temperature calculations are then averaged to determine the range of mean concrete temperatures if the slab is exposed to water on the upstream face and air on the downstream face. A detailed description and example of this calculation is available in the USBR Engineering Monograph No. 34 (Townsend 1965). This process has been automated and is available in the program TEMPER.

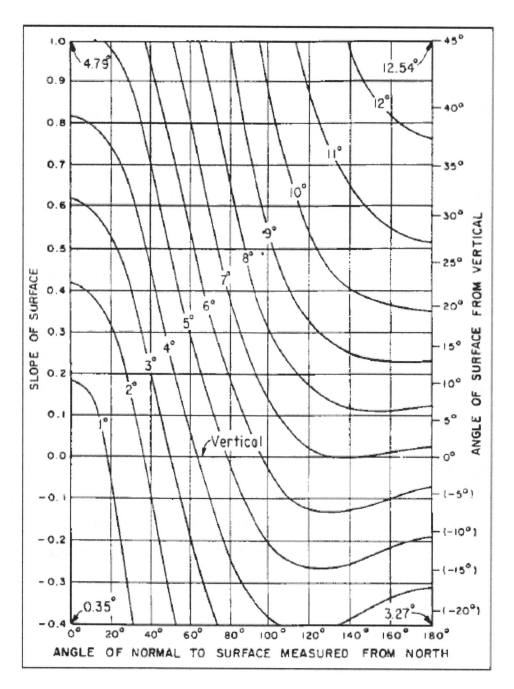

Figure 8-4

Increase in temperature due to solar radiation, latitudes 30° - 35°

Using the computer program will save a great deal of time; however, it would be very instructive to perform the calculation by hand at least once. The steps involved in this process are:

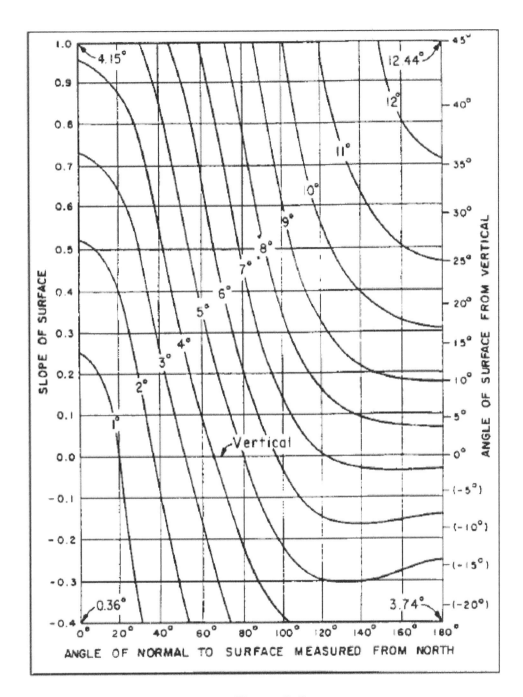

Figure 8-5

Increase in temperature due to solar radiation, latitudes 35° - 40°

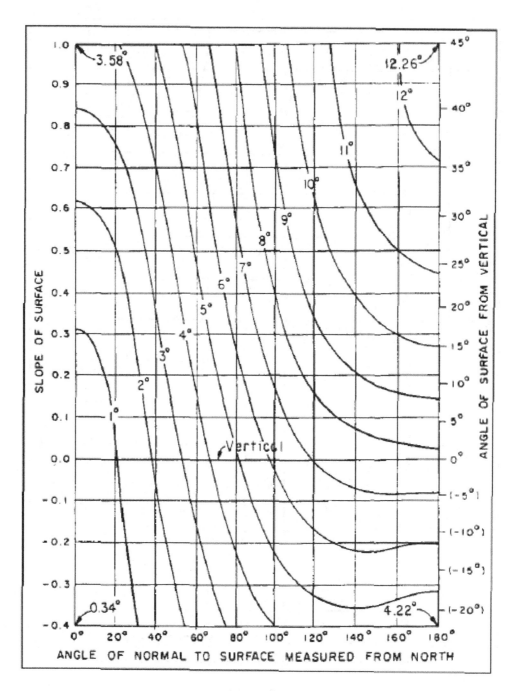

Figure 8-6

Increase in temperature due to solar radiation, latitudes 40° - 45°

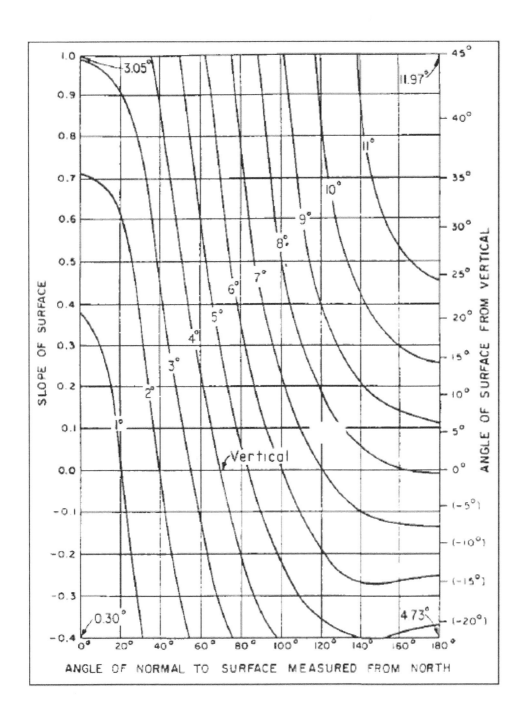

Figure 8-7

Increase in temperature due to solar radiation, latitudes 45° - 50°

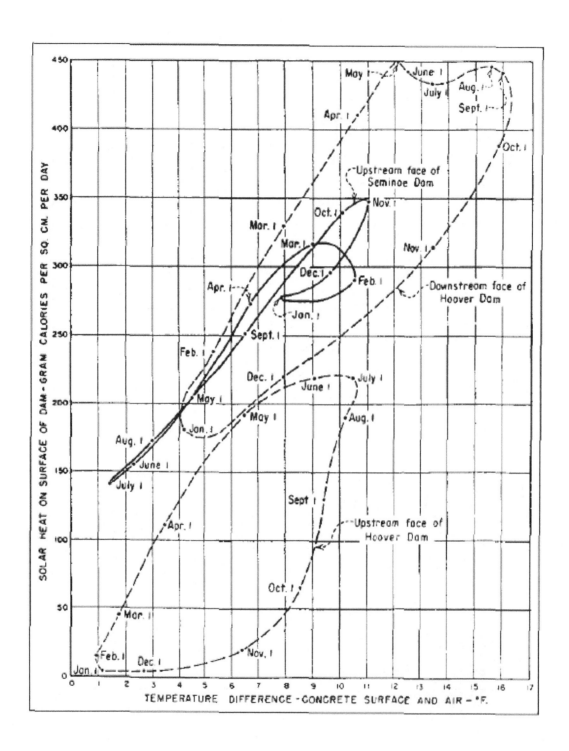

Figure 8-8

Variation of solar radiation during a typical year

12.2.5.2.1 DETERMINE THE INPUT TEMPERATURES. An explanation of the required data has already been given.

12.2.5.2.2 DETERMINE WHERE IN THE STRUCTURE TEMPERATURES ARE DESIRED. These locations should correspond to the "arch" elevations in a trial load analysis and element boundaries or nodal locations in a finite element analysis.

12.2.5.2.3 DETERMINE AIR AND WATER TEMPERATURE CYCLES. As previously mentioned, the reservoir temperatures may be assumed to experience only annual temperature cycles. At the elevations of interest, the reservoir cycle would be the average of the maximum water temperature and the minimum water temperature, plus or minus one-half the difference between the maximum and minimum water temperatures. As mentioned before, three air temperature cycles are required. Table 8-1 describes how these cycles are obtained.

12.2.5.2.4 PERFORM THE COMPUTATION. Only a general description will be presented in this discussion. The theory involved is that of heat flow through a flat slab of uniform thickness. The basis of the calculations is a curve of the thickness of the slab versus the ratio: variation of mean temperature of slab to variation of external temperature. To apply the curve, the thickness of the slab is an "effective" thickness related to the actual thickness of the dam, the diffusivity of the concrete, and the air cycle being utilized; yearly, 15-day, or daily cycle. Once the effective thickness is known, the graph is entered and the ratio is read from the ordinate. This is repeated for the three cycles and the ratios are noted. Then, using the cycles for air and then water, the maximum and minimum concrete temperatures for air on both faces and water on both faces are determined. These values are then averaged to determine the range of concrete temperatures for water on the upstream face and air on the downstream face.

12.2.5.2.5 CORRECT FOR THE EFFECTS OF SOLAR RADIATION.

12.2.5.2.6 APPLY RESULTS TO THE STRESS ANALYSIS.

12.2.5.3 ANOTHER METHOD TO DETERMINE concrete temperatures utilizes finite element techniques. Arch dams are truly 3-D structures from a stress standpoint; however, from a heat-flow standpoint, very little heat will be transmitted in a direction which is normal to vertical planes, i.e., longitudinally through the dam. This allows 2-D heat-flow analyses to be performed. Something to keep in mind is that the results from the heat-flow analyses must be applied to nodes of the 3-D stress model. Therefore, for ease of application, it may be worthwhile to use a 3-D heat-flow model. The benefits of ease of application must be weighed against an increase in computational costs and use of a "coarse" 3-D finite element mesh for the temperature calculations.

Period	Extreme Weather Condition		Usual Weather Condition	
	Above	Below	Above	Below
Annual	(1)	(2)	(1)	(2)
15-day	(4)	(5)	(6)	(7)
Daily	(3)	(3)	(3)	(3)

(1) The difference between the highest mean monthly and the mean annual

(2) The absolute difference between the lowest mean monthly and the mean annual

(3) One-half the minimum difference between any mean monthly maximum and the corresponding mean monthly minimum

(4) The difference between (1+3) and (the highest maximum recorded minus the mean annual)

(5) The difference between (2+3) and (the lowest minimum recorded difference from the mean)

(6) The difference between (1+3) and (the difference between the mean annual and the average of the highest maximum recorded and the highest mean monthly maximum)

(7) The difference between (2+3) and (the difference between the mean annual and the average of the minimum recorded and the lowest mean monthly minimum)

Example, °F

Month	Mean	Mean Max	Mean Min	Difference	High/Low
Jan	47.2	58.8	35.7	23.1	
Feb	51.4	63.5	39.3	24.2	21
Mar	57.1	70.6	43.6	27.0	
Apr	66.0	80.4	51.6	28.8	
May	74.8	89.5	60.0	29.5	
Jun	83.3	98.3	68.4	29.9	
Jul	89.1	102.4	75.6	26.8	116
Aug	86.6	99.7	73.6	26.1	
Sep	81.7	95.3	68.2	27.1	
Oct	70.4	84.1	56.7	27.4	
Nov	57.0	70.0	43.9	26.1	
Dec	49.3	60.4	38.3	22.1	
Annual	67.8	81.1	54.6		

TABLE 8-1

Amplitude of Air Temperatures

Mean annual	67.8
Highest mean monthly	89.1
Lowest mean monthly	47.2
Highest	116.0
Lowest	21.0
Highest mean max	102.4
Lowest mean min	35.7
Lowest difference	22.1

Period	Extreme Weather Condition		Usual Weather Condition	
	Above	Below	Above	Below
Annual	21.3	20.6	21.3	20.6
15-day	15.9	15.2	9.1	7.8
Daily	11.0	11.0	11.0	11.0

TABLE 8-1 (continued)

Amplitude of Air Temperatures

12.2.5.3.1 A FINITE ELEMENT MODEL of either the entire dam or of the crown cantilever should be prepared. The water and air cycles are applied around the boundaries of the model and the mean annual air temperature can be applied to the foundation.1 In most general-purpose finite element programs, steady state and transient solutions are possible. When performing these analyses, the transient solution is utilized. An initial temperature is required. By assuming the initial temperature to be the mean annual air temperature of the site, the transient solution will "settle" to a temperature distribution through the dam that is cyclic in nature. The key to this analysis is to let the solution run long enough for the cycle to settle down. The length of time necessary will be dependent on the thickness of the dam and the material properties. By plotting the response (temperature) of a node in the middle of the dam, a visual inspection can be made and a decision made as to whether or not the solution was carried out long enough. A cyclic response will begin at the initial temperature and the value about which the cycle is fluctuating will drift to a final stable value with all subsequent cycles fluctuating about this value. Based on these results, a solution time step can be chosen to represent the summer and winter concrete temperatures. Then the temperatures can be applied directly to the nodes of the 3-D stress model, if the

392

same model is used for the temperature calculations. If a different model is used for the temperature calculations, a procedure must be developed to spread the 2-D heat flow results throughout the 3-D stress model. If the dam site is in an area of geothermal activity, the mean annual air temperature may not be appropriate for the foundation temperature. In these cases, data should be collected from the site and foundation temperatures should be used based on this data.

12.2.6 SUMMARY. The data necessary to determine the operational temperature loads, the methods that can be used to estimate the data which may not be available at a new dam site, and the methods available to calculate the concrete temperatures have been described. It is necessary for the engineer to determine that the methods used are consistent with the level of evaluation being performed and the stress analysis technique to be employed. The thickness of the dam and, therefore, the resulting temperature distribution should be kept in mind while choosing the temperature analysis technique. The premise here is that thinner structures respond faster to environmental temperature changes than thicker structures. USBR Engineering Monograph No. 34 (Townsend 1965) is a good reference for both the techniques used and data that have been compiled for dams in the western portion of the United States. The program TEMPER may be used in determining the range of mean concrete temperatures. Finally, it is important to begin an instrumentation program early in the design process to verify the assumptions made during the temperature calculations.

12.3 CONSTRUCTION TEMPERATURES STUDIES.

12.3.1 GENERAL. Before the final stages of the design process it is necessary to begin considering how the dam will be constructed and what, if any, temperature control measures need to be implemented. Temperature controls are usually needed to minimize the possibility of thermally induced cracking, since cracking will affect the watertightness, durability, appearance, and the internal stress distribution in the dam. The most common temperature control measures include precooling, postcooling, using low heat cements and pozzolans, reducing cement content, reducing the water-cement ratio, placement in smaller construction lifts, and restricting placement to nighttimes (during hot weather conditions) or to warm months only (in areas of extreme cold weather conditions). This section will cover precooling methods, postcooling procedures, monolith size restrictions, and time of placing requirements. These items must be properly selected in order that a crack-free dam can be constructed with the desired closure temperature. This section also discusses how these variables influence the construction of the dam and how they can be determined.

12.3.2 THE TEMPERATURE CONTROL PROBLEM. The construction temperature control problem can be understood by looking at what happens to the mass concrete after it is placed.

12.3.2.1 DURING THE EARLY AGE OF THE concrete, as the cement hydrates, heat is generated and causes a rise in temperature in the entire mass. Under normal conditions some heat will be lost at the surface while the heat generated at the core is trapped. As the temperature in the core continues to increase, this concrete begins to expand; at the same time, the surface concrete is cooling and, therefore, contracting. In addition, the surface may also be drying which will cause additional shrinkage. As a result of the differential temperatures and shrinkage between the core and the surface, compression develops in the interior, and tensile stresses develop at the surface. When these tensile stresses exceed the tensile strength capacity, the concrete will crack.

12.3.2.2 OVER A PERIOD OF TIME THE COMPRESSIVE stresses that are generated in the core tend to be relieved as a result of the creep properties of the material. As this is happening, the massive core also begins to cool, and it contracts as it cools. This contraction, if restrained by either the foundation, the exterior surfaces, or the previously placed concrete, will cause tensile stresses to develop in the core. As with the previous case, once these tensile stresses exceed the tensile strength capacity of the concrete, the structure will crack.

12.3.3 THE IDEAL CONDITION. The ideal condition would be simply to eliminate any temperature gradient or temperature drop. This is possible only if the initial placement temperature of the concrete is set low enough so that the temperature rise due to hydration of the cement would just bring the concrete temperature up to its final stable state. For example, if the final stable temperature is determined to be 80 oF and the concrete is expected to have a 30 oF temperature rise, then the initial placement temperature could be set at 50 oF, and the designer could be assured that there would be little chance of thermally induced cracking. This example would result in no volumetric temperature shrinkage. However, it may not always be feasible or economical to place concrete at such a low temperature, especially where the final stable temperature falls below 70 oF. In most cases, it is more economical to set the initial placement temperature slightly above the value that would give the "ideal" condition, thereby accepting a slight temperature drop and a small amount of volumetric temperature shrinkage.

12.3.4 PRECOOLING.

12.3.4.1 PRECOOLING IS THE LOWERING OF THE placement temperature of the concrete and is one of the most effective and positive of the temperature control methods. Precooling can also improve the workability of the concrete as well as reduce the rate of heat generated during the hydration. The initial selection of the placement temperature can be achieved by assuming that a zero-stress condition will exist at the time of the initial peak temperature. A preliminary concrete placement temperature can be selected by using the following expression (American Concrete Institute (ACI) 1980):

$$T_i - T_f + (100*C)/(e*R) - dt \qquad (8-1)$$

where

T_i = placing temperature
T_f = final stable temperature
C = strain capacity (millionths)
e = coefficient of thermal expansion (millionths/degree of temperature)
R = degree of restraint (percent)
dt = initial temperature rise

In this expression, the final stable temperature is that temperature calculated as described. In the absence of that information, the final stable temperature can be assumed to be equal to the average annual air and water temperatures. By assuming 100 percent restraint (as would occur at the contact between the dam and the foundation), the equation becomes:

$$T_i = T_f + C/e - dt \qquad (8-2)$$

As an example, if the average annual air temperature is 45 oF, the slow load strain capacity is 120 millionths, the coefficient of thermal expansion is 5 millionths per °F, and the initial temperature rise is 20 °F, then the maximum placement temperature would be in the order of 49 °F. The material property values for these variables should be obtained from test results and from the other studies discussed in other parts of this chapter. Table 8-2 shows a comparison of the average annual temperature and the specified maximum placement temperature for various arch dams constructed in the United States.

Dam	Mean Annual	Placement
Swan Lake	45	50
Strontia Springs	52	55
Crystal	37	40-50
Mossyrock	50	60
Morrow Point	39	40-60
Glen Canyon	62	<50

TABLE 8-2

Comparison of Mean Annual and Placement Temperatures (°F)

(2) The method or methods of reducing concrete placement temperatures will vary depending upon the degree of cooling required, the ambient conditions, and the contractor's equipment. The typical methods of cooling concrete are listed in Table 8-3 in approximate order of increasing cost.

Method of Precooling Concrete	Approximate Temperature Reduction (°F)
Sprinkle coarse aggregate (CA) stockpiles	6
Chill mix water	3
Replace 80% of the mix water with ice	12
Vacuum cool CA to 35 or 38 °F	31
Cold-air cool CA to 40 °F	25
Cool CA by inundation to 40 °F	30
Vacuum cool fine aggregate to 34 °F	12
Contact cool cement to 80 °F	3

TABLE 8-3

Precooling Methods

12.3.5 POSTCOOLING. Postcooling is used both to reduce the peak temperature which occurs during the early stage of construction, and to allow for a uniform temperature reduction in the concrete mass to the point where the monolith joints can be grouted. Postcooling is accomplished by circulating water through cooling coils embedded between each lift of concrete. Following proper guidelines, concrete temperatures can safely be reduced to temperatures as low as 38 °F. Figure 8-9 shows a typical temperature history for postcooled concrete. Descriptions of the cooling periods and of the materials and procedures to be used in the postcooling operation are discussed in the following paragraphs.

12.3.5.1 INITIAL COOLING PERIOD. During the initial cooling period (see Figure 8-9) the initial rise in temperature is controlled and a significant amount of heat is withdrawn during the time when the concrete has a low modulus of elasticity. The total reduction in the peak temperature may be small (3 to 5 F), but it is significant. The initial cooling period will continue to remove a significant amount of heat during the early ages of the concrete when the modulus of elasticity is relatively low. It is preferable, however, not to remove more than 1/2 to 1 F per day and not to continue the initial cooling for more than 15 to 30 days. Rapid cooling could result in tensions developing in the area of the cooling coils which will exceed the tensile strength of the concrete.

12.3.5.2 INTERMEDIATE AND FINAL COOLING PERIODS. The intermediate and final cooling periods are used to lower the concrete temperature to the desired grouting temperature. In general, the same rules apply to the intermediate and final cooling periods as to the initial cooling period except that the cooling rate should not exceed 1/2 F per day. This lower rate is necessary because of the higher modulus of elasticity of the concrete. The need for the intermediate cooling period is dependent upon the need to reduce the vertical temperature gradient which occurs at the upper boundary of the grout lift. If an intermediate cooling period is needed, then the temperature drop occurring in the period is approximately half the total required. Each grout lift goes through this intermediate cooling period before the previous grout lift can go through its final cooling.

12.3.5.3 MATERIALS. The coils used in the postcooling process should be a thin-wall steel tubing. The diameter of the coils is selected as that which will most economically pass the required flow of water through the known length of coil. A small diameter may reduce the cost of the coil, but would increase the pumping cost. Coils with a 1-inch outside diameter are common for small flows. The water used in the postcooling operation must be free of silt which could clog the system. If cool river water is available year round, it usually will be cheaper than refrigerated water provided the required concrete temperature can be obtained within the desired time. The use of river water will usually require more and longer coils and a greater pumping capacity, but it could eliminate the need for a refrigeration plant.

12.3.5.4 LAYOUT. Individual coils can range in length from 600 to 1,300 feet. However, it is preferable to limit the length of each coil to 800 feet. Wherever possible, horizontal spacings equal to the vertical lift spacings give the most uniform temperature distribution during cooling. With lifts in excess of 7.5 feet, this may not be practical.

Horizontal spacings from 2 to 6 feet are most common. Coils are often spaced closer together near the foundation to limit the peak temperatures further in areas where the restraint is large.

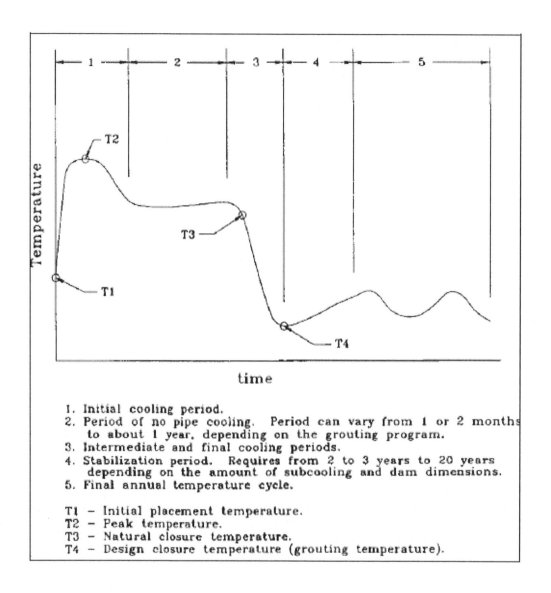

1. Initial cooling period.
2. Period of no pipe cooling. Period can vary from 1 or 2 months to about 1 year, depending on the grouting program.
3. Intermediate and final cooling periods.
4. Stabilization period. Requires from 2 to 3 years to 20 years depending on the amount of subcooling and dam dimensions.
5. Final annual temperature cycle.

T1 – Initial placement temperature.
T2 – Peak temperature.
T3 – Natural closure temperature.
T4 – Design closure temperature (grouting temperature).

Figure 8-9

Temperature history for artificially cooled concrete where monolith joints are grouted

12.3.5.5 PROCEDURES. The cooling coils should be fixed in position by the use of tie-down wires which were embedded in the lift surfaces prior to final set. Compression type connections should be used and the coil system should be pressure tested prior to placement of concrete. It should also undergo a pumping test at the design flow to check for friction losses. Each coil should include a visual flow indicator. Circulation of water through the cooling coils should be in process at the time that concrete placement begins. Since the water flowing through the coil is being warmed by the concrete, reversing the flow daily will give a more uniform reduction in temperature and help to prevent clogging. The cooling operations should be monitored by resistance-type thermometers embedded in the concrete at representative locations. When refrigerated systems are used, the flow seldom exceeds 4 gallons per minute (gpm). These are closed systems where the water is simply recirculated through the refrigeration plant. Systems using river water could have flow rates as high as 15 gpm. In these systems, the water is usually wasted after flowing through the system and new river water is supplied at the intake. Once the final cooling has been completed, the coils should be filled with grout.

12.3.6 CLOSURE TEMPERATURE ANALYSIS. One of the most important loadings on any arch dam is the temperature loading. The temperature loading is obtained by calculating the difference between the operational concrete temperature (paragraph 8-2) and the design closure temperature (Chapter 4). The design closure temperature is sometimes referred to as the grouting temperature, and is commonly obtained by cooling the concrete to the desired temperature and grouting the joints. However, grouting of the joints may not always be necessary, or possible. In some cases, it may be more desirable to select the placement temperature for the concrete so that the natural closure temperature of the structure corresponds to the design closure temperature. This is the "ideal condition" discussed in paragraph 8-3c. The purpose of the closure temperature analysis is to determine how the design closure temperature can be obtained while minimizing the possibility of cracking the structure.

12.3.6.1 BEFORE PERFORMING A DETAILED closure temperature analysis, a preliminary (simplified) analysis should be performed. The first step in the closure temperature study is to look at the typical temperature cycle for artificially cooled concrete. Artificially cooled concrete is concrete that incorporates the postcooling procedures discussed. Figure 8-9 shows a typical temperature cycle for artificially cooled concrete when the joints are to be grouted. The temperatures shown in this figure and those discussed in the next few paragraphs should be considered average temperatures. There are many factors that influence the temperature history including the placement temperature, the types and amounts of cementitious materials, the size of the monoliths, the placement rates, and the exposure conditions. As shown in the figure, there are five phases to the temperature history. Phase 1 begins as the concrete is being placed and continues while the cooling coils are in operation. Phase 2 covers the period between the initial postcooling operations and the intermediate and/or final cooling period. Phase 3 is the phase when the postcooling is restarted and continues until the joints are grouted. Phase 4 is the period after the grouting operation in which the concrete temperatures reach their final stable state. Phase 5 is the continuation of the final annual concrete temperature cycle, or the operating temperature of the structure, which is discussed.

12.3.6.2 THERE ARE FOUR IMPORTANT points along this temperature history line which are determined as part of the closure temperature analysis. Temperature T1 is the placement temperature of the concrete. Temperature T2 is the maximum or peak temperature. Temperature T3 is the natural closure temperature, or the temperature at which the joints begin to open. Temperature T4 is the design closure temperature, or the temperature of the concrete when the contraction joints are grouted. The preliminary analysis can be made to assure that the dam is constructable by evaluating each of these four temperatures. This is done by starting with temperature T4 and working back up the curve.

12.3.6.2.1 TEMPERATURE T4 IS SET BY the design analysis and is, therefore, fixed as far as the closure temperature analysis is concerned. For the example discussed in the next few paragraphs, a design closure temperature (T4) of 50 F is assumed.

12.3.6.2.2 TEMPERATURE T3 CAN be calculated by selecting a monolith width, using the coefficient of thermal expansion test results and assuming a required joint opening for grouting. An arch dam with a 50-foot monolith, a 5.0 × 10-6 inch/inch/oF coefficient of thermal expansion, and a joint opening of 3/32 inch would require a temperature drop of:

$$\Delta T = \frac{3/32 \text{ inch}}{50 \text{ feet} \times 12 \text{ inches/foot}} \times \frac{1}{5.0 \times 10^{-6}/\text{°F}} = 31.25 \text{ °F} \qquad (8-3)$$

For this type of analysis, temperatures can be rounded off to the nearest whole degree without a significant impact in the conclusions. Therefore, a DT of 31 F is acceptable and T3 becomes 81 F.

12.3.6.2.3 THE DIFFERENCE between T3 and T2 will vary according to the thickness of the lift and the placement temperature. This variation is usually small and is sometimes ignored for the preliminary closure temperature analysis. If included in the analysis, the following values can be assumed. For lift heights of 5 feet, a 3 oF difference can be assumed. For 10-foot lifts, a 5 F temperature difference is more appropriate. Therefore, for a 10-foot lift height the average peak temperature (T2) becomes 86 F (81 + 5 oF).

12.3.6.2.4 THE PLACEMENT TEMPERATURE (T1) can be calculated based on the anticipated temperature rise caused by the heat of hydration. There are many factors that influence the temperature rise such as the type and fineness of cement, the use of flyash to replace cement, the lift height, the cooling coil layout, the thermal properties of the concrete, the ambient condition, the construction procedures, etc. Because of the variety of factors affecting temperature rise, it is difficult to determine this quantity without specific information about the concrete materials, mix design, and ambient conditions. For the example discussed in this section, we will simply assume that a 25 F difference exists between T1 and T2, which is somewhat typical when Type II cement and flyash are used in the concrete mix and a 10-foot lift height is selected. This 25 F temperature rise will yield a placement temperature of 61 F. Allowing for some error in the analysis and some variation during the construction process, a temperature range of 60 + 5 F would be specified for this example.

12.3.6.3 USING THE PROCEDURE in the previous paragraphs, the temperatures along the temperature history curve can be estimated. The next step in this preliminary closure temperature analysis is to determine if any of the temperatures and/or changes in temperatures could result in thermal cracking, or if they represent conditions which are not constructible. Two aspects of the temperature history need to be closely evaluated:

12.3.6.3.1 THE PLACEMENT AND PEAK temperatures. To be economical, the placement temperature should be near the mean annual air temperature. If the calculated placement temperature from the preliminary analysis is less than 45 F or greater than 70 F, or if placement temperature is 10 F above the mean annual air temperature, then a more detailed closure temperature analysis should be performed. A detailed analysis should also be performed if the required peak temperature (T2) is above 105 F.

12.3.6.3.2 THE TEMPERATURE DROP from the peak to design closure temperature. The strain created during the final cooling period should not exceed the slow load strain capacity of the concrete as determined from test results. The maximum temperature drop can be determined by dividing the slow load strain capacity by the coefficient of thermal expansion. For example, if the slow load strain capacity is 120 millionths and the coefficient of thermal expansion is 5 millionths per F, the maximum temperature drop will be 24 F. Based on the values assumed (required temperature drop of 36 F), the monolith width would need to be increased, or a more detailed closure temperature analysis would be required. In this case, by increasing the monolith width to 80 feet the required temperature drop would be reduced to 24.5 F. The combination of the large monolith width and excessive temperature drop would usually require that a detailed closure analysis be performed to more accurately determine the construction parameters and temperature values.

12.3.6.4 IF EITHER OF THE CONDITIONS stated indicate a problem with obtaining the design closure temperature without jeopardizing the constructability of the dam, then a more detailed closure temperature analysis is required. The details of how to perform a detailed closure temperature analysis are presented in the next paragraphs.

12.3.6.5 TO PERFORM A DETAILED closure temperature analysis, the following assumptions are required:

12.3.6.5.1 THE PRINCIPLE OF SUPERPOSITION must apply. That is, the strains produced at any increment of time are independent of the effects of any strains produced at any previous increment of time.

12.3.6.5.2 WHEN THE MONOLITH joints are closed, the concrete is restrained from expanding by the adjacent monoliths and compressive stresses will develop in the monolith joints.

12.3.6.5.3 THE OPENING.

12.3.6.5.4 JOINT OPENING will occur only after all compressive stresses have been relieved.

12.3.6.5.5 CREEP IS APPLIED only to compressive stresses.

12.3.6.5.6 ONLY THE EFFECTS of thermal expansion or contraction and added weight are considered.

12.3.6.6 TO PERFORM THE CLOSURE temperature analysis, the time varying properties of coefficient of linear thermal expansion, rate of creep, modulus of elasticity, and Poisson's ratio will be needed. These material properties will be needed from the time of placement through several months. Chapter 9 furnishes more information on the material testing required.

12.3.6.7 THE FIRST STEP IN THE CLOSURE temperature analysis is to predict the temperature history of a "typical" lift within the dam. This can be done with a heat-flow finite element program. The main difference between the details discussed in the ETL and those discussed in this section is that the information needed for a closure temperature can be simplified such that the entire structure need not be modeled if a "typical" temperature history for each lift can be estimated. This can usually be done with a 2-D model with a limited number of lifts above the base of the dam. Ten lifts will usually be sufficient for most arch dam closure temperature analyses. If the thickness of the dam changes significantly near the crest, then additional heat-flow models may be necessary in that region.

12.3.6.8 ONCE THE TEMPERATURE HISTORY of a "typical" lift has been estimated, the next step is to calculate the theoretical strain caused by the change in temperature for each increment of time. This theoretical strain is calculated by:

$$\varepsilon_t = e_i * \Delta T = e_i * (T_i - T_{i-1}) \tag{8-4}$$

where

ε_t = incremental strain due to the change in temperature from time t_{i-1} to t_i

e_i = coefficient of linear thermal expansion at time t_i

ΔT = change in temperature

T_i = temperature at time t_i

T_{i-1} = temperature at time t_{i-1}

(9) In addition, the theoretical strain due to construction loads can be added by the following equation:

$$\varepsilon_{wt} = \frac{\mu_i * \Delta wt}{E_i} = \frac{\mu_i * (wt_i - wt_{i-1})}{E_i} \tag{8-5}$$

where

ε_{wt} = incremental strain due to added weight from time t_{i-1} to t_i

μ_i = Poisson's ratio at time t_i

Δwt = the incremental change in weight

E_i = modulus of elasticity at time t_i

wt_i = weight at time t_i

wt_{i-1} = weight at time t_{i-1}

(10)   The total incremental strain is the sum of the incremental strain
due to changes in temperature and added weight, as follows:

$$\varepsilon_i = \varepsilon_t + \varepsilon_{wt} = e_i * (T_i - T_{i-1}) + \frac{\mu_i * (wt_i - wt_{i-1})}{E_i} \qquad (8-6)$$

where

ε_i = total incremental strain at time t_i

(11)   The incremental stress can be calculated by:

$$\sigma_i = \varepsilon_i * E_i \qquad (8-7)$$

where

σ_i = total incremental stress at time t_i

(12) Once the stress has been determined for each time increment, creep
can be applied to the stress to determine how that incremental stress is
relaxed over time.   The following equation applies to stress relaxation under
constant strain:

$$\sigma_{i-n} = \frac{1}{1/E_i + [c_i * \ln(t_n - t_i + 1)]} \quad \text{per unit strain} \qquad (8-8)$$

where

σ_{i-n} = stress at time t_n due to an increment of strain at time t_i
   c_i = rate of creep at time t_i

(13)   To estimate the total stress at any time t_n, the following
equation can be used:

$$\sigma_n = \sum_{i=1}^{n} \sigma_{i-n} \qquad (8-9)$$

where

σ_n = total stress at time t_n

(14)   If the total stress in the monolith joint at the end of time t_n is
in compression ($\sigma_n \geq 0$), then the temperature drop necessary to relieve the
compressive stress can be determined by:

$$dT_n = T_n - T'_n = \frac{\sigma_n}{e_n \cdot E_n} \tag{8-10}$$

where

T'$_n$ = natural closure temperature of the structure at time t$_n$
T$_n$ = concrete temperature at time t$_n$.

(15) Under normal circumstances, T'$_n$ should not vary significantly after 20 to 30 days after concrete placement and can simply be referred to as T'. In the closure temperature analysis, the steady state value for T' is the critical value for estimating the monolith width. With T' and the design closure temperature, the minimum monolith width required to be able to grout the monolith joints can be determined by:

$$\ell_{min} = \frac{x}{e_n \cdot (T' - T_g)} \tag{8-11}$$

where

ℓ_{min} = minimum size (width) of monolith that will produce an acceptable joint opening for grouting
x = joint opening needed to be able to grout the joint
T$_g$ = temperature at which the joints are to be grouted (the design closure temperature)

12.3.7 THE UNGROUTED OPTION. If the preliminary and/or detailed closure temperature analysis indicates a problem with obtaining the design closure temperature because the required placement temperatures are higher than acceptable (greater than 70 oF), then the ungrouted option should be considered. The ungrouted option assumes that the "natural" closure temperature is the same as the "design" closure temperature. Figure 8-10 shows the temperature cycle for the ungrouted option. In this option, the concrete is placed at a low enough temperature such that the natural closure temperature falls within a specified value. A detailed closure temperature analysis is required in order to obtain adequate confidence that the dam will achieve the required closure temperature.

12.3.8 NONLINEAR, INCREMENTAL STRUCTURAL ANALYSIS. Once the closure temperature study has been satisfactorily completed, the next step is to perform a nonlinear, incremental structural analysis (NISA) using the construction parameters resulting from the closure temperature study. ETL 1110-2-324 provides guidance for performing a NISA. If the structural configuration or the construction sequence is modified as a result of the NISA, then a reanalysis of the closure temperature may be required.

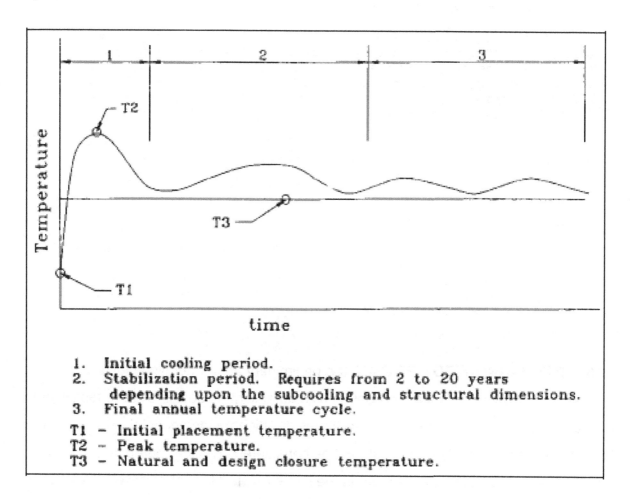

Figure 8-10

Temperature history for artificially cooled concrete where monolith joints are not grouted

CHAPTER 13: CONCRETE CONDUITS

13.1 LIFE CYCLE DESIGN

13.1.1 GENERAL. During the design process, selection of materials or products for conduits, culverts, or pipes should be based on engineering requirements and life cycle performance. This balances the need to minimize first costs with the need for reliable long-term performance and reasonable future maintenance costs.

13.1.2 PROJECT SERVICE LIFE. Economic analysis used as a part of project authorization studies usually calculates costs and benefits projected for a 50- or 75-year project life. However, many projects represent a major infrastructure for the Nation, and will likely remain in service indefinitely. For major infrastructure projects, designers should use a minimum project service life of 100 years when considering life cycle design.

13.1.3 PRODUCT SERVICE LIFE. Products made from different materials or with different protective coatings may exhibit markedly different useful lives. The service life of many products will be less than the project service life, and this must be considered in the life cycle design process. A literature search reported the following information on product service lives for pipe materials. In general, concrete pipe can be expected to provide a product service life approximately two times that of steel or aluminum. However, each project has a unique environment, which may either increase or decrease product service life. Significant factors include soil pH and resistivity, water pH, presence of salts or other corrosive compounds, erosion sediment, and flow velocity. The designer should investigate and document key environmental factors and use them to select an appropriate product service life.

13.1.3.1 CONCRETE. Most studies estimated product service life for concrete pipe to be between 70 and 100 years. Of nine state highway departments, three listed the life as 100 years, five states stated between 70 and 100 years, and one state gave 50 years.

13.1.3.2 STEEL. Corrugated steel pipe usually fails due to corrosion of the invert or the exterior of the pipe. Properly applied coatings can extend the product life to at least 50 years for most environments.

13.1.3.3 ALUMINUM. Aluminum pipe is usually affected more by soil-side corrosion than by corrosion of the invert. Long-term performance is difficult to predict because of a relatively short history of use, but the designer should not expect a product service life of greater than 50 years.

13.1.3.4 PLASTIC. Many different materials fall under the general category of plastic. Each of these materials may have some unique applications where it is suitable or unsuitable. Performance history of plastic pipe is limited. A designer should not expect a product service life of greater than 50 years.

13.1.4 FUTURE COSTS. The analysis should include the cost of initial construction and future costs for maintenance, repair, and replacement over the project service life. Where certain future costs are identical among all options, they will not affect the comparative results and may be excluded from the calculations. For example, costs might be identical for normal operation, inspection, and maintenance. In this case, the only future costs to consider are those for major repairs and replacement. Where replacement will be necessary during the project service life, the designer must include all costs for the replacement activities. This might include significant costs for construction of temporary levees or cofferdams, as well as significant disruptions in normal project operations.

13.2 CROSS-SECTION. The selection of the most economical conduit cross section must depend on the designer's judgment and the consideration of all design factors and site conditions for each application. For fills of moderate height, circular or rectangular openings will frequently be the most practicable because of the speed and economy obtainable in design and construction. For openings of less than about 5.6 m2 (60 ft2), a single rectangular box probably will be most economical for moderate fills up to about 18.3 m (60 ft). However, a rectangular conduit entrenched in rock to the top of the conduit may be economical for higher fills since the applied vertical load need be only the weight of the earth directly above with no increase for differential fill settlement. The ratio of height to width should be about 1.50 to accommodate the range of loading conditions economically. Where there is a battery of outlet gates, a multiple-box shape is sometimes economical where acceptable from a hydraulic standpoint.

13.2.1 SINGLE CONDUITS. For a single conduit of more than about 5.6-m2 (60-ft2) area and with a fill height over 18.3 m (60ft), it will generally be found economical to use a section other than rectangular for the embankment loading (Condition III). The circular shapes are more adaptable to changes in loadings and stresses that may be caused by unequal fill or foundation settlement. For cases in which the projection loading condition applies, no material stress reduction results from the provision of a variable cross section. These structures should be formed as shown in Figure 2-1 and should be analyzed as a ring of uniform thickness. While these sections show variations in thickness in the lower half of the conduit due to forming and other construction expedients, such variations may be disregarded in the design without appreciable error.

13.2.2 OBLONG SECTIONS. The oblong section shown in Figure 2-1 is formed by separating two semicircular sections by short straight vertical wall sections. The oblong section generally achieves maximum economy of materials by mobilizing more of the relieving fill pressure. The proportions should be selected carefully, and the tangent-length-to-radius ratio will usually be between 0.5 and 1.0. The conduit design should cover a range of possible loading conditions, from initial or construction condition to the long-time condition. Here also, a geologist or soils engineer should be consulted before final determination of the base shape of a conduit.

13.2.3 HORSESHOE SECTIONS. The "horseshoe" section in Figure 2-1 is generally less economical than the oblong and is therefore not often used. Its stress distribution is not as desirable as that of the circular or oblong section, and shear stirrups may be required in the base. It may be practicable, however, for some foundation conditions where the fill height is low.

13.2.4 INTERBEDDED FOUNDATIONS. It may be difficult to shape the foundation excavation when in closely bedded, flat-lying shale, or when in rock with frequent shale interbeds. For this condition, it may be economical to excavate the foundation level and backfill to the desired shape with a low-cement-content concrete. A geotechnical engineer should be consulted to help develop the excavation plan. Excavation drawings should show the pay excavation lines and not the actual excavation lines. For a conduit under a dam, the designer should show the actual excavation lines rather than the pay excavation lines and the contractor should limit excavation to the actual excavation lines.

13.3 MATERIALS

13.3.1 CONCRETE. Minimum compressive strength 28 MPa (4,000 psi) air entrained.

13.3.2 REINFORCEMENT. Minimum yield strength, Grade 400 MPa (60,000 psi).

13.3.3 INSTALLATION. Conduits through dams are cast directly against the soil or rock and, therefore, bedding is not a design consideration. When overexcavation of the foundation materials is required, concrete fill should be used to maintain proper conduit grade. All foundation materials for cast-in-place conduits should be reviewed by a geotechnical engineer.

13.4 LOADINGS. Typical conduit loads are shown in Figure 2-2. The conduit supports the weight of the soil and water above the crown. Internal and external fluid pressures and lateral soil pressures may be assumed as uniform loads along the horizontal axis of the conduit when the fluid head or fill height above the crown is greater than twice the conduit diameter or span. Foundation pressures are assumed to act uniformly across the full width of cast-in-place conduits. Uplift pressures should be calculated as uniform pressure at the base of the conduit when checking flotation.

13.4.1 GROUNDWATER AND SURCHARGE WATER. Because of the ratio of vertical to horizontal pressure, the most severe loading condition will generally occur when the reservoir is empty and the soil is in a natural drained condition. However, the following loads occur where there is groundwater and/or surcharge water.

13.4.1.1 VERTICAL PRESSURE. Use Equation 2-1 to determine vertical pressure due to the weight of the natural drained soil above the groundwater surface, the weight of the submerged soil below the groundwater surface, and the weight of the projected volume of water above the conduit, including any surcharge water above the fill surface.

$$W_w = \gamma_d\,H_d$$

or

$$
\begin{aligned}
W_w &= \gamma_d\,H_d + \gamma_s\,H_s \\
W_w &= \gamma_w\,H_w + (\gamma_s - \gamma_w)\,H_s
\end{aligned}
\tag{2-1}
$$

where

W_w = vertical pressure due to prism of soil above pipe, N/m^2 (psf)

γ = soil unit weight; d = dry, s = saturated, w = water, N/m^3 (pcf)

H = soil height; d = dry, s = saturated soil, m (ft)

H_w = water height above the point of interest, m (ft)

13.4.1.2 HORIZONTAL PRESSURE. Horizontal pressure from the lateral earth pressure is obtained by using soil weights for the appropriate moisture conditions and full hydrostatic pressure.

13.4.2 INTERNAL WATER PRESSURE. Internal water pressure should be considered but will seldom govern the design for the usual type of outlet works. However, internal pressures must be analyzed as indicated in Equation 2-2 for pressure conduits for interior drainage in local protection projects.

$$W_i = \gamma_w \, (H_G \pm r) \qquad\qquad (2\text{-}2)$$

where

W_i = internal pressure at point of interest, N/m^2 (psf)

γ_w = unit weight of water, 9.8 kN/m^3 (62.4 pcf)

H_G = hydraulic gradient above point of interest, m (ft)

r = inside radius of conduit, m (ft)

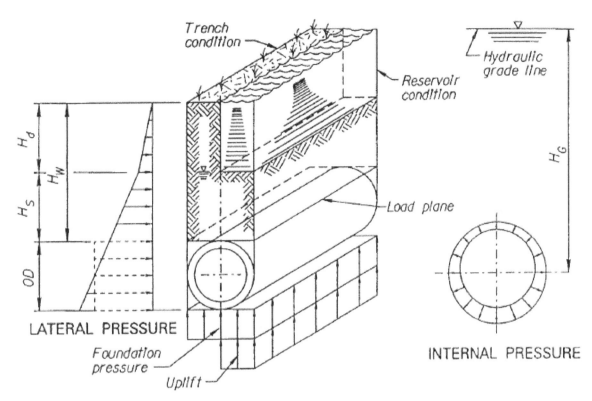

When the height of fill is greater than twice the trench width use
the average horizontal pressure computed at the pipe centerline

Figure 2-2

Typical conduit loadings

13.4.3 CONCENTRATED LIVE LOADS.

13.4.3.1 VERTICAL PRESSURE. Because soil conditions vary, designers can expect only a reasonable approximation when computing vertical pressures resulting from concentrated surface loads. The Boussinesq method is commonly used to convert surface point loads to vertical stress fields through the geometric relationship shown in Equation 2-3. This equation may be used for all types of soil masses including normally consolidated, overconsolidated, anisotropic, and layered soils. Stresses calculated by using this method are in close agreement with measured stress fields, and examples for using Equation 2-3 are shown in Figure 2-3.

$$W_c = \frac{3Pz^3}{2\pi R^5} \qquad (2\text{-}3)$$

where

W_c = vertical pressure due to concentrated load, N/m^2 (psf)

P = concentrated load, N (lb)

z = depth to pressure surface, m (ft)

R = radial distance to pressure surface, m (ft)

13.4.3.2 HORIZONTAL PRESSURE. Lateral loads caused by vehicles can be safely ignored due to their transient nature. However, a minimum lateral pressure of 0.005 of the wheel load for vehicles to a depth of 2.4 m (8 ft) should be considered in accordance with American Society for Testing and Materials (ASTM) C 857. For stationary surcharge loads, a lateral pressure can be calculated by using a Boussinesq equation such as Equation 2-4.

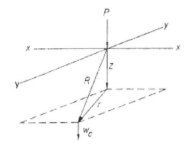

Given: $P = 44.5$ kN (10,000 lbs.)
$z = 0.92$ m (3 ft)
$r = 0.61$ m (2 ft)
$R = \sqrt{r^2 + z^2}$

$b = 0.52$ m (1.7 ft) radius of a circular area

$L = B = 0.92$ m (3 ft) dimension of a square area

$L =$ Long side of rectangular area

MINIMUM LATERAL PRESSURES

$p_e = 0.4\, w_c$ (AREA)
$p_e = 0.5$ or $1.0\, w_c$ (COE)

METHOD	FORMULA	LOAD PLANE PRESSURE	
		AT POINT Kpa (psf)	CENTER Kpa (psf)
BOUSSINEQ	$w_c = \frac{3Pz^3}{2\pi z R^5}$	10.3 (215)	25.1 (525)
SQUARE AREA	FIGURE 2-5	10.5 (220)	18.4 (384)
CIRCULAR	FIGURE 2-6	10.5 (220)	18.3 (383) 17.8 (372)
	$\frac{w_c}{w_o} = [1.0 - \frac{1}{[1+(b/z)^2]^{3/2}}]$		
SIMPLIFIED $\angle\overset{1}{\underset{}{}}^2$	$w_c = \frac{P}{(B + z)(L + z)}$	13.1 (275)	13.1 (275)
	$w_c = \frac{P}{(B + z)^2}$		
AASHTO $\frac{1}{1.75}^2$	Pyramid (2 ft) minimum cover	17.2 (359)	17.2 (359)

Figure 2-3

Typical live load stress distribution

$$p_c = \frac{P}{2\pi}\left(\frac{3zr^2}{R^5} - \frac{(1-2\mu)}{R(R+z)}\right) \qquad (2\text{-}4)$$

where

p_c = horizontal pressure from concentrated load, N/m^2 (psf)

r = surface radius from point load P, m (ft)

R = radial distance to point in question, m (ft)

μ = Poisson's ratio, 0.5 for saturated cohesive soils or 0.2 to 0.3 for other soils

Consult a geotechnical engineer for lateral loads from other surcharge conditions.

13.4.3.3 WHEEL LOADS. For relatively high fills, Equation 2-3 will give reasonably accurate results for highway and railroad wheel loads and the loads on relatively small footings. However, where the conduit is near the surface or where the contact area of the applied load is large, these loads must be divided into units for a more accurate analysis. The use of influence charts as developed by Newmark (1942) will be helpful in computing the stress due to loads on relatively large and irregular areas.

13.4.4 BACKFILL. The behavior of the soil pressures transmitted to a conduit or culvert by the overlying fill material is influenced by the physical characteristics and degree of compaction of the soil above and adjacent to the conduit or culvert as well as the degree of flexibility and the amount of settlement of the conduit or culvert. The effect of submergence in the backfill must also be considered as indicated in Figure 2-2. Direct measurements of such pressures have been made for small diameter pipes under relatively low fills. Until more data are available, the following loading should be used for rigid conduits and culverts for dams and levees and outlet conduits for interior drainage. The effect of submergence in the backfill must be considered. The three typical conduit installation conditions are trench, trench with superimposed fill, and embankment. Terms for these loading conditions are defined in Figure 2-4.

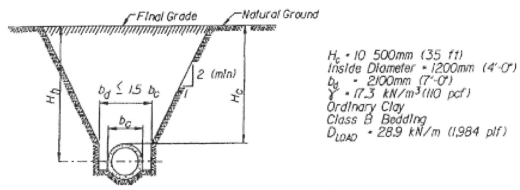

H_c = 10 500mm (35 ft)
Inside Diameter = 1200mm (4'-0")
b_d = 2100mm (7'-0")
γ = 17.3 kN/m³ (110 pcf)
Ordinary Clay
Class B Bedding
D_{LOAD} = 28.9 kN/m (1,984 plf)

TRENCH (CONDITION I)

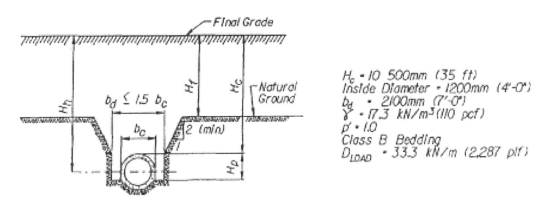

H_c = 10 500mm (35 ft)
Inside Diameter = 1200mm (4'-0")
b_d = 2100mm (7'-0")
γ = 17.3 kN/m³ (110 pcf)
p' = 1.0
Class B Bedding
D_{LOAD} = 33.3 kN/m (2,287 plf)

TRENCH WITH SUPERIMPOSED FILL (CONDITION II)

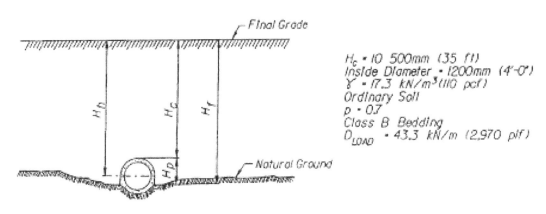

H_c = 10 500mm (35 ft)
Inside Diameter = 1200mm (4'-0")
γ = 17.3 kN/m³ (110 pcf)
Ordinary Soil
p = 0.7
Class B Bedding
D_{LOAD} = 43.3 kN/m (2,970 plf)

EMBANKMENT (CONDITION III)

Figure 2-4

Loading conditions for conduits

13.4.4.1 TRENCH WITH NO SUPERIMPOSED FILL (CONDITION I).

13.4.4.1.1 LOADS FROM THE TRENCH BACKFILL condition are applied to those structures that are completely buried in a trench with no superimposed fill above the top of the trench. To satisfy this condition, the width of the trench measured at the top of the conduit should be no greater than one and one-half times the overall width of the conduit, and the sides of the ditch above the top of the conduit should have a slope no flatter than one horizontal to two vertical. The total dead load of the earth at the top of the conduit should be computed as the larger of the two values obtained from Equations 2-5 through 2-7.

$$W_e = C_d \, \gamma \, B_d^2 \qquad\qquad (2\text{-}5)$$

or

$$W_e = \gamma \, B_c \, H \qquad\qquad (2\text{-}6)$$

$$C_d = \frac{1 - e^{-2 K\mu' \frac{H}{B_d}}}{2 K\mu'} \qquad\qquad (2\text{-}7)$$

where

W_e = total dead load of earth at top of conduit, N/m (lbf/ft)

C_d = trench coefficient, dimensionless

B_d = trench width at top of conduit < 1.5 b_c, m (ft)

B_c = outside diameter of conduit, m (ft)

H = variable height of fill, m (ft). When $H_c \geq 2B_d$, $H = H_h$. When $H_c < 2B_d$, H varies over the height of the conduit.

μ' = soil constant, dimensionless

Values for $K\mu'$ and C_d can be taken from Figure 2-5.

13.4.4.1.2 WHEN THE HEIGHT OF THE FILL ABOVE the top of the conduit (He) is less than twice the trench width, the horizontal pressure should be assumed to vary over the height of the conduit. When He is equal to or greater than 2sd• the horizontal pressure may be computed at the center of the conduit using an average value of H equal to H_h applied uniformly over the height of the conduit. When $H_e < 2B_d$, the horizontal pressure in N/m^2 (psf) at any depth should be computed using Equation 2-8.

$$p_e = \gamma H \tan^2 \left(45 - \frac{\phi}{2}\right) = K_a \gamma H \qquad (2\text{-}8)$$

where

p_e = horizontal earth pressure, N/m^2 (psf)

γ = unit weight of fill, N/m^3 (pcf)

ϕ = angle of internal friction of the fill material, degrees

K_a = active pressure coefficient, N (lb)

13.4.4.1.3 IN MOST CASES, the unit weight and the internal friction angle of the proposed backfill material in dry, natural drained, and submerged conditions should be determined by the laboratory and adapted to the design. However, where economic conditions do not justify the cost of extensive investigations by a soils laboratory, appropriate values of unit weight of the material and its internal friction angle should be determined by consultation with the soils engineer.

13.4.4.1.4 WHERE SUBMERGENCE and water surcharge are applicable, the loadings must be modified. To obtain the total vertical load, the weight of the projected volume of water above the conduit, including any surcharge water above the fill surface, is added to the larger value of we obtained by using the submerged weight of the material used in Equations 2-5 and 2-6. The horizontal pressure i~ obtained by adding the full hydrostatic pressure to the pressure found by Equation 2-8 using the submerged weight of material.

13.4.4.2 TRENCH WITH SUPERIMPOSED FILL (CONDITION II).

13.4.4.2.1 THIS LOADING condition applies to conduits that are completely buried in a trench with a superimposed fill H1 above the top of the trench. The trench width and side slopes have the same limitations as specified for the trench condition. The vertical and horizontal unit loads for this loading condition vary between the computed values for the Conditions I and III (trench and embankment conditions) in proportion to the ratio $H_f (H_c + H_p)$. The vertical load, in N/m (pounds per foot) of conduit length, for the Condition II (trench with superimposed fill) should be computed as the larger of the two values obtained from Equations 2-9 and 2-10.

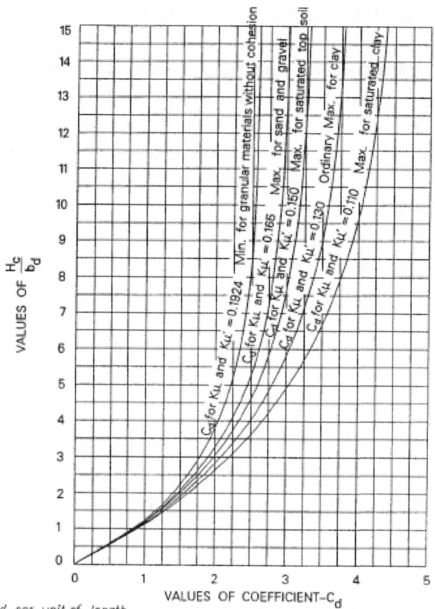

Load per unit of length,
$W_e = C_d \gamma b_d$ [M,KN²(ff. lb.)]
γ = unit weight of fill materials.
b_d = breadth of ditch at top of structure
H_c = height of fill over top of structure

u = the "coefficient of internal friction"
in the fill materials, abstract number.
u' = the "coefficient of sliding friction"
between the fill materials and the
sides of the ditch, abstract number.

$$K = \frac{\sqrt{u^2+1} - u}{\sqrt{u^2+1} + u}$$

Figure 2-5.

Earth loads trench condition

$$W_e = C_d \gamma b_d^2 +$$

$$\left(\frac{H_f}{H_c + H_p}\right)(1.5 \; \gamma b_c H_h - C_d \gamma b_d^2) \quad\quad (2\text{-}9)$$

or

$$W_e = \gamma b_c H_h +$$

$$\left(\frac{H_f}{H_c + H_p}\right)(1.5 \; \gamma b_c H_h - \gamma b_c H_h) \quad\quad (2\text{-}10)$$

where

γ = unit weight of fill, N/m^3 (pcf)

b_d = trench width, m (ft), $b_d \leq = 1.5 \; b_c$

H_f = height of superimposed fill above the top of the trench, m (ft)

H_c = height of fill above top of conduit, m (ft)

H_p = height of conduit above level adjacent foundation, m (ft)

b_c = outside dimension of conduit, m (ft)

H_h = height of fill above horizontal diameter of conduit, m (ft)

13.4.4.2.2 FOR LOW FILLS IT MAY be desirable to use an effective height slightly less than H_h. The horizontal pressure for Condition II loading IS determined using Equation 2-11.

$$p_e = \gamma H \tan^2\left(45° - \frac{\phi}{2}\right) + \left(\frac{H_f}{H_c + H_p}\right) \qquad (2\text{-}11)$$
$$\left[0.5\ \gamma H - \gamma H \tan^2\left(45° - \frac{\phi}{2}\right)\right]$$

where

$$H = \text{variable height of fill above conduit, m (ft)}$$

13.4.4.2.3 FOR LOADING CASES with submergence and water surcharge, the horizontal and vertical earth pressures should be similarly proportioned between the results obtained for Conditions I and III (trench and embankment conditions) with surcharge added to the hydrostatic pressure.

13.4.4.3 EMBANKMENTS (CONDITION III).

13.4.4.3.1 CONDITION III applies to conduits and culverts that project above an embankment subgrade and to conduits and culverts in ditches that do not satisfy the requirements of Condition I or II. For this condition, the design should cover a range of possible loading conditions from the initial condition to the long-time condition by satisfying two extreme cases: Case 1, with $p_e/W_e = 0.33$ ($W_e = 150$ percent vertical projected weight of fill material, lateral earth pressure coefficient $k = 0.50$); and Case 2, $p_e/W_e = 1.00$ ($W_e = 100$ percent vertical projected weight of fill material, $k = 1.00$). The total vertical load in N/m (lbf/ft) for this condition should be computed as shown in Equations 2-12 and 2-13:

For Case 1, $W_e = 1.5\ \gamma b_c H_h$ 　　　　　　　(2-12)

For Case 2, $W_e = \gamma b_c H_h$ 　　　　　　　(2-13)

or the unit vertical load N/m^2 (psf), W_e, as given by Equations 2-14 and 2-15:

For Case 1, $W_e = 1.5\ \gamma\ H_h$ 　　　　　　　(2-14)

For Case 2, $W_e = \gamma\ H_h$ 　　　　　　　(2-15)

The horizontal loading N/m^2 (psf) should be taken as shown in Equations 2-16 and 2-17:

For Case 1, $p_e = 0.5\ \gamma\ H$ 　　　　　　　(2-16)

For Case 2, $p_e = \gamma\ H$ 　　　　　　　(2-17)

Normal allowable working stresses should apply for both Case 1 and Case 2.

13.4.4.3.2 WHERE SUBMERGENCE and water surcharge are applicable, their effects must be considered as for Condition I. In such cases, the vertical load as computed by Equations 2-12 through 2-17, using the submerged weight of the material should be increased by the weight of the projected volume of water above the conduit including any surcharge water above the fill surface. When a clay blanket is applied to the face of the darn, the weight of water above the blanket must be included but the soil weight below the blanket and above the phreatic line (or the line of saturation where capillarity exists) is that for the natural drained condition. The horizontal unit pressure is found by adding full hydrostatic pressure to the value of p, obtained from Equation 2-16 or 2-17 using the submerged weight of the material.

13.5 SPECIAL CONDITIONS

13.5.1 GENERAL. If conditions are encountered that warrant deviation from the loading criteria discussed above, justification for the change should be submitted with the analysis of design. However, the designer must first select the most economical method of installation. Where the rock surface occurs above the elevation of the bottom of the conduit, the designer should investigate the relative costs of excavating away from the conduit and backfilling between the conduit and the excavation line, allowing sufficient space between the conduit and the excavation line for operation of compaction rollers, and placing the conduit directly against rock as indicated for the following conditions.

13.5.2 WALLS CAST AGAINST ROCK. Where the conduit walls are placed directly against rock and the rock surface is at or above the top of the crown, the soil weight should be taken as 1.0 times the weight of material above, rather than 1.5, and the lateral pressure should be hydrostatic only, where applicable. Where the rock surface is at an intermediate level between crown and invert, use judgment to select a value between 1.0 and 1.5 to multiply by the weight of material above to obtain the correct soil design load. Lateral soil pressure should be applied only above the rock level and hydrostatic pressure as applicable over the full height of conduit. For either of these cases, the condition with no hydrostatic pressure should also be considered.

13.6 METHODS OF ANALYSIS. Cast-in-place conduits can be designed using simplified elastic analysis or with finite element codes. Specialized finite element codes are available that feature nonlinear soil elements. These specialized codes provide the most accurate analysis. If these codes are not available, general finite element codes can be used, but they may need to be calibrated to the actual soil conditions. The finite element approach lends itself to parametric studies for rapid analysis of various foundation, bedding, and compaction conditions. Consult a geotechnical engineer for determination of soil spring constants to be used in the finite element model. Both concrete thickness and reinforcing steel area should be varied to obtain the best overall economy.

13.6.1 FINITE ELEMENT ANALYSIS. Finite element analysis is a useful method to design sections with unique shape for various field stresses. This method can be used to approximate the soil-structure interaction using spring foundations and friction between elements. These models calculate flexure and shear loads on the design section directly from soil-structure interaction relationships. The design of reinforcement for flexure and shear should be in accordance with the technical literature. When the inside face steel is in tension, the area of steel needs to be limited to reduce the effects of radial tension. Therefore, limits on the amount of inside face steel that can be developed are necessary to prevent interior face concrete spalls or "slabbing failures." If more steel is required to develop the flexural capacity of the section, use radial ties. They should be designed in accordance with American Concrete Institute (ACI) 318 for shear reinforcement.

13.6.2 CURVILINEAR CONDUITS AND CULVERTS (CURCON). This Computer-Aided Structural Engineering (CASE) program performs a structural analysis for conduit shapes including horseshoe, arch, modified oblong, and oblong sections with constant thickness, base fillets, or a square base. Loads that can be analyzed include groundwater and surcharge water in embankment backfills.

13.7 REINFORCEMENT

13.7.1 MINIMUM LONGITUDINAL. Longitudinal reinforcement should be placed in both faces of the conduit as shown in Figure 2-6. The minimum required area of reinforcement should not be less than 0.0028 times the gross area of concrete, half in each face, with a maximum of #30M at 300 mm (#9 at 12 in.) in each face. Generally, the same reinforcement will be in each face. Maximum spacing of bars should not exceed 450 mm (18 in).

13.7.2 MINIMUM TRANSVERSE. Minimum transverse reinforcement should be placed in both inside and outside faces. Minimum required area of transverse steel, even when not carrying computed stresses, should not be less than 0.002 times the nominal area of concrete in each face, but not more than #25M at 300 mm (#8 at 12 in.) in each face, unless required to carry the computed stresses. Compression reinforcement in excess of this minimum should not be used.

13.7.3 MINIMUM COVER. Minimum concrete cover of reinforcement should not be less than 100 mm (4 in.).

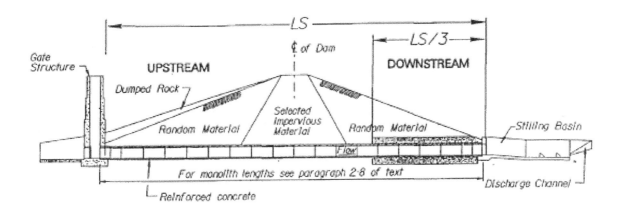

RESERVOIR OUTLET WORKS–LONGITUDINAL SECTION THROUGH CONDUIT ON ROCK

Notes:
1. Conduit strength should vary roughly in accordance with height of overburden or other loading conditions so the overall structure will have essentially a constant safety factor throughout its length. Prefabricated conduit can usually be varied for strength class commercially available. For cast-in-place conduit both concrete thickness and reinforcing steel area should be varied to obtain the best overall economy.
2. The "Corps EM 1110-2-2102, Waterstops and other Joint Materials", illustrates various shapes of rubber and polyvinylchloride commercially available.

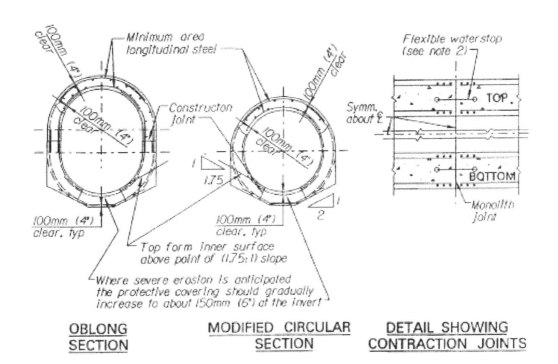

OBLONG SECTION

MODIFIED CIRCULAR SECTION

DETAIL SHOWING CONTRACTION JOINTS

Figure 2-6

Typical conduit details (large dams)

13.8 JOINTS

13.8.1 TRANSVERSE MONOLITH JOINTS. Maximum contraction joint spacing should not exceed 6 m (20 ft) on earth foundations and 9 m (30 ft) on rock, as shown in Figure 2-6. When large settlements are expected, these maximum spacings should be reduced to allow for more movement in the joint. A geotechnical engineer should be consulted for soil settlements.

13.8.2 LONGITUDINAL CONSTRUCTION JOINTS. The position of the longitudinal construction joints indicated in Figure 2-6 can be varied to suit the construction methods used. When circular and oblong conduits are used, the concrete in the invert section should be top-formed above the point where the tangent to the invert is steeper than 1 vertical on 1.75 horizontal.

13.9 WATERSTOPS. Flexible-type waterstops should be used in all transverse contraction joints, as shown in Figure 2-6. Where large differential movement is expected, a center-bulb-type waterstop and a joint separation of approximately 13 mm (112 in.) should be used. When the conduit rests on a rather firm foundation, a two-bulb or equivalent type waterstop should be used with a joint separation of approximately 6 mm (1/4 in.). For conduit on rock foundations with little expected deformation, the joint should be coated with two coats of mastic and an appropriate waterstop should be used.

13.10. CAMBER. When conduits are cast-in-place, large settlements are usually not a major concern. However, where considerable foundation settlements are likely to occur, camber should be employed to ensure positive drainage.

CHAPTER 14: ANALYSIS OF CONCRETE GRAVITY DAMS

14.1 STABILITY ANALYSIS

14.1.1 INTRODUCTION

14.1.1.1 THIS DISCUSSION PRESENTS information on the stability analysis of concrete gravity dams. The basic loading conditions investigated in the design and guidance for the dam profile and layout are discussed. The forces acting on a structure are determined as outlined.

14.1.1.2 FOR NEW PROJECTS, the design of a gravity dam is performed through an interative process involving a preliminary layout of the structure followed by a stability and stress analysis. If the structure fails to meet criteria then the layout is modified and reanalyzed. This process is repeated until an acceptable cross section is attained.

14.1.1.3 ANALYSIS OF THE STABILITY and calculation of the stresses are generally conducted at the dam base and at selected planes within the structure. If weak seams or planes exist in the foundation, they should also be analyzed.

14.1.2 BASIC LOADING CONDITIONS

14.1.2.1 THE FOLLOWING BASIC LOADING conditions are generally used in concrete gravity dam designs (see Figure 4-1). Loadings that are not indicated should be included where applicable. Power intake sections should be investigated with emergency bulkheads closed and all water passages empty under usual loads.

14.1.2.1.1 LOAD CONDITION NO. 1 - unusual loading condition - construction.
- Dam structure completed.
- No headwater or tailwater.

14.1.2.1.2 LOAD CONDITION NO. 2 - usual loading condition - normal operating.

- Pool elevation at top of closed spillway gates where spillway is gated, and at spillway crest where spillway is ungated.
- Minimum tailwater.
- Uplift.
- Ice and silt pressure, if applicable.

14.1.2.1.3 LOAD CONDITION NO. 3 - unusual loading condition - flood discharge.

- Pool at standard project flood (SPF).
- Gates at appropriate flood-control openings and
- tailwater at flood elevation.
- Tailwater pressure.
- Uplift.
- Silt, if applicable.
- No ice pressure.

14.1.2.1.4 LOAD CONDITION NO. 4 - extreme loading condition - construction with operating basis earthquake (OBE).

- Operating basis earthquake (OBE).
- Horizontal earthquake acceleration in upstream direction.
- No water in reservoir.
- No headwater or tailwater.

14.1.2.1.5 LOAD CONDITION NO. 5 - Unusual loading condition - normal operating with operating basis earthquake.

- Operating basis earthquake (OBE).
- Horizontal earthquake acceleration in downstream direction.
- Usual pool elevation.

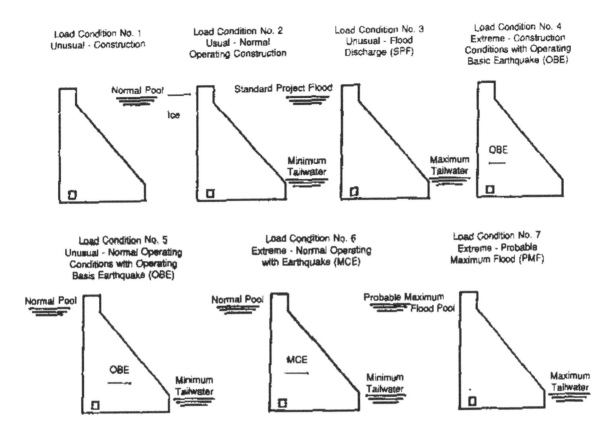

Figure 4-1

Basic loading conditions in concrete gravity dam design

- Minimum tailwater.

- Uplift at pre-earthquake level.

- Silt pressure, if applicable.

- No ice pressure.

14.1.2.1.6 LOAD CONDITION NO. 6 - extreme loading condition - normal operating with maximum credible earthquake.

- Maximum credible earthquake (MCE).

- Horizontal earthquake acceleration in downstream direction.

- Usual pool elevation.

- Minimum tailwater.

- Uplift at pre-earthquake level.

- Silt pressure, if applicable.

- No ice pressure.

14.1.2.1.7 LOAD CONDITION NO. 7 - Extreme loading condition - probable maximum flood.

- Pool at probable maximum flood (PMF).

- All gates open and tailwater at flood elevation.

- Uplift.

- Tailwater pressure.

- Silt, if applicable.

- No ice pressure.

14.1.2.2 IN LOAD CONDITION NOS. 5 AND 6, the selected pool elevation should be the one judged likely to exist coincident with the selected design earthquake event. This means that the pool level occurs, on the average, relatively frequently during the course of the year.

14.1.3 DAM PROFILES

14.1.3.1 NONOVERFLOW SECTION.

14.1.3.1.1 THE CONFIGURATION OF THE nonoverflow section is usually determined by finding the optimum cross section that meets the stability and stress criteria for each of the loading conditions. The design cross section is generally established at the maximum height section and then used along the rest of the nonoverflow dam to provide a smooth profile. The upstream face is generally vertical, but may include a batter to increase sliding stability or in existing projects provided to meet prior stability criteria for construction requiring the resultant to fall within the middle third of the base. The downstream face will usually be a uniform slope transitioning to a vertical face near the crest. The slope will usually be in the range of 0.7H to 1V, to 0.8H to 1V, depending on uplift and the seismic zone, to meet the stability requirements.

14.1.3.1.2 IN THE CASE OF RCC DAMS not using a downstream forming system, it is necessary for construction that the slope not be steeper than 0.8H to 1V and that in applicable locations, it include a sacrificial concrete because of the inability to achieve good compaction at the free edge. The thickness of this sacrificial material will depend on the climatology at the project and the overall durability of the mixture. The weight of this material should not be included in the stability analysis. The upstream face will usually be vertical to facilitate construction of the facing elements. When overstressing of the foundation material becomes critical, constructing a uniform slope at the lower part of the downstream face may be required to reduce foundation pressures. In locations of slope changes, stress concentrations will occur. Stresses should be analyzed in these areas to assure they are within acceptable levels.

14.1.3.1.3 THE DAM CREST SHOULD have sufficient thickness to resist the impact of floating objects and ice loads and to meet access and roadway requirements. The freeboard at the top of the dam will be determined by wave height and runup. In significant seismicity areas, additional concrete near the crest of the dam results in stress increases. To reduce these stress concentrations, the crest mass should be kept to a minimum and curved transitions provided at slope changes.

14.1.3.2 OVERFLOW SECTION. The overflow or spillway section should be designed in a similar manner as the nonoverflow section, complying with stability and stress criteria. The upstream face of the overflow section will have the same configuration as the nonoverflow section. The required downstream face slope is made tangent to the exponential curve of the crest and to the curve at the junction with the stilling basin or flip bucket. Piers may be included in the overflow section to support a bridge crossing the spillway and to support spillway gates. Regulating outlet conduits and gates are generally constructed in the overflow section.

14.1.4 STABILITY CONSIDERATIONS

14.1.4.1 GENERAL REQUIREMENTS. The basic stability requirements for a gravity dam for all conditions of loading are:

14.1.4.1.1 THAT IT BE SAFE AGAINST OVERTURNING at any horizontal plane within the structure, at the base, or at a plane below the base.

14.1.4.1.2 THAT IT BE SAFE AGAINST SLIDING on any horizontal or near-horizontal plane within the structure at the base or on any rock seam in the foundation.

14.1.4.1.3 THAT THE ALLOWABLE UNIT STRESSES in the concrete or in the foundation material shall not be exceeded. Characteristic locations within the dam in which a stability criteria check should be considered include planes where there are dam section changes and high concentrated loads. Large galleries and openings within the structure and upstream and downstream slope transitions are specific areas for consideration.

14.1.4.2 STABILITY CRITERIA. The stability criteria for concrete gravity dams for each load condition are listed in Table 4-1. The stability analysis should be presented in the design memoranda in a form similar to that shown on Figure 4-1. The seismic coefficient method of analysis, as outlined in Chapter 3, should be used to determine resultant location and sliding stability for the earthquake load conditions. The seismic coefficient used in the analysis should be no less than that given by code. Any deviation from the criteria in Table 4-1 shall be accomplished only with the approval of the appropriate authority, and should be justified by comprehensive foundation studies of such nature as to reduce uncertainties to a minimum.

14.1.5 OVERTURNING STABILITY

14.1.5.1 RESULTANT LOCATION. The overturning stability is calculated by applying all the vertical forces (SV) and lateral forces for each loading condition to the dam and, then, summing moments (SM) caused by the consequent forces about the downstream toe. The resultant location along the base is:

Load Condition	Resultant Location at Base	Minimum Sliding FS	Foundation Bearing Pressure	Concrete Stress	
				Compressive	Tensile
Usual	Middle 1/3	2.0	≤ allowable	0.3 f_c'	0
Unusual	Middle 1/2	1.7	≤ allowable	0.5 f_c'	0.6 $f_c'^{2/3}$
Extreme	Within base	1.3	≤ 1.33 × allowable	0.9 f_c'	1.5 $f_c'^{2/3}$

Note: f_c' is 1-year unconfined compressive strength of concrete. The sliding factors of safety (FS) are based on a comprehensive field investigation and testing program. Concrete allowable stresses are for static loading conditions.

$$Resultant\ location = \frac{\Sigma M}{\Sigma V} \qquad (4\text{-}1)$$

14.1.5.2 CRITERIA. When the resultant of all forces acting above any horizontal plane through a dam intersects that plane outside the middle third, a non-compression zone will result. The relationship between the base area in compression and the location of the resultant is shown in Figure 4-2. For usual loading conditions, it is generally required that the resultant along the plane of study remain within the middle third to maintain compressive stresses in the concrete. For unusual loading conditions, the resultant must remain within the middle half of the base. For the extreme load conditions, the resultant must remain sufficiently within the base to assure that base pressures are within prescribed limits.

14.1.6 SLIDING STABILITY

14.1.6.1 GENERAL. The sliding stability is based on a factor of safety (FS) as a measure of determining the resistance of the structure against sliding. The multiple-wedge analysis is used for analyzing sliding along the base and within the foundation.

14.1.6.2 DEFINITION OF SLIDING FACTOR OF SAFETY.

14.1.6.2.1 THE SLIDING FS IS CONCEPTUALLY related to failure, the ratio of the shear strength (τ_F), and the applied shear stress (τ) along the failure planes of a test specimen according to Equation 4-2:

$$FS = \frac{\tau_F}{\tau} = \frac{(\sigma \tan \phi + c)}{\tau} \tag{4-2}$$

where $t_F = s \tan f + c$, according to the Mohr-Coulomb Failure Criterion (Figure 4-3). The sliding FS is applied to the material strength parameters in a manner that places the forces acting on the structure and rock wedges in sliding equilibrium.

14.1.6.2.2 THE SLIDING FS IS DEFINED AS THE RATIO of the maximum resisting shear (TF) and the applied shear (T) along the slip plane at service conditions:

$$FS = \frac{T_F}{T} = \frac{(N \tan \phi + cL)}{T} \tag{4-3}$$

where

N = resultant of forces normal to the assumed sliding plane

ϕ = angle of internal friction

c = cohesion intercept

L = length of base in compression for a unit strip of dam

14.1.6.2.3 LIMIT EQUILIBRIUM. Sliding stability is based on a limit equilibrium method. By this method, the shear force necessary to develop sliding equilibrium is determined for an assumed failure surface. A sliding mode of failure will occur along the presumed failure surface when the applied shear (T) exceeds the resisting shear (T_F).

14.1.6.2.4 FAILURE SURFACE. The analyses are based on failure surfaces that can be any combination of planes and curves; however, for simplicity all failure surfaces are assumed to be planes. These planes form the bases of the wedges. It should be noted that for the analysis to be realistic, the assumed failure planes have to be kinematically possible. In rock the slip planes may be predetermined by discontinuities in the foundation. All the potential planes of failure must be defined and analyzed to determine the one with the least FS.

14.1.6.3 TWO-DIMENSIONAL ANALYSIS. The principles presented for sliding stability are based on a two-dimensional analysis. These principles should be extended to a three dimensional analysis if unique three-dimensional geometric features and loads critically affect the sliding stability of a specific structure.

14.1.6.4 FORCE EQUILIBRIUM ONLY. Only force equilibrium is satisfied in the analysis. Moment equilibrium is not used. The shearing force acting parallel to the interface of any two wedges is assumed to be negligible; therefore, the portion of the failure surface at the bottom of each wedge is loaded only by the forces directly above or below it. There is no interaction of vertical effects between the wedges. The resulting wedge forces are assumed horizontal.

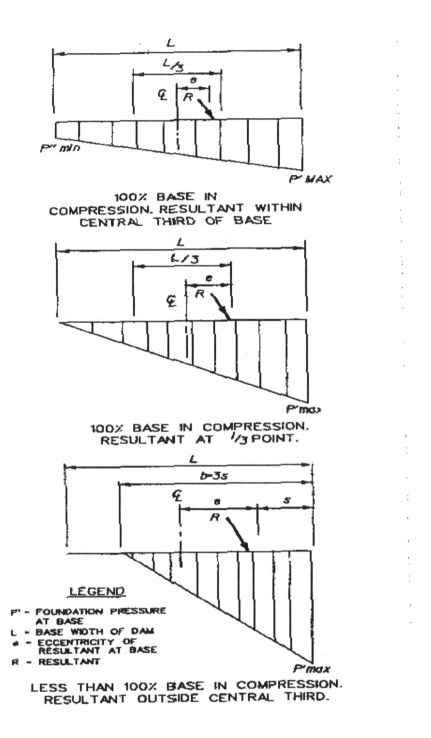

Figure 4-2

Relationship between base area in compression and resultant location

14.1.6.5 DISPLACEMENTS. Considerations regarding displacements are excluded from the limit equilibrium approach. The relative rigidity of different foundation materials and the concrete structure may influence the results of the sliding stability analysis. Such complex structure-foundation systems may require a more intensive sliding investigation than a limit-equilibrium approach. The effects of strain compatibility along the assumed failure surface may be approximated in the limit equilibrium approach by selecting the shear strength parameters from in situ or laboratory tests according to the failure strain selected for the stiffest material. (6) Relationship between shearing and normal forces. A linear relationship is assumed between the resisting shearing force and the normal force acting on the slip plane beneath each wedge. The Coulomb-Mohr Failure Criterion defines this relationship.

14.1.6.6 MULTIPLE WEDGE ANALYSIS.

14.1.6.6.1 GENERAL. This method computes the sliding FS required to bring the sliding mass, consisting of the structural wedge and the driving and resisting wedges, into a state of horizontal equilibrium along a given set of slip planes.

14.1.6.6.2 ANALYSIS MODEL. In the sliding stability analysis, the gravity dam and the rock and soil acting on the dam are assumed to act as a system of wedges. The dam foundation system is divided into one or more driving wedges, one structural wedge, and one or more resisting wedges, as shown in Figures 4-4 and 4-5.

14.1.6.6.3 GENERAL WEDGE EQUATION. By writing equilibrium equations normal and parallel to the slip plane, solving for N_i and T_i, and substituting the expressions for N_i and T_i into the equation for the factor of safety of the typical wedge, the general wedge and wedge interaction equation can be written as shown in Equation 4-5.

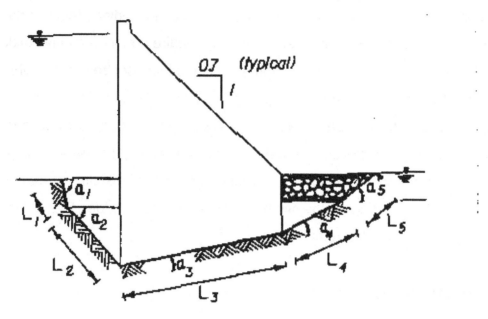

Figure 4-4

Geometry of structure foundation system

$$FS = \left\{ \left[(W_i + V_i) \cos \alpha_i + (H_{Li} - H_{Ri}) \sin \alpha_i + (P_{i-1} - P_i) \sin \alpha_i - U_i \right] \tan \phi_i \right.$$

$$\left. + C_i L_i \right\} / \left[(H_{Li} - H_{Ri}) \cos \alpha_i + (P_{i-1} - P_i) \cos \alpha_i - (W_i + V_i) \sin \alpha_i \right]$$

(4-5)

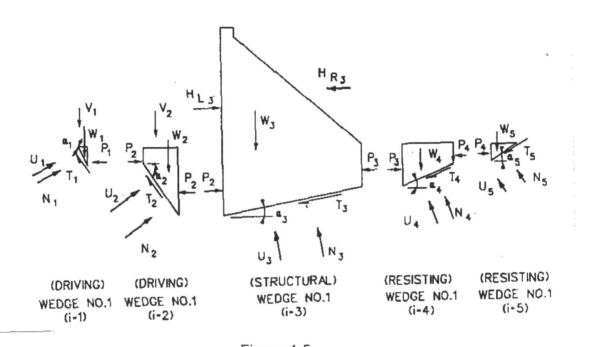

Figure 4-5

Dam foundation system, showing driving, structural, and resisting wedges

Solving for $(P_{i-1} - P_i)$ gives the general wedge equation.

$$(P_{i-1} - P_i) = \left[(W_i + V_i)(\tan \phi_{di} \cos \alpha_i + \sin \alpha_i) - U_i \tan \phi_{di} (H_{Li} - H_{Ri}) \right.$$
$$\left. (\tan \phi_{di} \sin \alpha_i - \cos \alpha_i) + c_{di} L_i \right] / (\cos \alpha_i - \tan \phi_{di} \sin \alpha_i)$$

(4-6)

where

i = number of wedge being analyzed

$(P_{i-1} - P_i)$ = summation of applied forces acting horizontally on the ith wedge. (A negative value for this term indicates that the applied forces acting on the ith wedge exceed the forces resisting sliding along the base of the wedge. A positive value for the term indicates that the applied forces acting on the ith wedge are less than the forces resisting sliding along the base of that wedge.)

W_i = total weight of water, soil, rock, or concrete in the ith wedge

V_i = any vertical force applied above top of ith wedge

$\tan \phi_{di}$ = $\tan \phi_i / FS$

α_i = angle between slip plane of ith wedge and horizontal. Positive is counterclockwise

U_i = uplift force exerted along slip plane of the ith wedge

H_{Li} = any horizontal force applied above top or below bottom of left side adjacent wedge

H_{Ri} = any horizontal force applied above top or below bottom of right side adjacent wedge

c_{di} = c_i / FS

L_i = length along the slip plane of the ith wedge

This equation is used to compute the sum of the applied forces acting horizontally on each wedge for an assumed FS. The same FS is used for each wedge.

14.1.6.6.4 FAILURE PLANE ANGLE. For the initial trial, the failure plane angle alpha for a driving wedge can be approximated by:

$$\alpha = 45° + \frac{\phi_d}{2}$$

$$\text{where} \quad \phi_d = \tan^{-1}\left(\frac{\tan \phi}{FS}\right)$$

For a resisting wedge, the slip plane angle can be approximated by:

$$\alpha = 45° - \frac{\phi_d}{2}$$

These equations for the slip plane angle are the exact solutions for wedges with a horizontal top surface with or without a uniform surcharge.

14.1.6.6.5 PROCEDURE FOR A MULTIPLE-wedge analysis. The general procedure for analyzing multi-wedge systems includes:

14.1.6.6.5.1 ASSUMING A POTENTIAL failure surface based on the stratification, location and orientation, frequency and distribution of discontinuities of the foundation material, and the configuration of the base.

14.1.6.6.5.2 DIVIDING THE ASSUMED slide mass into a number of wedges, including a single-structure wedge.

14.1.6.6.5.3 DRAWING FREE BODY DIAGRAMS that show all the forces assuming to be acting on each wedge.

14.1.6.6.5.4 ESTIMATE THE FS FOR THE FIRST TRIAL.

14.1.6.6.5.5 COMPUTE THE CRITICAL SLIDING angles for each wedge. For a driving wedge, the critical angle is the angle that produces a maximum driving force. For a resisting wedge, the critical angle is the angle that produces a minimum resisting force.

14.1.6.6.5.6 COMPUTE THE UPLIFT PRESSURE, if any, along the slip plane. The effects of seepage and foundation drains should be included.

14.1.6.6.5.7 COMPUTE THE WEIGHT of each wedge, including any water and surcharges.

14.1.6.6.5.8 COMPUTE THE SUMMATION of the lateral forces for each wedge using the general wedge equation. In certain cases where the loadings or wedge geometries are complicated, the critical angles of the wedges may not be easily calculated. The general wedge equation may be used to iterate and find the critical angle of a wedge by varying the angle of the wedge to find a minimum resisting or maximum driving force.

14.1.6.6.5.9 SUM THE LATERAL FORCES FOR ALL THE WEDGES.

14.1.6.6.510 IF THE SUM OF THE LATERAL FORCES is negative, decrease the FS and then recompute the sum of the lateral forces. By decreasing the FS, a greater percentage of the shearing strength along the slip planes is mobilized. If the sum of the lateral forces is positive, increase the FS and recompute the sum of the lateral forces. By increasing the FS, a smaller percentage of the shearing strength is mobilized.

14.1.6.6.5.11 CONTINUE THIS TRIAL AND ERROR process until the sum of the lateral forces is approximately zero for the FS used. This procedure will determine the FS that causes the sliding mass in horizontal equilibrium, in which the sum of the driving forces acting horizontally equals the sum of the resisting forces that act horizontally.

14.1.6.6.5.12 IF THE FS IS LESS THAN THE minimum criteria, a redesign will be required by sloping or widening the base. e. Single-plane failure surface. The general wedge equation reduces to Equation 4-7 providing a direct solution for FS for sliding of any plane within the dam and for structures defined by a single plane at the interface between the structure and foundation material with no embedment. Figure 4-6 shows a graphical representation of a single-plane failure mode for sloping and horizontal surfaces.

$$FS = \frac{[W \cos \alpha - U + H \sin \alpha] \tan \phi + CL}{H \cos \alpha - W \sin \alpha} \quad (4\text{-}7)$$

where

H = horizontal force applied to dam

C = cohesion on slip plane

L = length along slip plane

For the case of sliding through horizontal planes, generally the condition analyzed within the dam, Equation 4-7 reduces to Equation 4-8:

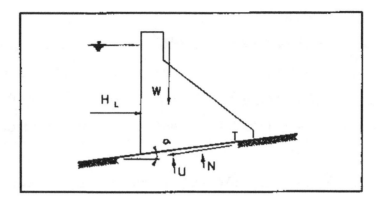

a. Upslope sliding, α > 0

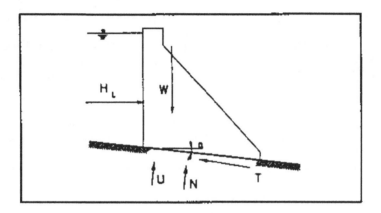

b. Downslope slid

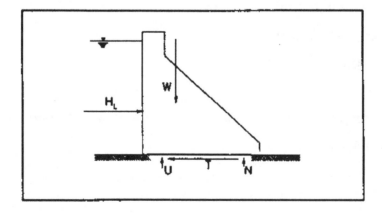

c. Horizontal sliding, α = 0

Figure 4-6

Single plane failure mode

452

$$FS = \frac{(W - U)\tan\phi + CL}{H_L} \qquad (4\text{-}8)$$

14.1.6.6 DESIGN CONSIDERATIONS.

14.1.6.6.1 DRIVING WEDGES. The interface between the group of driving wedges and the structural wedge is assumed to be a vertical plane that is located at the heel of the structural wedge and extends to its base. The magnitudes of the driving forces depend on the actual values of the safety factor and the inclination angles of the slip path. The inclination angles, corresponding to the maximum active forces for each potential failure surface, can be determined by independently analyzing the group of driving wedges for a trial safety factor. In rock, the inclination may be predetermined by discontinuities in the foundation. The general equation applies directly only to driving wedges with assumed horizontal driving forces.

14.1.6.6.2 STRUCTURAL WEDGE. The general wedge equation is based on the assumption that shearing forces do not act on the vertical wedge boundaries; hence there can be only one structural wedge because concrete structures transmit significant shearing forces across vertical internal planes. Discontinuities in the slip path beneath the structural wedge should be modeled by assuming an average slip plane along the base of the structural wedge.

14.1.6.6.3 RESISTING WEDGES. The interface between the group of resisting wedges and the structural wedge is assumed to be a vertical plane that is located at the toe of the structural wedge and extends to its base. The magnitudes of the resisting forces depend on the actual values of the safety factor and the inclination angles of the slip path. The inclination angles, corresponding to the minimum passive forces for each potential failure mechanism, can be determined by independently analyzing the group of resisting wedges for a trial safety factor. The general wedge equation applies directly only to resisting wedges with assumed horizontal passive forces. If passive resistance is used, then rock that may be subjected to high velocity water scouring should not be used unless adequately protected. Also, the compressive strength of the rock layers must be sufficient to develop the wedge resistance. In some cases, wedge resistance should not be included unless rock anchors are installed to stabilize the wedge.

14.1.6.6.4 EFFECTS OF CRACKS IN FOUNDATION. Sliding analyses should consider the effects of cracks on the driving side of the structural wedge in the foundation material resulting from differential settlement, shrinkage, or joints in a rock mass. The depth of cracking in massive strong rock foundations should be assumed to extend to the base of the structural wedge. Shearing resistance along the crack should be ignored, and full hydrostatic pressure should be assumed to act at the bottom of the crack. The hydraulic gradient across the base of the structural wedge should reflect the presence of a crack at the heel of the structural wedge.

14.1.6.6.5 UPLIFT. The effects of uplift forces should be included in the sliding analysis. Uplift pressures on the wedges and within any plane within the structure should be determined as described.

14.1.6.6.6 RESULTANT OUTSIDE KERN. As previously stated, requirements for rotational equilibrium are not directly included in the general wedge equation. For some load cases, the normal component of the resultant applied loads will lie outside the kern of the base area, and not all of the structural wedge will be in contact with the foundation material. The sliding analysis should be modified for these load cases to reflect the following secondary effects due to coupling of the sliding and rational behavior.

14.1.6.6.6.1 THE UPLIFT PRESSURE ON THE portion of the base not in contact with the foundation material should be a uniform value that is equal to the maximum value of the hydraulic pressure across the base (except for instantaneous load cases such as those resulting from seismic forces).

14.1.6.6.6.2 THE COHESIVE COMPONENT OF the sliding resistance should include only the portion of the base area in contact with the foundation material.

14.1.6.7 SEISMIC SLIDING STABILITY. The sliding stability of a structure for an earthquake-induced base motion should be checked by assuming the specified horizontal earthquake and the vertical earthquake acceleration, if included in the analysis, to act in the most unfavorable direction. The earthquake-induced forces on the structure and foundation wedges may then be determined by the seismic coefficient method as outlined.

14.1.6.8 STRAIN COMPATIBILITY. Shear resistance in a dam foundation is dependent on the strength properties of the rock. Slide planes within the foundation rock may pass through different materials, and these surfaces may be either through intact rock or along existing rock discontinuities. Less deformation is required for intact rock to reach its maximum shear resistance than for discontinuity surfaces to develop their maximum frictional resistances. Thus, the shear resistance developed along discontinuities depends on the amount of displacement on the intact rock part of the shear surface. If the intact rock breaks, the shear resistance along the entire length of the shear plane is the combined frictional resistance for all materials along the plane.

14.1.7 BASE PRESSURES

14.1.7.1 COMPUTATIONS OF BASE PRESSURES. For the dam to be in static equilibrium, the resultant of all horizontal and vertical forces including uplift must be balanced by an equal and opposite reaction of the foundation consisting of the total normal reaction and the total tangential shear. The location of this force is such that the summation of moments is equal to zero.

14.1.7.2 ALLOWABLE BASE PRESSURE. The maximum computed base pressure should be equal to or less than the allowable bearing capacity for the usual and unusual load conditions. For extreme loading condition, the maximum bearing pressure should be equal to or less than 1.33 times the allowable bearing capacity.

14.1.8 COMPUTER PROGRAMS

14.1.8.1 PROGRAM FOR SLIDING STABILITY ANALYSIS OF CONCRETE STRUCTURES (CSLIDE).

14.1.8.1.1 THE COMPUTER PROGRAM CSLIDE has the capability of performing a two-dimensional sliding stability analysis of gravity dams and other concrete structures. It uses the principles of the multi-wedge system of analysis as discussed.

14.1.8.1.2 THE POTENTIAL FAILURE PLANES and the associated wedges are chosen for input and, by satisfying limit equilibrium principles, the FS against sliding failure is computed for output. The results also give a summary of failure angles and forces acting on the wedges.

14.1.8.1.3 THE PROGRAM CONSIDERS the effects of:
- Multiple layers of rock with irregular surfaces.
- Water and seepage effects. The line-of-creep and seepage factor/gradient are provided.
- Applied vertical surcharge loads including line, uniform, strip, triangular, and ramp loads.
- Applied horizontal concentrated point loads.
- Irregularly shaped structural geometry with a horizontal or sloped base.
- Percentage of the structure base in compression because of overturning effects.
- Single and multiple-plane options for the failure surfaces.
- Horizontal and vertical induced loads because of earthquake accelerations.
- Factors requiring the user to predetermine the failure surface.

14.1.8.1.4 IT WILL NOT ANALYZE CURVED SURFACES or discontinuities in the slip surface of each wedge. In those cases, an average linear geometry should be assumed along the base of the wedge.

14.1.8.2 THREE-DIMENSIONAL STABILITY analysis and design program (3DSAD), special purpose modules for dams (CDAMS).

14.1.8.2.1 GENERAL. The computer program called CDAMS performs a three-dimensional stability analysis and design of concrete dams. The program was developed as a specific structure implementation of the three-dimensional stability analysis and design (3DSAD) program. It is intended to handle two cross-sectional types:

14.1.8.2.1.1 AN OVERFLOW MONOLITH with optional pier.

14.1.8.2.1.2 A NONOVERFLOW MONOLITH. The program can operate in either an analysis or design mode. Load conditions outlined in paragraph 4-1 can be performed in any order. A more detailed description and information about the use of the program can be found in the technical literature

14.1.8.2.2 ANALYSIS. In the analysis mode, the program is capable of performing resultant location, bearing, and sliding computations for each load condition. A review is made of the established criteria and the results outputted.

14.1.8.2.3 DESIGN. In the design mode, the structure is incrementally modified until a geometry is established that meets criteria. Different geometric parameters may be varied to achieve a stable geometry. A design memorandum plate option is also available.

14.2 STATIC AND DYNAMIC STRESS ANALYSES

14.2.1 STRESS ANALYSIS

14.2.1.1 GENERAL.

14.2.1.1.1 A STRESS ANALYSIS OF GRAVITY dams is performed to determine the magnitude and distribution of stresses throughout the structure for static and dynamic load conditions and to investigate the structural adequacy of the substructance and foundation.

14.2.1.1.2 GRAVITY DAM STRESSES are analyzed by either approximate simplified methods or the finite element method depending on the refinement required for the particular level of design and the type and configuration of the dam. For preliminary designs, simplified methods using cantilever beam models for two-dimensional analysis or the trial load twist method for three-dimensional analysis are appropriate as described in the US Bureau of Reclamation (USBR), "Design of Gravity Dams" (1976). The finite element method is ordinarily used for the feature and final design stages if a more exact stress investigation is required.

14.2.1.2 FINITE ELEMENT ANALYSIS.

14.2.1.2.1 FINITE ELEMENT MODELS are used for linear elastic static and dynamic analyses and for nonlinear analyses that account for interaction of the dam and foundation. The finite element method provides the capability of modeling complex geometries and wide variations in material properties. The stresses at corners, around openings, and in tension zones can be approximated with a finite element model. It can model concrete thermal behavior and couple thermal stresses with other loads. An important advantage of this method is that complicated foundations involving various materials, weak joints on seams, and fracturing can be readily modeled. Special purpose computer programs designed specifically for analysis of concrete gravity dams are CG-DAMS which performs static, dynamic, and nonlinear analyses and includes a smeared crack model, and MERLIN which includes a discrete cracking fracture mechanics model.

14.2.1.2.2 TWO-DIMENSIONAL, FINITE element analysis is generally appropriate for concrete gravity dams. The designer should be aware that actual structure response is three dimensional and should review the analytical and realistic results to assure that the two-dimension approximation is acceptable and realistic. For long conventional concrete dams with transverse contraction joints and without keyed joints, a two-dimensional analysis should be reasonably correct. Structures located in narrow valleys between steep abutments and dams with varying rock moduli which vary across the valley are conditions that necessitate three-dimensional modeling.

14.2.1.2.3 THE SPECIAL PURPOSE PROGRAMS Earthquake Analysis of Gravity Dams including Hydrodynamic Interaction (EADHI) and Earthquake Response of Concrete Gravity Dams Including Hydrodynamic and Foundation Interaction Effects (EAGD84) are available for modeling the dynamic response of linear two-dimensional structures. Both programs use acceleration time records for dynamic input. The program SDOFDAM is a two-dimensional finite element model (Cole and Cheek 1986) that computes the hydrodynamic loading using Chopra's simplified procedure. The finite element programs such as GTSTRUDL, SAP, ANSYS, ADINA, and ABAQUS provide general capabilities for modeling static and dynamic responses.

14.2.2 DYNAMIC ANALYSIS. The structural analysis for earthquake loadings consists of two parts: an approximate resultant location and sliding stability analysis using an appropriate seismic coefficient (see Chapter 4) and a dynamic internal stress analysis using site-dependent earthquake ground motions if the following conditions exist:

14.2.2.1 THE DAM IS 100 FEET OR MORE in height and the peak ground acceleration (PGA) at the site is greater than 0.2 g for the maximum credible earthquake.

14.2.2.2 THE DAM IS LESS THAN 100 feet high and the PGA at the site is greater than 0.4 g for the maximum credible earthquake.

14.2.2.3 THERE ARE GATED SPILLWAY monoliths, wide roadways, intake structures, or other monoliths of unusual shape or geometry.

14.2.2.4 THE DAM IS IN A WEAKENED condition because of accident, aging, or deterioration. The requirements for a dynamic stress analysis in this case will be decided on a project-by-project basis in consultant and approved by the Owner.

14.2.3 DYNAMIC ANALYSIS PROCESS. The procedure for performing a dynamic analysis include the following:

- Review the geology, seismology, and contemporary tectonic setting.
- Determine the earthquake sources.
- Select the candidate maximum credible and operating basis earthquake magnitudes and locations.
- Select the attenuation relationships for the candidate earthquakes.
- Select the controlling maximum credible and operating basis earthquakes from the candidate earthquakes based on the most severe ground motions at the site.
- Select the design response spectra for the controlling earthquakes.
- Select the appropriate acceleration-time records that are compatible with the design response spectra if acceleration-time history analyses are needed.
- Select the dynamic material properties for the concrete and foundation.
- Select the dynamic methods of analysis to be used.
- Perform the dynamic analysis.
- Evaluate the stresses from the dynamic analysis.

14.2.4 INTERDISCIPLINARY COORDINATION. A dynamic analysis requires a team of engineering geologists, seismologists, and structural engineers. They must work together in an integrated approach so that elements of conservatism are not unduly compounded. An example of undue conservatism includes using a rare event as the

MCE, upper bound values for the PGA, upper bound values for the design response spectra, and conservative criteria for determining the earthquake resistance of the structure. The steps in performing a dynamic analysis should be fully coordinated to develop a reasonably conservative design with respect to the associated risks. The structural engineers responsible for the dynamic structural analysis should be actively involved in the process of characterizing the earthquake ground motions in the form required for the methods of dynamic analysis to be used.

14.2.5 PERFORMANCE CRITERIA FOR RESPONSE TO SITE-DEPENDENT EARTHQUAKES

14.2.5.1 MAXIMUM CREDIBLE EARTHQUAKE. Gravity dams should be capable of surviving the controlling MCE without a catastrophic failure that would result in loss of life or significant damage to property. Inelastic behavior with associated damage is permissible under the MCE.

14.2.5.2 OPERATING BASIS EARTHQUAKE. Gravity dams should be capable of resisting the controlling OBE within the elastic range, remain operational, and not require extensive repairs.

14.2.6 GEOLOGICAL AND SEISMOLOGICAL INVESTIGATION A geological and seismological investigation of all damsites is required for projects located in seismic zones 2 through 4. The objectives of the investigation are to establish controlling maximum and credible operating basis earthquakes and the corresponding ground motions for each and to assess the possibility of earthquake induced foundation dislocation at the site. Selecting the controlling earthquakes is discussed below.

14.2.7 SELECTING THE CONTROLLING EARTHQUAKES

14.2.7.1 MAXIMUM CREDIBLE EARTHQUAKE. The first step for selecting the controlling MCE is to specify the magnitude and/or modified Mercalli (MM) intensity of the MCE for each seismotectonic structure or source area within the region examined around the site. The second step is to select the controlling MCE based on the most severe vibratory ground motion within the predominant frequency range of the dam and determine the foundation dislocation, if any, capable of being produced at the site by the candidate MCE's. If more than one candidate MCE produce the largest ground motions in different frequency bands significant to the response of the dam, each should be considered a controlling MCE.

14.2.7.2 OPERATING BASIS EARTHQUAKE.

14.2.7.2.1 THE SELECTION OF THE OBE is based upon the desired level of protection for the project from earthquake- induced damage and loss of service project life. The project life of new dams is usually taken as 100 years. The probability of exceedance of the OBE during the project life should be no greater than 50 percent unless the cost savings in designing for a less severe earthquake outweighs the risk of incurring the cost of repairs and loss of service because of a more severe earthquake.

14.2.7.2.2 THE PROBABILISTIC ANALYSIS for the OBE involves developing a magnitude frequency or epicentral intensity frequency (recurrence) relationship of each seismic source; projecting the recurrence information from regional and past data into forecasts concerning future occurrence; attenuating the severity parameter, usually either PGA of MM intensity, to the site; determining the controlling recurrence relationship for the site; and finally, selecting the design level of earthquake based upon the probability of exceedance and the project life.

14.2.8 CHARACTERIZING GROUND MOTIONS

14.2.8.1 GENERAL. After specifying the location and magnitude (or epicentral intensity) of each candidate earthquake and an appropriate regional attenuation relationship, the characteristics of vibratory ground motion expected at the site can be determined. Vibratory ground motions have been described in a variety of ways, such as peak ground motion parameters, acceleration-time records (accelerograms), or response spectra. For the analysis and design of concrete dams, the controlling characterization of vibratory ground motion should be a site-dependent design response spectra.

14.2.8.2 SITE-SPECIFIC DESIGN RESPONSE SPECTRA.

14.2.8.2.1 WHEREVER POSSIBLE, site-specific design response spectra should be developed statistically from response spectra of strong motion records of earthquakes that have similar source and propagation path properties as the controlling earthquake(s) and are recorded on a foundation similar to that of the dam. Important source properties include magnitude and, if possible, fault type and tectonic environment. Propagation path properties include distance, depth, and attenuation. As many accelerograms as possible that are recorded under comparable conditions and have a predominant frequency similar to that selected for the design earthquake should be included in the development of the design response spectra. Also, accelerograms should be selected that have been corrected for the true baseline of zero acceleration, for errors in digitization, and for other irregularities.

14.2.8.2.2 WHERE A LARGE ENOUGH ensemble of site-specific strong motion records is not available, design response spectra may be approximated by scaling that ensemble of records that represents the best estimate of source, propagation path, and site properties. Scaling factors can be obtained in several ways. The scaling factor may be determined by dividing the peak or effective peak acceleration specified for the controlling earthquake by the peak acceleration of the record being rescaled. The peak velocity of the record should fall within the range of peak velocities specified for the controlling earthquake, or the record should not be used. Spectrum intensity can be used for scaling by using the ratio of the spectrum intensity determined for the site and the spectrum intensity of the record being rescaled (USBR 1978). Acceleration attenuation relationships can be used for scaling by dividing the acceleration that corresponds to the source distance and magnitude of the controlling earthquake by the acceleration that corresponds to the source distance and magnitude of the record being rescaled (Guzman and Jennings 1970). Because the scaling of accelerograms is an approximate operation at best, the closer the characteristics of the actual earthquake are to those of the controlling earthquake, the more reliable the results. For this reason, the scaling factor should be held to within a range of 0.33 to 3 for gravity dam.

14.2.8.2.3 GUIDANCE FOR DEVELOPING design response spectra, statistically, from strong motion records is given in Vanmarcke (1979).

14.2.8.2.4 SITE-DEPENDENT RESPONSE spectra developed from strong motion records, as described in paragraphs 5-8b, should have amplitudes equal to or greater than the mean response spectrum for the appropriate foundation given by Seed, Ugas, and Lysmer (1976), anchored by the PGA determined for the site. This minimum response spectrum may be anchored by an effective PGA determined for the site, but supporting documentation for determining the effective PGA will be required (Newmark and Hall 1982).

14.2.8.2.5 A MEAN SMOOTH RESPONSE spectrum of the response spectra of records chosen should be presented for each damping value of interest. The statistical level of response spectra used should be justified based on the degree of conservatism in the preceding steps of the seismic design process and the thoroughness of the development of the design response spectra. If a rare event is used as the controlling earthquake and the earthquake records are scaled by upper bound values of ground motions, then use a response spectrum corresponding to the mean of the amplification factors if the response spectrum is based on five or more earthquake records.

14.2.8.3 ACCELEROGRAMS FOR ACCELERATION-TIME HISTORY ANALYSIS.
Accelerograms used for dynamic input should be compatible with the design response spectrum and account for the peak ground motions parameters, spectrum intensity, and duration of shaking. Compatibility is defined as the envelope of all response spectra derived from the selected accelerograms that lie below the smooth design response spectrum throughout the frequency range of structural significance.

14.2.9 DYNAMIC METHODS OF STRESS ANALYSIS

14.2.9.1 GENERAL. A dynamic analysis determines the structural response based on the characteristics of the structure and the nature of the earthquake loading. Dynamic methods usually employ the modal analysis technique. This technique is based on the simplifying assumption that the response in each natural mode of vibration can be computed independently and the modal responses can be combined to determine the total response (Chopra 1987). Modal techniques that can be used for gravity dams include a simplified response spectrum method and finite element methods using either a response spectrum or acceleration-time records for the dynamic input. A dynamic analysis should begin with the response spectrum method and progress to more refined methods if needed. A time-history analysis is used when yielding (cracking) of the dam is indicated by a response spectrum analysis. The time-history analysis allows the designer to determine the number of cycles of nonlinear behavior, the magnitude of excursion into the nonlinear range, and the time the structure remains nonlinear.

14.2.9.2 SIMPLIFIED RESPONSE SPECTRUM METHOD.

14.2.9.2.1 THE SIMPLIFIED RESPONSE spectrum method computes the maximum linear response of a nonoverflow section in its fundamental mode of vibration due to the horizontal component of ground motion (Chopra 1987). The dam is modeled as an elastic mass fully restrained on a rigid foundation. Hydrodynamic effects are modeled as an added mass of water moving with the dam. The amount of the added water mass depends on the fundamental frequency of vibration and mode shape of the dam and the effects of interaction between the dam and reservoir. Earthquake loading is computed directly from the spectral acceleration, obtained from the design earthquake response spectrum, and the dynamic properties of the structural system.

14.2.9.2.2 THIS SIMPLIFIED METHOD can be used also for an ungated spillway monolith that has a section similar to a nonoverflow monolith. A simplified method for gated spillway monoliths is presented in WES Technical Report SL-89-4 (Chopra and Tan 1989).

14.2.9.2.3 THE PROGRAM SDOFDAM IS AVAILABLE to easily model a dam using the finite element method and Chopra's simplified procedure for estimating the hydrodynamic loading. This analysis provides a reasonable first estimate of the tensile stress in the dam. From that estimate, one can decide if the design is adequate or if a refined analysis is needed.

14.2.9.3 FINITE ELEMENT METHODS.

14.2.9.3.1 GENERAL. The finite element method is capable of modeling the horizontal and vertical structural deformations and the exterior and interior concrete, and it includes the response of the higher modes of vibrations, the interaction effects of the foundation and any surrounding soil, and the horizontal and vertical components of ground motion.

14.2.9.3.2 FINITE ELEMENT RESPONSE SPECTRUM METHOD.

14.2.9.3.2.1 THE FINITE ELEMENT response spectrum method can model the dynamic response of linear two- and three dimensional structures. The hydrodynamic effects are modeled as an added mass of water moving with the dam using Westergaard's formula (Westergaard 1933). The foundations are modeled as discrete elements or a half space.

14.2.9.3.2.2 SIX GENERAL PURPOSE finite element programs are compared by Hall and Radhakrishnan (1983).

14.2.9.3.2.3 A FINITE ELEMENT program computes the natural frequencies of vibration and corresponding mode shapes for specified modes. The earthquake loading is computed from earthquake response spectra for each mode of vibration induced by the horizontal and vertical components of ground motion. These modal responses are combined to obtain an estimate of the maximum total response. Stresses are computed by a static analysis of the dam using the earthquake loading as an equivalent static load.

14.2.9.3.2.4 THE COMPLETE QUADRATIC combination (CQC) method (Der Kiureghian 1979 and 1980) should be used to combine the modal responses. The CQC method degenerates to the square root of the sum of squares (SRSS) method for two-dimensional structures in which the frequencies are well separated. Combining modal maxima by the SRSS method can dramatically overestimate or significantly underestimate the dynamic response for three-dimensional structures.

14.2.9.3.2.5 THE FINITE ELEMENT response spectrum method should be used for dam monoliths that cannot be modeled two dimensionally or if the maximum tensile stress from the simplified response spectrum method (paragraph 5-9b) exceeds 15 percent of the unconfined compressive strength of the concrete.

14.2.9.3.2.6 NORMAL STRESSES should be used for evaluating the results obtained from a finite element response spectrum analysis. Finite element programs calculate normal stresses that, in turn, are used to compute principal stresses. The absolute values of the dynamic response at different time intervals are used to combine the modal responses. These calculations of principal stress overestimate the actual condition. Principal stresses should be calculated using the finite element acceleration-time history analysis for a specific time interval.

14.2.9.3.3 FINITE ELEMENT ACCELERATION-TIME HISTORY METHOD.

14.2.9.3.3.1 THE ACCELERATION-time history method requires a general purpose finite element program or the special purpose computer program called EADHI. EADHI can model static and dynamic responses of linear two-dimensional dams. The hydrodynamic effects are modeled using the wave equation. The compressibility of water and structural deformation effects are included in computing the hydrodynamic pressures. EADHI was developed assuming a fixed base for the dam. The most comprehensive two-dimensional earthquake analysis program available for gravity dams is EAGD84, which can model static and dynamic responses of linear two-dimensional dams, including hydrodynamic and foundation interaction. Dynamic input for EADHI and EAGD84 is an acceleration time record.

14.2.9.3.3.2 THE ACCELERATION-TIME HISTORY method computes the natural frequencies of vibration and corresponding mode shapes for specified modes. The response of each mode, in the form of equivalent lateral loads, is calculated for the entire duration of the earthquake acceleration-time record starting with initial conditions, taking a small time interval, and computing the response at the end of each time interval. The modal responses are added for each time interval to yield the total response. The stresses are computed by a static analysis for each time interval.

14.2.9.3.3.3 AN ACCELERATION-TIME HISTORY analysis is appropriate if the variation of stresses with time is required to evaluate the extent and duration of a highly stressed condition.

CHAPTER 15: MISCELLANEOUS CONSIDERATIONS FOR GRAVITY DAMS

15.1 TEMPERATURE CONTROL OF MASS CONCRETE

15.1.1 INTRODUCTION. Temperature control of mass concrete is necessary to prevent cracking caused by excessive tensile strains that result from differential cooling of the concrete. The concrete is heated by reaction of cement with water and can gain additional heat from exposure to the ambient conditions. Cracking can be controlled by methods that limit the peak temperature to a safe level, so the tensile strains developed as the concrete cools to equilibrium are less than the tensile strain capacity.

15.1.2 THERMAL PROPERTIES OF CONCRETE

15.1.2.1 GENERAL. The properties of concrete used in thermal studies for the design of gravity dams are thermal diffusivity, thermal conductivity, specific heat, coefficient of thermal expansion, heat of hydration of the cement, tensile strain capacity, and modulus of elasticity. The most significant factor affecting the thermal properties is the composition of the aggregates. The selection of suitable aggregates is based on other considerations, so little or no control can be exercised over the thermal properties of the aggregates. Type II cement with optional low heat of hydration limitation and a cement replacement are normally specified. Type IV low-heat cement has not been used in recent years, because in most cases heat development can be controlled by other measures and type IV cement is not generally available.

15.1.2.2 THERMAL CONDUCTIVITY. The thermal conductivity of a material is the rate at which it transmits heat and is defined as the ratio of the flux of heat to the temperature gradient. Water content, density, and temperature significantly influence the thermal conductivity of a specific concrete. Typical values are 2.3, 1.7, and 1.2 British thermal units (Btu)/hour/foot/Fahrenheit degree (°F) for concrete with quartzite, limestone, and basalt aggregates, respectively.

15.1.2.3 THERMAL DIFFUSIVITY. Diffusivity is described as an index of the ease or difficulty with which concrete undergoes temperature change and, numerically, is the thermal conductivity divided by the product of specific heat and density. Typical diffusivity values for concrete range from 0.03 square foot/hour for basalt concrete to 0.06 square foot/hour for quartzite concrete.

15.1.2.4 SPECIFIC HEAT. Specific heat or heat capacity is the heat required to raise a unit weight of material 1 degree. Values for various types of concrete are about the same and vary from 0.22 to 0.25 Btu's/pound/°F.

15.1.2.5 COEFFICIENT OF THERMAL EXPANSION. The coefficient of thermal expansion can be defined as the change in linear dimension per unit length divided by the temperature change expressed in millionths per °F. Basalt and limestone concretes have values from 3 to 5 millionths/°F; quartzite concretes range up to 8 millionths/°F.

15.1.2.6 HEAT OF HYDRATION. The reaction of water with cement is exothermic and generates a considerable amount of heat over an extended period of time. Heats of hydration for various cements vary from 60 to 95 calories/gram at 7 days and 70 to 110 calories/gram at 28 days.

15.1.2.7 TENSILE STRAIN CAPACITY. Design is based on maximum tensile strain. The modulus of rupture test (CRD-C 16) is done on concrete beams tested to failure under third-point loading. Tensile strain capacity is determined by dividing the modulus of rupture by the modulus of elasticity. Typical values range from 50 to 200 million depending on loading rate and type of concrete.

15.1.2.8 CREEP. Creep of concrete is deformation that occurs while concrete is under sustained stress. Specific creep is creep under unit stress. Specific creep of mass concrete is in the range of $1.4 \times 10\text{-}6$/pounds per square inch (psi).

15.1.2.9 MODULUS OF ELASTICITY. The instantaneous loading modulus of elasticity for mass concrete ranges from about 1.5 to 6×10^6 psi and under sustained loading from about 0.5 to 4×10^6 psi.

15.1.3 THERMAL STUDIES

15.1.3.1 GENERAL. During the design of gravity dams, it is necessary to assess the possibility that strain induced by temperature changes in the concrete will not exceed the strain capacity of the concrete. Detailed design procedures for control of the generation of heat and volume changes to minimize cracking may be found in the ACI Manual of Concrete Practice, Section 207. The following concrete parameters should be determined by a division laboratory: heat of hydration (CRD-C 229), adiabatic temperature rise (CRD-C 38), thermal conductivity (CRD-C 44), thermal diffusivity (CRD-C 37), specific heat (CRD-C 124), coefficient of thermal expansion (CRD-C 397, 125, and 126), creep (CRD-C 54), and tensile strain capacity (CRD-C 71). Thermal properties testing should not be initiated until aggregate investigations have proceeded to the point that the most likely aggregate sources are determined and the availability of cementitious material is known.

15.1.3.2 ALLOWABLE PEAK TEMPERATURE. The peak temperature for the interior mass concrete must be controlled to prevent cracking induced by surface contraction. The allowable peak temperature commonly used to prevent serious cracking in mass concrete structures is the mean annual ambient temperature plus the number of degrees Fahrenheit determined by dividing the tensile strain capacity by the coefficient of linear expansion. This assumes that the concrete will be subjected to 100-percent restraint against contraction. When the potential temperature rise of mass concrete is reduced to this level, the temperature drop that causes tensile strain and cracking is reduced to an acceptable level.

15.1.4 TEMPERATURE CONTROL METHODS The temperature control methods available for consideration all have the basic objective of reducing increases in temperature due to heat of hydration, reducing thermal differentials within the structure, and reducing exposure to cold air at the concrete surfaces that would create cracking. The most common techniques are the control of lift thickness, time interval between lifts, maximum allowable placing temperature of the concrete, and surface insulation. Postcooling may be economical for large structures. Analysis should be made to determine the most economic method to restrict temperature increases and subsequent temperature drops to levels just safely below values that could cause undesirable cracking. For structures of limited complexity, such as conventionally shaped gravity dams, satisfactory results may be obtained by use of the design procedures in ACI 207 "Mass Concrete for Dams and Other Massive Structures." Roller compacted concrete thermal control options include the installation of contraction joints, winter construction, mixture design, and increased heat dissipation. Contraction joints can be created by inserting a series of cuts or metal plates into each lift to produce a continuous vertical joint. Using very high production and placement rates, RCC construction can be limited to colder winter months without excessive schedule delays. The normal lift height of 1 to 2 feet provides for an increased rate of heat dissipation during cool weather.

15.2 STRUCTURAL DESIGN CONSIDERATIONS

15.2.1 INTRODUCTION. This section discusses the layout, design, and construction considerations associated with concrete gravity dams. These general considerations include contraction and construction joints, waterstops, spillways, outlet works, and galleries.

15.2.2 CONTRACTION AND CONSTRUCTION JOINTS

15.2.2.1 TO CONTROL THE FORMATION OF CRACKS in mass concrete, vertical transverse contraction (monolith) joints will generally be spaced uniformly across the axis of the dam about 50 feet apart. Where a powerhouse forms an integral part of a dam and the spacing of the units is in excess of this dimension, it will be necessary to increase the joint spacing in the intake block to match the spacing of the joints in the powerhouse. In the spillway section, gate and pier size and other requirements are factors in the determination of the spacing of the contraction joints. The location and spacing of contraction joints should be governed by the physical features of the damsite, details of the appurtenant structures, results of temperature studies, placement rates and methods, and the probable concrete mixing plant capacity. Abrupt discontinuities along the dam profile, material changes, defects in the foundation, and the location of features such as outlet works and penstock will also influence joint location. In addition, the results of thermal studies will provide limitations on monolith joint spacing for assurance against cracking from excessive temperature-induced strains. The joints are vertical and normal to the axis, and they extend continuously through the dam section. The joints are constructed so that bonding does not exist between adjacent monoliths to assure freedom of volumetric change of individual monoliths. Reinforcing should not extend through a contraction joint. At the dam faces, the joints are chamfered above minimum pool level for appearance and for minimizing spalling. The monoliths are numbered, generally sequentially, from the right abutment.

15.2.2.2 HORIZONTAL OR NEARLY HORIZONTAL CONSTRUCTION joints (lift joints) will be spaced to divide the structure into convenient working units and to control construction procedure for the purpose of regulating temperature changes. A typical lift will usually be 5 feet consisting of three 20-inch layers, or 7-1/2 feet consisting of five 18-inch layers. Where necessary as a temperature control measure, lift thickness may be limited to 2-1/2 feet in certain areas of the dam. The best lift height for each project will be determined from concrete production capabilities and placing methods.

15.2.3 WATERSTOPS. A double line of waterstops should be provided near the upstream face at all contraction joints. The waterstops should be grouted 18 to 24 inches into the foundation or sealed to the cutoff system and should terminate near the top of the dam. For gated spillway sections, the tops of the waterstops should terminate near the crest of the ogee. A 6- to 8-inch-diameter formed drain will generally be provided between the two waterstops. In the nonoverflow monolith joints, the drains extend from maximum pool elevation and terminate at about the level of, and drain into, the gutter in the grouting and drainage gallery. In the spillway monolith joints, the drains extend from the gate sill to the gallery. A single line of waterstops should be placed around all galleries and other openings crossing monolith joints.

15.2.4 SPILLWAY

15.2.4.1 THE PRIMARY FUNCTION of a spillway is to release surplus water from reservoirs and to safely bypass the design flood downstream in order to prevent overtopping and possible failure of the dam. Spillways are classified as controlled (gate) or uncontrolled (ungated). The overflow (ogee) spillway is the type usually associated with concrete gravity dams. Other less common spillway types such as chute, side channel, morning glory, and tunnel are not addressed in this manual.

15.2.4.2 AN OVERFLOW SPILLWAY PROFILE is governed in its upper portions by hydraulic considerations rather than by stability requirements. The downstream face of the spillway section terminates either in a stilling basin apron or in a bucket type energy dissipator, depending largely upon the nature of the site and upon the tailwater conditions. The design of the spillway shall include the stability and internal stress analysis and the structural performance. Operating equipment should be designed to be operational following a maximum credible earthquake.

15.2.4.3 DISCHARGE OVER THE SPILLWAY or flip bucket section must be confined by sidewalls on either side, terminating in training walls extending along each side of the stilling basin or flip bucket. Height and length of training walls are usually determined by model tests or from previous tests of similar structures. Sidewalls should be of sufficient height to contain the spillway design flow, with a 2-foot freeboard. Negative pressures due to flowing water should be considered in the design of the sidewalls, with the maximum allowance being made at the stilling basin, decreasing uniformly to no allowance at the crest. Sidewalls are usually designed as cantilevers projecting out of the monolith. A wind load of 30 pounds/square foot or earthquake loading should be assumed for design of reinforcing in the outer face of the walls. The spillway section surfaces should be designed to withstand the high flow velocities expected during peak discharge and reduced pressures resulting from the hydrodynamic effects.

15.2.4.4 THE DYNAMIC LOADS occurring in the energy dissipators will include direct impact, pulsating loads from turbulence, multidirectional and deflected hydraulic flows, surface erosion from high velocities and debris, and cavitation. The downstream end of the dissipator should include adequate protection against undermining from turbulence and eddies. Concrete apron, riprap, or other measures have been used for stabilization.

15.2.4.5 SPILLWAY BRIDGE

15.2.4.5.1 BRIDGES ARE PROVIDED across dam spillways to furnish a means of access for pedestrian and vehicular traffic between the nonoverflow sections; to provide access or support for the operating machinery for the crest gates; or, usually, to serve both purposes. In the case of an uncontrolled spillway and in the absence of vehicular traffic, access between the nonoverflow sections may be provided by a small access bridge or by stair shafts and a gallery beneath the spillway crest.

15.2.4.5.2 THE DESIGN OF A DECK-type, multiple-span spillway bridge should generally conform to the following criteria. The class of highway design loading will normally not be less than HS-20. Special loadings required for performing operation and maintenance functions and those that the bridge is subjected to during construction should be taken into account, including provisions for any heavy concentrated loads. Heavy loadings for consideration should include those due to powerhouse equipment transported during construction, mobile cranes used for maintenance, and gantry cranes used to operate the regulating outlet works and to install spillway stoplogs. If the structure carries a state or county highway, the design will usually conform to the standard specification for highway bridges adopted by the American Association of State Highway and Transportation Officials (AASHTO).

15.2.4.5.3 MATERIALS USED IN THE DESIGN and construction of the bridge should be selected on the basis of life cycle costs and functional requirements. Floors, curbs, and parapets should be reinforced concrete. Beams and girders may be structural steel, precast or cast-in-place reinforced concrete, or prestressed concrete. Prestressed concrete is often used because it combines economy, simple erection procedures, and low maintenance.

15.2.6 SPILLWAY PIERS

15.2.6.1 FOR UNCONTROLLED SPILLWAYS, the piers function as supports for the bridge. On controlled spillways, the piers will also contain the anchorage or slots for the crest gates and may support fixed hoists for the gates. The piers are generally located in the middle of the monolith, and the width of pier is usually determined by the size of the gates, with the average width being between 8 and 10 feet. The spillway piers in RCC dams are constructed with conventional concrete.

15.2.6.2 SINCE EACH PIER SUPPORTS a gate on each side, the following pier loading conditions should be investigated:

- Case 1--both gates closed and water at the top of gates.
- Case 2--one gate closed and the other gate wide open with water at the top of the closed gate.
- Case 3--one gate closed and the other open with bulkheads in place and water at the top of the closed gate.

15.2.6.3 CASES 1 AND 3 RESULT IN MAXIMUM horizontal shear normal to the axis of the dam and the largest overturning moment in the downstream direction.

15.2.6.4 CASE 2 RESULTS IN LOWER horizontal shear and downstream overturning moment, but in addition the pier will have a lateral bending moment due to the water flowing through the open gate and to the hoisting machinery when lifting a closed gate. A torsional shear in the horizontal plane will also be introduced by the reaction of the closed gate acting on one side of the pier. When tainter gates with inclined end frames are used, Cases 2 and 3 introduce the condition of the lateral component of the thrust on the trunion as a load on only one side of the pier in addition to the applicable loads indicated above.

15.2.7 OUTLET WORKS

15.2.7.1 THE OUTLET WORKS FOR CONCRETE dams are usually conduits or sluices through the mass with an intake structure on the upstream face, gates or valves for regulation control, and an energy dissipator on the downstream face. Multiple conduits are normally provided because of economics and operating flexibility in controlling a wide range of releases. The conduits are frequently located in the center line of the overflow monoliths and discharge into the spillway stilling basin. Outlet works located in nonoverflow monoliths will require a separate energy dissipator. All conduits may be at low level, or some may be located at one or more higher levels to reduce the head on the gates, to allow for future reservoir silting, or to control downstream water quality and temperature. The layout, size, and shape of the outlet works are based on hydraulic and hydrology requirements, regulation plans, economics, site conditions, operation and maintenance needs, and interrelationship to the construction plan and other appurtenant structures. Conduits may be provided for reservoir evacuation, regulation of flows for flood control, emergency drawdown, navigation, environmental (fish), irrigation, water supply, maintaining minimum downstream flows and water quality, or for multiple purposes. Low-level conduits are used to aid water quality reservoir evacuation and are sometimes desirable for passage of sediment. These openings are generally unlined except for short sections adjacent to the control gates. For lined conduits, it is assumed that the liner is designed for the full loading. In conduits where velocities will be 40 feet/second or higher, precautions will be taken to ensure that the concrete in the sidewalls and inverts will be of superior quality.

15.2.7.2 THE EFFECT OF PROJECT FUNCTIONS upon outlet works design and hydraulic design features, including trashrack design and types for sluice outlets, are discussed in the technical literature. A discussion of the structural features of design for penstocks and trashracks for power plant intakes is included in the technical literature.

15.2.8 FOUNDATION GROUTING AND DRAINAGE It is good engineering practice to grout and drain the foundation rock of gravity dams. A well-planned and executed grouting program should assist in disclosing weaknesses in the foundation and improving any existing defects. The program should include area grouting for foundation treatment and curtain grouting near the upstream face for seepage cutoff through the foundation. Area grouting is generally done before concrete placement. Curtain grouting is commonly done after concrete has been placed to a considerable height or even after the structure has been completed. A line of drainage holes is drilled a few feet downstream from the grout curtain to collect seepage and reduce uplift across the base.

15.2.9 GALLERIES. A system of galleries, adits, chambers, and shafts is usually provided within the body of the dam to furnish means of access and space for drilling and grouting and for installation, operation, and maintenance of the accessories and the utilities in the dam. The primary considerations in the arrangement of the required openings within the dam are their functional usefulness and efficiency and their location with respect to maintaining the structural integrity.

15.2.9.1 GROUTING AND DRAINAGE GALLERY. A gallery for grouting the foundation cutoff will extend the full length of the dam. It will also serve as a collection main for seepage from foundation drainage holes and the interior drainage holes. The location of the gallery should be near the upstream face and as near the rock surface as feasible to provide the maximum reduction in overall uplift. A minimum distance of 5 feet should be maintained between the foundation surface and gallery floor and between the upstream face and the gallery upstream wall. It has been standard practice to provide grouting galleries 5 feet wide by 7 feet high. Experience indicates that these dimensions should be increased to facilitate drilling and grouting operations. Where practicable, the width should be increased to 6 or 8 feet and the height to 8 feet. A gutter may be located along the upstream wall of the gallery where the line of grout holes is situated to carry away drill water and cuttings. A gutter should be located along the downstream gallery wall to carry away flows from the drain pipes. The gallery is usually arranged as a series of horizontal runs and stair flights. The stairs should be provided with safety treads or a nonslip aggregate finish. Metal treads are preferable where it is probable that equipment will be skidded up or down the steps since they provide protection against chipping of concrete. Where practicable, the width of tread and height of riser should be uniform throughout all flights of stairs and should never change in any one flight.

15.2.9.2 GATE CHAMBERS AND ACCESS GALLERIES. Gate chambers are located directly over the service and emergency sluice gates. These chambers should be sized to accommodate the gate hoists along with related mechanical and electrical equipment and should provide adequate clearances for maintenance. Access galleries should be sufficient size to permit passage of the largest component of the gates and hoists and equipment required for maintenance. Drainage gutters should be provided and the floor of the gallery sloped to the gutter with about 1/4 inch/foot slope.

15.2.10 INSTRUMENTATION. Structural behavior instrumentation programs are provided for concrete gravity dams to measure the structural integrity of the structure, check design assumptions, and monitor the behavior of the foundation and dam during construction and the various operating phases. The extent of instrumentation at projects will vary between projects depending on particular site conditions, the size of the dam, and needs for monitoring critical sections. Instrumentation can be grouped into those that either directly or indirectly measure conditions related to the safety of the structure. Plumbing, alignment, uplift, and seismic instruments fall into the category of safety instruments. In the other group, the instruments measure quantities such as stress and strain, length change, pore pressure, leakage, and temperature change.

15.3 REEVALUATION OF EXISTING DAMS

15.3.1 GENERAL. Existing gravity dams and foundations should be reevaluated for integrity, strength, and stability when:

15.3.1.1 IT IS EVIDENT THAT distress has occurred because of an accident, aging, or deterioration.

15.2.1.2 DESIGN CRITERIA HAVE become more stringent.

15.3.1.3 EXCAVATION IS TO be performed near existing structures.

15.3.1.4 STRUCTURAL DEFICIENCIES have been detected.

15.3.1.5 ACTUAL LOADINGS ARE, or anticipated loadings will be, greater than those used in the original design. Loadings can increase as a result of changed operational procedures or operational deficiencies, an increase in dam height, or an increase in the maximum credible earthquake as a result of seismological investigations. Conditions such as excessive uplift pressures, unusual horizontal or vertical displacements, increased seepage through the concrete or foundation, and structural cracking are indications that a reevaluation should be performed.

15.3.2 REEVALUATION. The reevaluation should be based on current design criteria and prevailing geological, structural, and hydrological conditions. If the investigations indicate a fundamental deficiency, then the initial effort should concentrate on restoring the dam to a safe and acceptable operating condition. Efforts could include measures to reduce excessive uplift pressures, reduce leak, repair cracks, or restore deteriorated concrete. Should restoration costs be unreasonable or should the fundamental deficiency be due to changes in load or stability criteria, a detailed analysis should be performed in accordance with the following procedures. Reevaluation of structures not designed to current standards should be in accordance with the appropriate requirements.

15.3.3 PROCEDURES. The following procedures shall be used in evaluating current structural conditions and determining the necessary measures for rehabilitation of existing concrete gravity dams.

15.3.3.1 EXISTING DATA. Collect and review all the available information for the structure including geologic and foundation data, design drawings, as-built drawings, periodic inspection reports, damage reports, repair and maintenance records, plans of previous modifications to the structure, measurements of movement, instrumentation data, and other pertinent information. Any unusual structural behavior that may be an indication of an unsafe condition or any factor that may contribute to the weakening of the structure's stability should be noted and investigated further.

15.3.3.2 SITE INSPECTION. To be evaluated.

15.3.3.3 PRELIMINARY ANALYSES. Perform the preliminary analyses based on current structural criteria and available data. If the structure does not meet the current criteria, list the possible remedial schemes and prepare a preliminary cost estimate for each scheme.

15.3.3.4 DESIGN MEETING. Schedule a meeting when the preliminary analyses indicate that the structure does not meet current criteria. The meeting should include appropriate representatives to decide on a plan for proposed analyses, the extent of the sampling and testing program, the remedial schemes to be studied, and the proposed schedule. This meeting will facilitate the design effort and should obviate the need for major revisions or additional studies when the results are submitted for review and approval.

15.3.3.5 PARAMETRIC STUDY. Perform a parametric study to determine the effect of each parameter on the structure's safety. The parameters to be studied should include, but not be limited to, unit weight of concrete, groundwater levels, uplift pressures, and shear strength parameters of rock fill material, rock foundation, and structure foundation interface. The maximum variation of each parameter should be considered in determining its effect.

15.3.3.6 FIELD INVESTIGATIONS. Develop an exploration, sampling, testing, and instrumentation program, if needed, to determine the magnitude and reasonable range of variation for the parameters that have significant effects on the safety of the structure as determined by the parametric study.

15.3.3.7 DETAILED STRUCTURAL ANALYSES. Perform detailed analyses using data obtained from studies, field investigations, and procedures outlined. Three-dimensional modeling should be used as appropriate to more accurately predict the structural behavior.

15.3.3.8 REFINED STRUCTURAL ANALYSIS. The conventional methods described may be more conservative than necessary, especially when making a determination as to the need for remedial strengthening to improve the stability of an existing dam. If the conventional analyses indicate remedial strengthening is required, then a refined finite element analysis should be performed. This refined analysis should accurately model the strength and stiffness of the dam and foundation to determine the following:

15.3.3.8.1 THE EXTENT OF TENSILE cracking at the dam foundation interface.

15.3.3.8.2 THE BASE AREA in compression.

15.3.3.8.3 THE ACTUAL MAGNITUDE and distribution of foundation pressures.

15.3.3.8.5 THE MAGNITUDE AND distribution of concrete stresses.

15.3.3.9 REVIEW AND APPROVAL. Present the results of detailed structural analyses and cost estimates for remedial measures to the appropriate office for review and approval. If a deviation from current structural criteria was made in the analyses, the results should be forwarded to the appropriate authority for approval.

15.3.3.10 PLANS AND SPECIFICATIONS. Develop design plans, specifications, and a cost estimate for proposed remedial measures.

15.3.4 CONSIDERATIONS OF DEVIATION FROM STRUCTURAL CRITERIA

15.3.4.1 THE PURPOSE OF INCORPORATING a factor of safety in structural design is to provide a reserve capacity with respect to failure and to account for strength variability of the dam and foundation materials. The required margin depends on the consequences of failure and on the degree of uncertainties due to loading variations, analysis simplifications, design assumptions, variations in material strengths, variations in construction control, and other factors. For evaluation of existing structures, a higher degree of confidence may be achieved when the critical parameters can be determined accurately at the site. Therefore, deviation from the current structural criteria for an existing structure may be allowed under the conditions listed.

15.3.4.2 IN ADDITION TO THE DETAILED analyses and cost estimates as listed, the following information should also be presented with the request for a deviation from the current structural criteria:

15.3.4.2.1 PAST PERFORMANCE of the structure, including instrumentation data and a description of the structure condition such as cracking, spalling, displacements, etc.

15.3.4.2.2 THE ANTICIPATED remaining life of the structure.

15.3.4.2.3 A DESCRIPTION OF consequences in case of failure.

15.3.4.3 APPROVAL OF THE DEVIATION depends upon the degree of confidence in the accuracy of design parameters determined in the field, the remaining life of the structure, and the potential adverse effect on lives, property, and services in case of failure.

15.3.5 STRUCTURAL REQUIREMENTS FOR REMEDIAL MEASURE. When it is determined that remedial measures are required for the existing structure, they should be designed to meet the structural criteria.

15.3.6 METHODS OF IMPROVING STABILITY IN EXISTING STRUCTURES

15.3.6.1 GENERAL. Several methods are available for improving the rotational and sliding stability of concrete gravity dams. In general, the methods can be categorized as those that reduce loadings, in particular uplift, or those that add stabilizing forces to the structure and increase overturning or shear-frictional resistance. Stressed foundation anchor systems are considered one of the most economical methods of increasing rotational and sliding resistance along the base of the dam. Foundation grouting and drainage may also be effective in reducing uplift, reducing foundation settlements and displacements, thereby increasing bearing capacity. Regrouting the foundation could adversely affect existing foundation drainage systems unless measures are taken to prevent plugging the drains; otherwise, drain redrilling will be required. Various methods of transferring load to more competent adjacent structures or foundation material through shear keys, buttresses, underpinning, etc., are also possible ways of improving stability.

15.3.6.2 REDUCING UPLIFT FORCES. In many instances, measured uplift pressures are substantially less than those used in the original design. These criteria limit drain efficiency to a maximum of 50 percent. Many designs are based on efficiencies less than 50 percent. Existing drainage systems can produce efficiencies of 75 percent or more if they extend through the most pervious layers of the foundation, if the elevation of the drainage gallery is at or near tailwater, and if the drains are closely spaced and effectively maintained. If measured uplift pressures are substantially less than design values, then parametric studies should determine what benefit it may have towards improving stability. Uplift pressures less than design allowables should be data from reliable instrumentation which assures that the measured uplift is indicative of pressures within the upper zones and along the entire foundation. Uplift pressures can be reduced by additional foundation grouting and re-establishing drains. Uplift also be reduced by increasing the depth of existing drains, adding new drains, or rehabilitating existing drains by reaming and cleaning.

15.3.6.3 PRESTRESSED ANCHORS. Prestressed anchors with double corrosion protection may be used to stabilize existing concrete monoliths, but generally should not be used in the design of new concrete gravity dams. They are effective in improving sliding resistance, resultant location, and excessive foundation pressure. Anchors may be used to secure thrust blocks or stilling basins for the sole purpose of improving sliding stability. The anchor force required to stabilize a dam will depend largely on the orientation of the anchors. Anchors should be oriented for maximum efficiency subject to constraints of access, embedded features, galleries, and stress concentrations they induce in the dam. Analyses of tensile stresses under anchor heads should be made, and reinforcing should be provided as required. Tendon size, spacing, and embedment length should be based on the required anchor force, and should be provided the geotechnical engineer for determination of the required embedment length. Design, installation, and testing of anchors and anchorages should be guided by information in "Recommendations for Prestressed Rock and Soil Anchors" (Post- Tensioning Institute (PTI) 1985). Allowable bond stresses used to determine the length of embedment between grout and rocks are recommended to be one half of the ultimate bond stress determined by tests. The typical values of bond strength given in the above referenced PTI publication may be used in lieu of test values during design, but the design value should be verified by test before or during construction. The first three anchors installed and a minimum of 2 percent of the remaining anchors selected by the engineer should be performance tested. All other anchors must be proof tested upon installation in accordance with the PTI recommendations. Additionally, initial lift-off readings should be taken after the anchor is seated and before the jack is removed. Lift-off tests of random anchors selected by the engineer should be made 7 days after lock-off and prior to secondary grouting. Long-term monitoring of selected anchors using load cells and unbonded tendons should be employed where unusual conditions exist or the effort and expense can be justified by the importance of the structure. In addition to stability along the base of the dam, prestressed anchors may be required for deep-seated stability problems as discussed in the following paragraph. Non-prestressed anchors shall not be used to improve the stability of dams.

15.3.7 STABILITY ON DEEP-SEATED FAILURE PLANES. A knowledge of the rock structure of a foundation is crucial to a realistic stability analysis on deep-seated planes. If instability is to occur, it will take place along zones of weakness within the rock mass. A team effort between the geotechnical and structural engineers is important in evaluating the foundation and its significance to the design of the dam. Deep-seated sliding is of primary interest as it is the most common problem encountered. Significant foundation features are: rock surface joint patterns that admit water to potential deep-seated sliding planes; inclination of joints and fracturing that affect passive resistance; relative permeability of foundation materials that affect uplift; and discontinuities such as gouge zones and faulting which affect both strength and uplift along failure planes. Strength values for failure planes are required for design. As these values are often difficult to define with a high level of confidence, they should be described in terms of expected values and standard deviations. Analyses of resultant location and maximum bearing pressure will also be required.

15.3.7.1 METHOD AND ASSUMPTIONS. Stability on deep-seated planes is similar to methods described in Chapter 4 for the dam. Tensile strength within the foundation is neglected except where it can be demonstrated by exploration and testing. Vertical and near vertical joints are assumed to be fully pressurized by the pool to which they are exposed. Normally a pressurized vertical joint will be assumed to exist at or near the heel of the dam. Uplift on flat and inclined bedding planes will be dependent on their state of compression and the presence of drains passing through these planes as described for dams. Passive resistance will be based on the rock conditions downstream of the dam. Adversely inclined joints, faults, rock fracturing, or damage from excavation by blasting will affect available passive resistance.

15.3.7.2 ANCHOR PENETRATION. Required anchor penetration depends on the purpose of the anchor. Anchors provided to resist uplift of the heel must have sufficient penetration to develop the capacity of the anchors. Anchors provided to resist sliding must be fully developed below the lowest critical sliding plane. Critical sliding planes are those requiring anchors to meet minimum acceptable factors of safety against sliding.

15.3.7.3 ANCHOR RESISTANCE. The capacity of the anchor to resist uplift should be limited to the force that can be developed by the submerged weight of the rock engaged by the anchor. Rock engaged will either be shaped as cones or intersecting cones depending on the length and spacing of the anchors. The anchor force that can be developed should be based on the pullout resistance of a cone with an apex angle of 90 deg. Tensile stresses will occur in the anchorage zone of prestressed anchors. The possibility of foundation cracking as a result of these tensile stresses must be considered. It is possible that cracks in the foundation could open at the lower terminal points of the anchors and propagate downstream. To alleviate this potential problem, a sufficient weight of submerged rock should be engaged to resist the anchor force, and the anchor depths should be staggered.

15.3.8 EXAMPLE PROBLEM. The following example is a gated outlet structure for an earth fill dam. The existing gated spillway monoliths are deficient in sliding resistance along a weak seam in the foundation which daylights in the stilling basin. A crosssection of the spillway monoliths is shown in Figure 8-1. The spillway monoliths are founded at elevation 840 on moderately hard silty shale. A continuous soft, plastic clay shale seam approximately 1/2 inch in thickness exists at elevation 830. A free body diagram showing forces acting on the gravity structure and foundation above the weak seam is shown in Figure 8-2. Even though the foundation drains penetrate the potential sliding plane, the drains are assumed ineffective as they are insufficient to drain a thin clay seam. The sliding plane is in full compression, and uplift is assumed to vary uniformly from upper pool head to zero in the stilling basin. A drained shear strength of 20° 30¢ has been assigned to this potential sliding surface, and a sliding factor of safety of 0.49 has been calculated for loading condition No. 2, i.e., pool to top of closed spillway gates. The tailwater is below the level of the sliding surface. A summary of loads and the resulting factor of safety for this critical loading condition is shown in Table 8-1. The design of anchors to provide a required factor of safety of 1.70 is summarized in Table 8-2. The anchors are located as shown in Figure 8-3. Details of the anchors are shown in Figure 8-4. The 45-deg angle for the anchors was selected to minimize drilling and to provide a large component of resisting force without creating a potential upstream sliding problem during low pools. Tips of anchors are staggered to avoid tensile stress concentrations in the foundation. The anchors are embedded below the lowest sliding plane requiring anchors to meet required safety factors. Reinforcement similar to that used in post-tensioned beams is provided under the bearing plates to resist the high tensile bursting stresses associated with large capacity anchors. The anchors were tensioned in the sequence shown in Figure 8-4 to avoid unacceptable stress concentrations in the concrete monoliths. The anchors were designed, installed, and tested in accordance with PTI (1985). The anchors are designed for a working load of 826 kips and were locked-off at 910 kips (i.e., working load plus 10 percent) to allow for calculated relaxation of the anchors, creep in the concrete structure, and consolidation of the foundation. Proof testing of all anchors to 80 percent of ultimate strength confirmed the adequacy of the anchors for a working load of 826 kips per anchor (approximately 60 percent of ultimate strength). Each anchor successfully passed a 14th day lift-off test, secondary grouting was accomplished, and anchor head recesses were filled with concrete to restore the spillway profile.

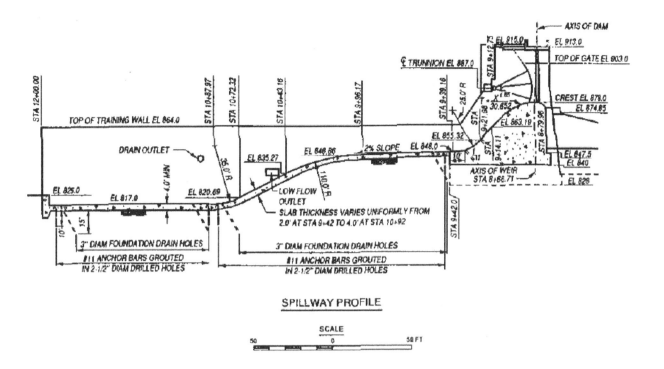

SPILLWAY PROFILE

SCALE

50   0   50 FT

Figure 8-1

Cross-section of spillway monoliths

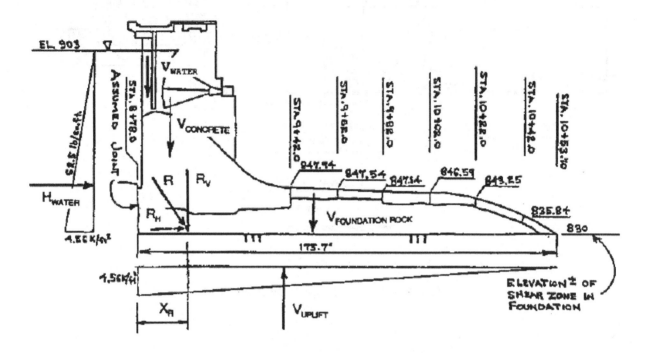

Figure 8-2

Free body diagram, R_y = resultant of vertical forces, R_H = resultant of horizontal forces,

and X_R = distance from heel to resultant location on sliding plane

	Σ Vert, kips	Σ Horz, kips
Concrete	11,910	
Rock (Saturated Weight)	13,160	
Machinery	10	
Gates	70	
Water Down	870	
Water Up	- 90	
Uplift	- 16,830	
Horizontal Water		6,990
Totals, Loading Condition No. 2	9,100	6,990

$$Sliding\ FS,\ Without\ Anchors = \frac{TAN\ 20.5° \times 9,100}{6,990} = 0.49$$

Table 8-1

Summary of Forces on the Sliding Plane. Loading Condition No. 2 (Pool at Top of Gates, Tailwater Below Sliding Surface)

	Σ Vert, kips	Σ Horz, kips
Concrete	11,910	
Rock	13,160	
Machinery	10	
Gates	70	
Water Down	870	
Water Up	- 90	
Uplift	- 16,830	
Horizontal Water		6,990
Anchors (Vertical) 7 x 826 x 0.707	4,088	
Anchors (Horizontal)		- 4,088
Totals, Loading Condition No. 2	13,188	2,902

$$Sliding\ FS,\ With\ Anchors = \frac{TAN\ 20.5° \times 13,188}{6,990 - 4,088} = 1.70$$

Table 8-2

Summary of Forces on the Sliding Plane. Loading Condition No. 2, With Anchors

ANCHORING OF SPILLWAY MONOLITHS

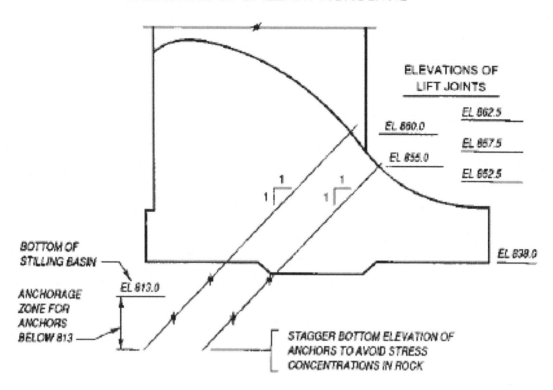

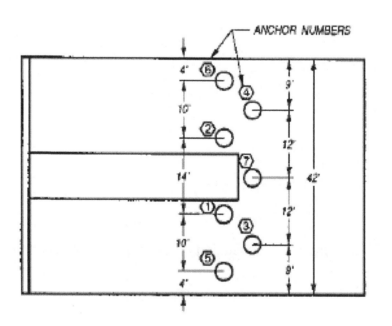

Figure 8-3

Location of anchors

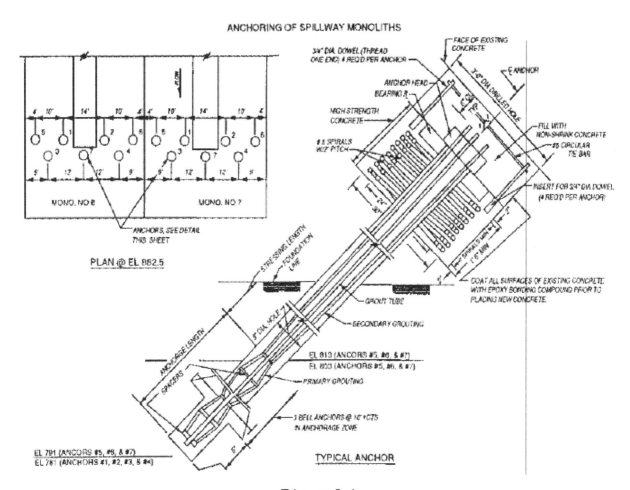

Figure 8-4

Detail of anchors

15.4 ROLLER COMPACTED CONCRETE GRAVITY DAMS

15.4.1 INTRODUCTION. Gravity dams built using the RCC construction method, afford economies over conventional concrete through rapid placement techniques. Construction procedures associated with RCC require particular attention be given in the layout and design to watertightness and seepage control, horizontal and transverse joints, facing elements, and appurtenant structures. The designer should take advantage of the latitude afforded by RCC construction and use engineering judgment to balance cost reductions and technical requirements related to safety, durability, and long-term performance. A typical cross section of an RCC dam is shown in Figure 9-1. RCC mix design and construction should be in accordance with appropriate technical guidance.

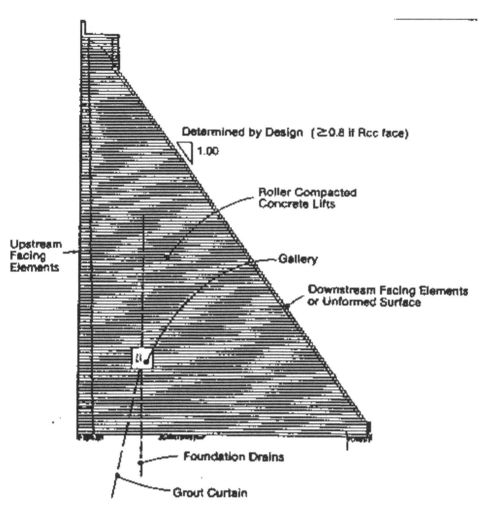

Determined by Design (≥ 0.8 if Rcc face)

1.00

Roller Compacted
Concrete Lifts

Upstream
Facing
Elements

Gallery

Downstream Facing Elements
or Unformed Surface

B

Foundation Drains

Grout Curtain

Figure 9-1. Typical RCC dam section

15.4.2 CONSTRUCTION METHOD. Construction techniques used for RCC placement often result in a much lower unit cost per cubic yard compared with conventional concrete placement methods. The dry, nonflowable nature of RCC makes the use of a wide range of equipment for construction and continuous placement possible. End and bottom dump trucks and/or conveyors can be used for transporting concrete from the mixer to the dam. Mechanical spreaders, such as caterpillars and graders, place the material in layers or lifts. Self-propelled, vibratory, steel-wheeled, or pneumatic rollers along with the dozers perform the compaction. The thickness of the placement layers, ranging from 8 to 24 inches, is established by the compaction capabilities. With the flexibility of using the above equipment and continuous placement, RCC dams can be constructed at significantly higher rates than those achievable with conventional mass concrete. A typical work layout for the RCC placement spreading operation is illustrated in Figure 9-2.

15.4.3 ECONOMIC BENEFITS. RCC construction techniques have made gravity dams an economically competitive alternative to embankment structures. The following factors tend to make RCC more economical than other dam types:

15.4.3.1 MATERIAL SAVINGS. Construction cost histories of RCC and conventional concrete dams show the unit cost per cubic yard of RCC is considerably less. The unit cost of concrete for both types of dam varies with the volume of the material in the dam. As the volume increases, the unit cost decreases. The cost savings of RCC increase as the volume decreases. RCC dams have considerably less volume of construction material than embankments of the same height. As the height increases, the volume versus height for the embankment dam increases almost exponentially in comparison to the RCC dam. Thus, the higher the structure, the more likely the RCC dam will be less costly than the embankment alternative.

15.4.3.2 RAPID CONSTRUCTION. The rapid construction techniques and reduced concrete volume account for the major cost savings in RCC dams. Maximum placement rates of 5,800 to 12,400 cubic yards/day have been achieved. These production rates make dam construction in one construction season readily achievable. When compared with embankment dams, construction time is reduced by 1 to 2 years. Other benefits from rapid construction include reduced construction administration costs, earlier project benefits, and possible selection of sites with limited construction seasons. Basically, RCC construction offers economic advantages in all aspects of dam construction that are related to time.

15.4.3.3 SPILLWAYS AND APPURTENANT STRUCTURES. The location and layout alternatives for spillways, outlet and hydropower works, and other appurtenant

structures in RCC dams provide additional economic advantages compared with
embankment dams.

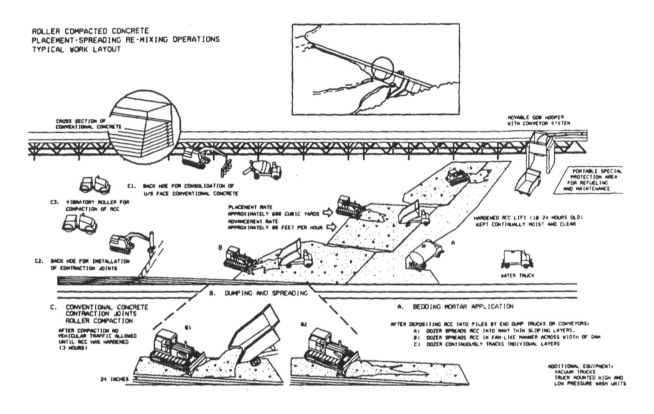

Figure 9-2

Typical work layout for RCC placement spreading operation

The arrangements of these structures is similar to conventional concrete dams, but with
certain modifications to minimize costly interference to the continuous RCC placement
operation. Gate structures and intakes should be located outside the dam mass.
Galleries, adits, and other internal openings should be minimized. Details on the layout
and design of spillways and appurtenant structures are discussed in paragraph

15.4.3.4 SPILLWAYS FOR RCC dams can be directly incorporated into the structure.
The layout allows discharging flows over the dam crest and down the downstream face.
In contrast, the spillway for an embankment dam is normally constructed in an abutment
at one end of the dam or in a nearby natural saddle. Generally, the embankment dam

spillway is more costly. For projects that require a multiple-level intake for water quality control or for reservoir sedimentation, the intake structure can be readily anchored to the upstream face of the dam. For an embankment dam, the same type of intake tower is a freestanding tower in the reservoir or a structure built into or on the reservoir side of the abutment. The economic savings for an RCC dam intake is considerably cheaper, especially in high seismic areas. The shorter base dimension of an RCC dam compared with an embankment dam reduces the size and length of the conduit and penstock for outlet and hydropower works.

15.4.4 DIVERSION AND COFFERDAM. RCC dams provide cost advantages in river diversion during construction and reduce damages and risks associated with cofferdam overtopping. The diversion conduit will be shorter compared with embankment dams. With a shorter construction period, the size of the diversion conduit and cofferdam height can be reduced. These structures may need to be designed only for a seasonal peak flow instead of annual peak flows. With the high erosion resistance of RCC, if overtopping of the cofferdam did occur, the potential for a major failure would be minimal and the resulting damage would be less.

15.4.5 OTHER ADVANTAGES. The smaller volume of an RCC dam makes the construction material source less of a driving factor in site selection of a dam. Furthermore, the borrow source will be considerably smaller and more environmentally acceptable. The RCC dam is also inherently safer against internal erosion, overtopping, and seismic ground motions.

15.4.5 DESIGN AND CONSTRUCTION CONSIDERATIONS

15.4.5.1 WATERTIGHTNESS AND SEEPAGE CONTROL. Achieving watertightness and controlling seepage through RCC dams are particularly important design and construction considerations. Excessive seepage is undesirable from the aspect of structural stability and because of the adverse appearance of water seeping on the downstream dam face, the economic value associated with lost water, and possible long-term adverse impacts on durability. RCC that has been properly proportioned, mixed, placed, and compacted should be as impermeable as conventional concrete. The joints between the concrete lifts and interface with structural elements are the major pathways for potential seepage through the RCC dam. This condition is primarily due to segregation at the lift boundaries and discontinuity between successive lifts. It can also be the result of surface contamination and excessive time intervals between lift placements. Seepage can be controlled by incorporating special design and construction procedures that include contraction joints with waterstops making the upstream face watertight, sealing the interface between RCC layers, and draining and collecting the seepage.

15.4.5.2 UPSTREAM FACING. RCC cannot be compacted effectively against upstream forms without the forming of surface voids. An upstream facing is required to produce a surface with good appearance and durability. Many facings incorporate a watertight barrier. Facings with barriers include the following:

- Conventional form work with a zone of conventional concrete placed between the forms and RCC material.
- Slip-formed interlocking conventional concrete elements. RCC material is compacted against the cured elements.
- Precast concrete tieback panels with a flexible waterproof membrane placed between the RCC and the panels.

A waterproof membrane sprayed or painted onto a conventional concrete face is another method; however, its use has been limited since such membranes are not elastic enough to span cracks that develop and because of concerns about moisture developing between the membrane and face and subsequent damage by freezing.

15.4.5.3 HORIZONTAL JOINT TREATMENT. Bond strength and permeability are major concerns at the horizontal lift joints in RCC. Good sealing and bonding are accomplished by improving the compactability of the RCC mixture, cleaning the joint surface, and placing a bedding mortar (a mixture of cement paste and fine aggregate) between lifts. When the placement rate and setting time of RCC are such that the lower lift is sufficiently plastic to blend and bond with the upper layer, the bedding mortar is unnecessary; however, this is rarely feasible in normal RCC construction. Compactibility is improved by increasing the amount of mortar and fines in the RCC mixture. The lift surfaces should be properly moist cured and protected. Cleanup of the lift surfaces prior to RCC placement is not required as long as the surfaces are kept clean and free of excessive water. Addition of the bedding mortar serves to fill any voids or depressions left in the surface of the previous lift and squeezes up into the voids in the bottom of the new RCC lift as it is compacted. A bedding mix consisting of a mixture of cement paste and fine and 3/8-in.-MSA aggregate is also applied at RCC contacts with the foundation, abutment surfaces, and any other hardened concrete surfaces.

15.4.5.4 SEEPAGE COLLECTION. A collection and drainage system is a method for stopping unsightly seepage water from reaching the downstream face and for preventing excessive hydrostatic pressures against conventional concrete spillway or downstream facing. It will also reduce uplift pressures within the dam and increase stability. Collection methods include vertical drains with waterstops at the upstream face and vertical drain holes drilled from within the gallery near the upstream or downstream face. Collected water can be channeled to a gallery or the dam toe.

15.4.5.5 NONOVERFLOW DOWNSTREAM FACING. Downstream facing systems for nonoverflow sections may be required for aesthetic reasons, maintaining slopes steeper than the natural repose of RCC, and freeze-thaw protection in severe climate locations. Facing is necessary when the slope is steeper than 0.8H to 1.0V when lift thickness is limited to 12 inches or less. Thicker lifts require a flatter slope. Experience has demonstrated that these are the steepest uncompacted slopes that can be practically controlled without special equipment or forms. The exposed edge of an uncompacted slope will have a rough stair-stepped natural gravel appearance with limited strength within 12 inches of the face. Downstream facing systems include conventional vertical slipforming placement and horizontal slipforming similar to that used on the upstream face. When this type of slope is used, the structural cross section should include a slight

overbuild to account for deterioration and unraveling of material loosened from severe
weather exposure over the project life (see Figure 9-3). Several recent projects have
compacted downstream faces using a tractor-mounted vibrating plate.

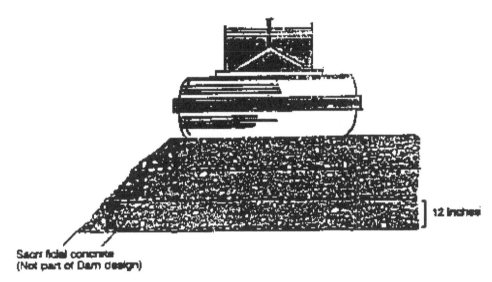

Figure 9-3

Compaction of RCC at downstream face

15.4.5.6 TRANSVERSE CONTRACTION JOINTS. Transverse contraction joints are
required in most RCC dams. The potential for cracking may be slightly lower in RCC
because of the reduction in mixing water and reduced temperature rise resulting from
the rapid placement rate and lower lift heights. In addition, the RCC characteristic of
point-topoint aggregate contact decreases the volume shrinkage. Thermal cracking
may, however, create a leakage path to the downstream face that is aesthetically
undesirable. Thermal studies should be performed to assess the need for contraction
joints. Contraction joints may also be required to control cracking if the site configuration
and foundation conditions may potentially restrain the dam. If properly designed and
installed, contraction joints will not interfere or complicate the continuous placement
operation of RCC. At Elk Creek Dam, contraction joints were installed with no impact to
RCC placement operations by inserting galvanized steel sheeting into the uncompacted
RCC for the entire thickness and height of the dam. The sheets were pushed vertically
into the RCC by means of a tractor-mounted vibratory blade, as shown in Figure 9-4.

Figure 9-4

Contract joint placement using a vibrating blade to insert galvanized steel sheeting

15.4.5.7 WATERSTOPS. Standard waterstops may be installed in an internal zone of conventional concrete placed around the joint near the upstream face. Waterstops and joint drains are installed in the same manner as for conventional concrete dams. Typical internal waterstops and joint drain construction in RCC dams are shown in Figure 9-5. Around galleries and other openings crossing joints, waterstop installation will require a section of conventional concrete around the joint.

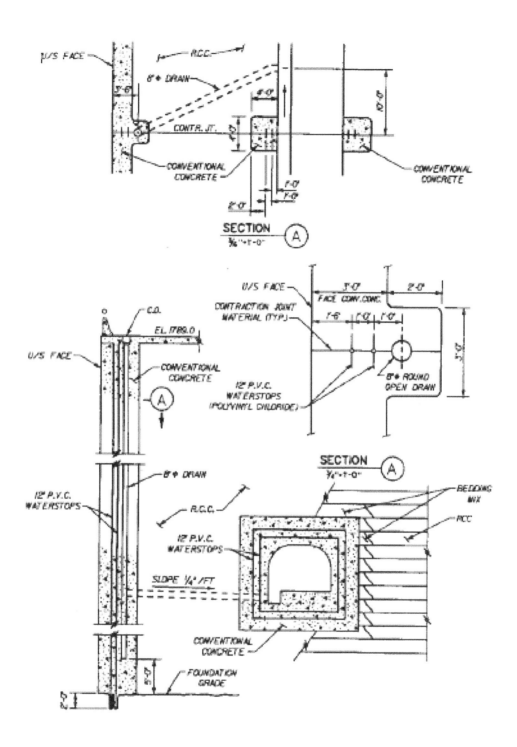

Figure 9-

Typical internal waterstops and joint drain construction in RCC dams

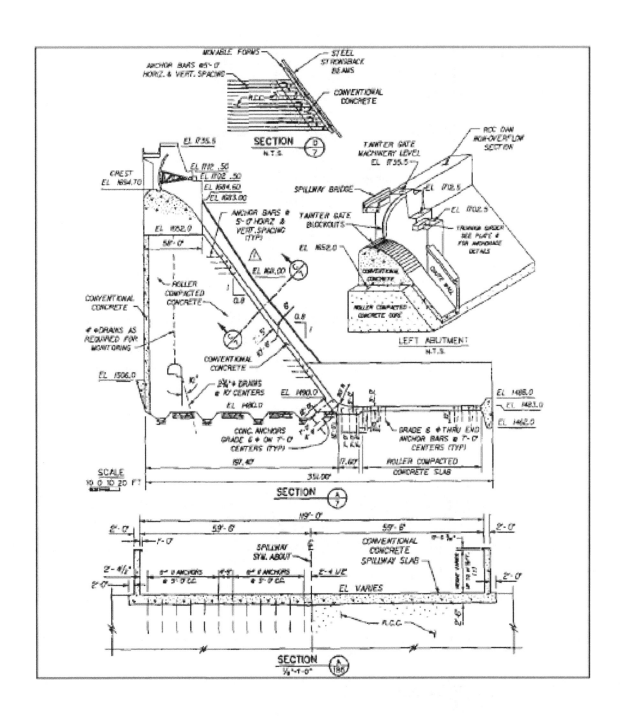

Figure 9-6

RCC spillway details

Made in the USA
Las Vegas, NV
28 September 2023

78236587R00280